Life
Science
Ethics

Life Science Ethics

Edited by

Gary L. Comstock

Iowa State Press
A Blackwell Publishing Company

Gary L. Comstock is director of the North Carolina State University research ethics program and professor of philosophy. The author of *Vexing Nature? On the Ethical Case Against Agricultural Biotechnology*, he is perhaps best known as the director of the ISU Bioethics Institutes, an international faculty development workshop that has helped five hundred life scientists to integrate discussions of ethics into their courses. Comstock has published more than fifty articles and book chapters and edited the volumes *Is There A Moral Obligation to Save the Family Farm?* and *Religious Autobiographies*.

© 2002 Iowa State Press
A Blackwell Publishing Company

Iowa State Press
2121 State Avenue, Ames, Iowa 50014

Orders:	1-800-862-6657
Office:	1-515-292-0140
Fax:	1-515-292-3348
Web site:	www.iowastatepress.com

Authorization to photocopy items for internal or personal use, or the internal or personal use of specific clients, is granted by Iowa State Press, provided that the base fee of $.10 per copy is paid directly to the Copyright Clearance Center, 222 Rosewood Drive, Danvers, MA 01923. For those organizations that have been granted a photocopy license by CCC, a separate system of payments has been arranged. The fee code for users of the Transactional Reporting Service is 0-8138-2835-X/2002 $.10.

♾   Printed on acid-free paper in the United States of America

First edition, 2002

Library of Congress Cataloging-in-Publication Data

Life science ethics / Gary L. Comstock, editor.--1st ed.
       p. cm.
Includes bibliographical references and index.
   ISBN 0-8138-2835-X (acid-free paper)
   1. Bioethics. 2. Life sciences. I. Comstock, Gary, 1954-
QH332 .L54 2002
174'.957--DS21

                                  2002003327

The last digit is the print number: 9 8 7 6 5 4 3 2

Dedicated to my brother Doug and his beloved wife,

Emily Goyer Comstock,

whose grace is exceeded only by her courage

CONTENTS

CONTRIBUTORS

Gary L. Comstock is director of the North Carolina State University research ethics program and professor of philosophy. The author of *Vexing Nature? On the Ethical Case Against Agricultural Biotechnology* (Kluwer, 2000), he is perhaps best known as the director of the ISU Bioethics Institutes, an international faculty development workshop that has helped five hundred life scientists to integrate discussions of ethics into their courses. Comstock has published more than fifty articles and book chapters and edited the volumes *Is There A Moral Obligation to Save the Family Farm?* (ISU Press, 1987) and *Religious Autobiographies* (Wadsworth, 1995).

Lilly-Marlene Russow is professor of philosophy at Purdue University, with an appointment in the Department of Veterinary Pathobiology. She is co-author with Martin Curd of *Principles of Reasoning* (St. Martin's Press, 1989) and she wrote the seminal, often reprinted article, "Why Do Species Matter?" She has also authored numerous articles on philosophy of mind, environmental ethics, and ethics and animals. A special interest in the treatment of laboratory animals is reflected in several articles in scientific journals and in her work with national groups such as the Scientist's Center for Animal Welfare (SCAW). She directed three Bioethics Institutes at Purdue University, two of which were sponsored by NSF Grant SBR–9601759. Parts of her contributions here were results of the Institutes.

Hugh LaFollette is professor of philosophy at East Tennessee State University. He is author of *Personal Relationships: Love, Identity, and Morality* (Blackwell, 1995), co-author (with Niall Shanks) of *Brute Science: Dilemmas of Animal Experimentation* (Routledge, 1996), and author of numerous essays in ethics and political philosophy. He is also editor or co-editor of seven volumes, most recently *Blackwell Guide to Ethical Theory* (Blackwell, 2000) and *Ethics in Practice*, 2nd edition (Blackwell, 2001). LaFollette is also currently editing *The Oxford Handbook of Practical Ethics* and writing a book entitled *The Practice of Ethics*, which seeks to integrate the discussion of practical ethics and ethical theory.

Larry May is professor of philosophy at Washington University in St. Louis. He is the author of *The Morality of Groups* (Notre Dame, 1987), *Sharing Responsibility* (Chicago, 1992), *The Socially Responsive Self* (Chicago, 1996), *Masculinity and Morality* (Cornell, 1998), and co-author of *Praying for a Cure* (Rowman & Littlefield, 1999). He has just completed the penultimate draft of *Humanitarian Crimes*. He has also co-edited nine books, including *Collective Responsibility* (Rowan & Littlefield, 1991), *Mind and Morals* (MIT, 1996), and *Rights and Reason* (Kluwer, 2000). He has a law degree as well as a doctorate in philosophy.

Gary Varner is associate professor of philosophy at Texas A&M University. His book, *In Nature's Interests?* (Oxford University Press, 1998), provides an original analysis of what it means to have morally significant interests and examines the alleged divide between animal rights views and sound environmental policy. His published papers cover related topics in medical research, cloning, animal agriculture, and human nutrition, as well as philosophical issues associated with the National Environmental Policy Act, the Endangered Species Act, and the property takings debate. He has spoken on related topics to the American Bar Association's National Judicial College, the American Veterinary Medicine Association, and the Wildlife Society.

Paul B. Thompson holds the Joyce and Edward E. Brewer Chair in Applied Ethics in the Department of Philosophy at Purdue University. He has published dozens of articles on the ethics of food and agricultural technology, and several books, including *The Spirit of the Soil: Agriculture and Environmental Ethics*. He was twice the recipient of the American Agricultural Economics Association award for professional excellence in communication. Most recently, he has edited a collection of essays with Thomas Hilde entitled *The Agrarian Roots of Pragmatism* (Vanderbilt University Press, 2000), to which he contributed a chapter on the agrarian philosophy of Thomas Jefferson.

Fred Gifford is associate professor of philosophy at Michigan State University and the on-site coordinator for the 1995 and 1996 Bioethics Institutes at MSU. The author of numerous articles on medical ethics and philosophy of science, including "Outcomes Research and Practice Guidelines: Upstream Issues for Downstream Users," *Hastings Center Report* 26 (1996), Gifford is also associate faculty member of the Center for Ethics and Humanities in the Life Sciences at MSU. He developed and co-teaches a course in the Department of Horticulture entitled "Biotechnology in Agriculture: Applications and Ethical Issues." His research and publications focus largely on issues concerning scientific research.

Charles Taliaferro is associate professor of philosophy and a member of the Environmental Studies Concentration Faculty at St. Olaf College. He has contributed to the journals *Environmental Ethics* and *Agriculture and Human Values*, and addressed environmental concerns in a series of articles as well as in his book *Consciousness and the Mind of God* (Cambridge University Press). He co-edited *A Companion to Philosophy of Religion* and *An Introduction to Philosophy of Religion* (both forthcoming with Basil Blackwell, UK). Taliaferro is a member of an Agricultural Concerns Group in Northfield, Minnesota, and is active in the Land Stewardship Project, based in Boston, Massachusetts.

Case Study Authors

Rare Plants

Lynn G. Clark, Botany, Iowa State University

Marine Mammal Protection

Donald J. Orth, Fisheries and Wildlife Sciences, Virginia Polytechnic Institute (Written after the 1996 Bioethics Institute at Michigan State University)

Infant Deaths in Developing Countries

Lois Banta, Biology, Haverford College
Jeffrey Beetham, Veterinary Pathology and Entomology, Iowa State University
Donald Draper, Veterinary Anatomy, Iowa State University
Nolan Hartwig, Veterinary Medicine, Iowa State University
Marvin Klein, Food Marketing and Agribusiness Management, California State Polytech
Grace Marquis, Food Science and Human Nutrition, Iowa State University
(Written at the 1999 Bioethics Institute at Iowa State University)

Edible Antibiotics in Food Crops

Mike Zeller, Office of Biotechnology, Iowa State University
Terrance Riordan, Agronomy and Horticulture, University of Nebraska
Halina Zaleski, Human Nutrition, Food and Animal Science, University of Hawaii
Dean Herzfeld, Plant Pathology, Extension, University of Minnesota
Kathryn Orvis, 4-H Youth Development and Horticulture, Purdue University
(Written at the 2001 Bioethics Institute at Iowa State University)

Beef, Milk, and Eggs

Gary Varner, Philosophy, Texas A&M
(Written for the 1998 Bioethics Institute at Iowa State University)

Veterinary Euthanasia

Bernard Rollin, Philosophy, Colorado State University
Jerrold Tannenbaum, Population Health and Reproduction, University of California, Davis
Courtney Campbell, Philosophy, Oregon State University
Kathleen Moore, Philosophy, Oregon State University
Gary L. Comstock, Philosophy and Religious Studies, Iowa State University

Hybrid Corn

Jochum Wiersma, Agronomy and Plant Genetics, University of Minnesota
Deon Stuthman, Agronomy and Plant Genetics, University of Minnesota
David Fan, Genetics, Cell Biology and Development, University of Minnesota
Donald Duvick, Johnston, IA
Victor Konde, Biochemistry, University of Zambia, and Center for International Development, Harvard University
(Written at the 2001 Bioethics Institute at Iowa State University)

Trait Protection System

Thomas Peterson, Agronomy, and Zoology and Genetics, Iowa State University
Bryony Bonning, Entomology, Iowa State University
(Written at the 1998 Bioethics Institute at Iowa State University)

Golden Rice

Kristen Hessler, Philosophy and Religious Studies, Iowa State University
Ross Whetten, Forestry, North Carolina State University
Carol Loopstra, Forest Science, Texas A&M University
Sharon Shriver, Biology, Pennsylvania State University
Karen Pesaresi Penner, Food Science, Kansas State University
Robert Zeigler, Plant Pathology, Kansas State University
Jacqueline Fletcher, Entomology and Plant Pathology, Oklahoma State University
Melanie Torrie, undergraduate student, Iowa State University
Gary L. Comstock, Philosophy and Religious Studies, Iowa State University
(Written at the 1999 Bioethics Institute at North Carolina State University)

Organ Transplantation

Christopher Baldwin, Chemical Engineering, Iowa State University
David Bristol, Food Animal and Equine Medicine, North Carolina State University
Emily Deaver, Environmental Biology, Chowan College
Bruce Hammerberg, Microbiology, Pathology, and Parasitology, North Carolina State University
Carole A. Heath, Chemical Engineering, Iowa State University
Surya Mallapragada, Chemical Engineering, Iowa State University
Gavin J. Naylor, Zoology and Genetics, Iowa State University
Elaine Richardson, Animal and Veterinary Sciences, Clemson University
Jim Wilson, Industrial Engineering, North Carolina State University
(Written by two teams, one at the 1998 Bioethics Institute at North Carolina State University, and the second at the 1998 Bioethics Institute at Iowa State University)

Lost in the Maize

Isabel Lopez-Calderon, Genetics, University of Seville
Steven Hill, Plant Sciences, University of Oxford
L. Horst Grimme, Biology and Chemistry, University of Bremen
Michael Lawton, Biotechnology, Rutgers University
Anabela M. L. Romano, Engineering and Natural Resources, University of Algarve
(Written at the 2001 Bioethics Institute at the Foundation for Luso-American Development in Lisbon, Portugal)

Magnanimous Iowans

Ricardo Salvador, Agronomy, Iowa State University
Stephen Moose, Crop Sciences, University of Illinois
Bruce Chassy, Food Science and Human Nutrition, University of Illinois
Kathie Hodge, Plant Pathology, Cornell University
(Written at the 2001 Bioethics Institute at Iowa State University)

ACKNOWLEDGMENTS

This book is one of the products of National Science Foundation (NSF) grants SBR–9254504 and SES–9906244, grants that supported the "Iowa State University Model Bioethics Institutes." The Iowa State University Model Institutes are faculty development workshops for life scientists that began in 1991 at Iowa State. In the last decade, the Institutes have visited five other U.S. campuses: the University of Illinois, Michigan State, Purdue, North Carolina State, and Oregon State. In the year 2000, they expanded to reach an international audience, with Institutes at the Foundation for Luso-American Development (FLAD), in Lisbon, Portugal. We are grateful for FLAD's support and specifically for the contributions of FLAD's forward-looking director, Dr. Charles Buchanan.

The majority of this book's chapters grew out of papers originally presented at one ISU Model Institute or another. Almost all the case studies in Part III began their lives as drafts produced at an Institute.

Deeply grateful to NSF for its support, we hasten to offer special recognition to Dr. Rachelle Hollander, Director of NSF's Ethics and Values Studies Program. Dr. Hollander has exhibited extraordinary vision and courage in her efforts to stimulate the integration of discussions of ethics into the life science curriculum. Were it not for her efforts in championing a broad range of highly successful research and teaching projects in the area of science and values, the gap between the so-called "two cultures" would, alas, be wider than it is.

The editor was able to finish the project thanks to grant support from the Cooperative State Research, Education, and Extension Service, U.S. Department of Agriculture, under agreement 00–52100–9617. The USDA requires us to add: "Any opinions, findings, conclusions, or recommendations expressed in this publication are those of the author(s) and do not necessarily reflect the view of the U.S. Department of Agriculture."

Finally, speaking personally, I wish to acknowledge the active support of Patricia B. Swan, formerly Vice-Provost for Research at Iowa State University. Not only did Dr. Swan rigorously guard the autonomy and health of the ISU Bioethics Program during her tenure as Vice-Provost from 1990–2000, she also powerfully shaped the future of biological education by first suggesting the outlines of what would become the ISU Bioethics Institute.

INTRODUCTION

Life science ethics is the normative evaluation of human actions affecting living things. We affect living things in virtually everything we do, from drinking water to cooking dinner and from sending e-mail to flushing the toilet. Sometimes we pause to reflect about these activities and, when we do, we may ask ourselves some basic philosophical questions. Does nature have intrinsic value? Should we be doing more to save wilderness and ocean ecosystems? What are our duties to future generations of humans? Do animals have rights? Should scientists sign agreements that prevent them, for a time, from making the results of their experiments known to anyone except the private industry that has funded their research? These are some of the questions we find in life science ethics.

The book is a work of applied ethics and is intended to fill a gap in the ethics literature. The gap concerns moral issues that arise when humans use what Aldo Leopold called "the land."[1] The book has three parts. Part I introduces ethics, the relationship of religion to ethics, how we assess ethical arguments, and a method that ethicists use to reason about ethical theories.

Part II demonstrates the relevance of ethical reasoning to six topics:

- The relative moral standing of ecosystems, nonhuman animals, and future human generations
- Our duties to aid the hungry in developing countries
- Obligations to animals used to produce food, fiber, and knowledge
- Public policies to adjudicate conflicting rights-claims among urban consumers, environmentalists, and farmers over the use of water and land
- The moral justifiability of genetic engineering as a whole and the patenting of life forms in particular
- The virtues traditionally associated with family farms

Part III offers twelve case studies, two cases for each of the six topics. We have found the cases useful in promoting reasoned discussion of fundamental questions in life science ethics.

A word about our title: One of the branches of life science ethics is *bioethics*, a term that has come to mean the normative evaluation of actions affecting *humans*. Is the fetus a person? Should physicians be permitted to help patients commit suicide? Who should pay for health care for the poor? These are profound and urgent matters, and a veritable bioethics industry has grown up to reckon with them during the last four decades.

Yet, the prefix *bio* derives from the Greek word *bios*, meaning *all* life, so why restrict our attention to humans? Could *bioethics* not be used in a broader way, to encompass more than human, medical ethics? The etymological origins of *bio* focus on life in all its myriad forms, including animal, plant, microorganismic and ecosystemic life.[2]

To date, professional ethicists have not been inclined to use the term in its original, more inclusive, sense. Representative of the debate, for example, is this call for grant applications written by a well-respected private foundation:

> Through its Interdisciplinary Program in Bioethics, the Foundation provides funding for physicians, lawyers, philosophers, economists, theologians, and other professionals to address micro and macro issues in bioethics, providing guidance for those engaged in decision making at the bedside as well as those responsible for shaping institutional and public policy.

The terms *bedside* and *physicians* clearly convey the assumption of the granting agency: Proposals should focus on the care of humans. Proposals from agronomists and animal scientists focused on ethical issues having to do with the care of plants and animals, endangered species, and farm animal welfare are not likely to be considered, much less funded.

A recent experience of the editor of this volume is also telling. There is a widely respected international academic journal with a title that sounds very much like *bioethics*. I suggested to its editor that the journal consider reviewing a new book on the ethical dimensions of agricultural biotechnology. The proposal was rejected on the grounds that the subject matter of the book was not within the scope of the journal.

The narrower understanding of bioethics as medical clinical ethics is currently dominant. Hence, a new phrase is needed to convey the original meaning of the word. We have adopted *life science ethics*.

How should undergraduates be introduced to life science ethics? We have kept this question in front of us, hoping to create a text that will assist its users in sharpening their critical reasoning skills while also providing essential background concepts in moral theory. We intend our essays to be accessible to first-year college students while also introducing cutting-edge philosophical ideas. Authors, therefore, were selected because of their original contributions to ethics scholarship and on the basis of their ability to explain difficult philosophical concepts to novices.

A significant feature of the collection is its case study approach, an innovative pedagogical structure that should make the book particularly appealing to nonspecialists. The book begins with a brief narrative introducing a student, Emily, who must decide whether to cheat. The readers are invited to assess the case for themselves, look into the facts of the case, and reach their own decision about the permissibility of cheating. Emily's case should not only prove entertaining but also provoke energetic and reflective classroom discussion of topics such as the following: What is ethics? How does ethics differ from custom, law, science, and religion? Is there anything objective about ethics? Succeeding chapters introduce the concepts of moral reasoning and argumentation, providing students with exercises that they can complete in order to help them master the skills of critical reasoning (in Parts I and II, these exercises are located in the book's appendix), and follow Emily's story as she confronts other critical questions.

The contributors shared five objectives in writing the cases that introduce each chapter in Part II:

Accessibility: The majority of students encountering this book will not have had a course in philosophy. The cases and essays are written in such a way that these students will be able to understand them.

Plausibility: The cases are not factual because they involve imaginary characters, but they are plausible, with a high degree of verisimilitude. Ideally, they are based on actual incidents and describe situations students may face.

Philosophical fecundity: With the right tools and careful guidance, philosophical novices can be led to discuss ethical issues with a high degree of sophistication. The cases provide an introduction to key terms and ideas by which instructors can lead classes in in-depth discussions. Discussion of the cases that open the chapters in Part II (with help from the discussion questions found in the book's appendix) may be further developed by close reading of the essays that follow.

Drama: We have constructed the cases to appeal to the imagination, using narrative and dialog to heighten interest.

Coherence: We introduce a cast of characters taking a university course called "Agricultural Ethics." We follow them throughout the book, thus presenting a single narrative plot that builds on previous cases, lending coherence to the whole.

Each case study in Part II is accompanied by a set of discussion questions located in the book's appendix. These questions are meant to elicit conversations about the issues taken up in the essays that follow.

Each essay begins with the author discussing a new development in the Ag Ethics class and returns to the case at the end. Each essay surveys the philosophical literature, introduces different answers that have been given to the discussion questions, and leads the student through relevant philosophical topics. Each author also suggests the outlines of his or her own position on the central questions.

Our over-arching goal is to improve the students' skills in analyzing ethical arguments and to help them discover which argument they have the best reasons to believe and act upon. Is it possible to achieve this goal? Research suggests that students can "make substantial gains in moral reasoning skills."[3] Teachers of critical thinking have created and tested various methods to improve these ethical capacities.[4] And there is some reason to think, perhaps a bit optimistically, that as we improve our reasoning abilities in the area of ethics, we also improve our behavior.[5]

What should university students be able to do when they have completed a science curriculum enriched with an ethics component? They should be able to speak and write with discernment and good reasoning. We will know they are discerning if their discourse evidences the ability to recognize issues as moral issues; articulate and apply moral principles, values, and approaches; and analyze cases in a self-reflective way. We will know they can reason well if their discourse evidences knowledge of the accepted moral standards within their field; knowledge of key ethical arguments, figures, and texts; the ability to speak and write in a way that is logical, complete, consistent, and clear and that can recognize potential objections to one's position.

Students need to be able to discern and reason if they are to live a good life, the life of a reflective, mature citizen and a morally responsible professional. Science graduates will enter a workplace in which many issues once thought to be purely technical, scientific, or legal now clearly have an ethical dimension. The challenges they face as professionals will increasingly be challenges their mentors have not faced. Consequently, students may find themselves having to say something intelligent, perhaps with a television camera in their face, without having had a chance to discuss the question with peers.

We can assist these students by helping them to recognize, organize, and evaluate moral arguments; by creating a learning environment that fosters cooperation, analysis, and criticism; by introducing them to moral arguments relevant to their disciplines; by modeling proper scientific conduct; and by providing them with case studies that raise relevant ethical issues. We hope this book will help achieve at least some of these goals.

Notes

1. Aldo Leopold, *A Sand County Almanac* (New York: Oxford University Press, 1949).
2. It seems that Van Renssalaer Potter II coined the word *bioethics* in 1971 in his book *Bioethics, Bridge to the Future* (Englewood Cliffs, NJ: Prentice-Hall, Inc.). As Potter has taken pains to point out, he did not intend the word to refer narrowly to human clinical ethics but rather to the wide range of problems associated with the global survival of all life forms. Cf. Potter, "What Does Bioethics Mean?" *The Ag Bioethics Forum* 8 (June 1996), available on-line at: http://www.biotech.iastate.edu/Bioethics.html.
3. A. Garrod, ed., *Approaches to Moral Development: New Research and Emerging Themes* (New York Teachers College Press, 1993).
4. M. J. Bebeau and S. J. Thoma, "The Impact of a Dental Ethics Curriculum on Moral Reasoning," *Journal of Dental Education* 58 (1994): 684–692.
5. "The link between moral reasoning and moral behavior is well established." Thoma, "Moral Judgments and Moral Actions," in J. R. Rest and D. Narvaez, eds., *Moral Development in the Professions: Psychology and Applied Ethics* (Hillsdale: Erlbaum Associates, 1994).

Life
Science
Ethics

Part

I

ETHICAL REASONING

Chapter 1

ETHICS

Gary L. Comstock

CASE STUDY: EMILY THE STUDENT

Emily is starting her sophomore year at a large state university. Having not yet decided on a major and still trying to decide which courses to take during her second year, she consults her friend Doug. He tells her that his courses are largely determined by the Department of Agricultural Economics in the College of Agriculture. Thinking that she might like to take a class with Doug, she asks him what he's registered for. On his list is "Agricultural Ethics." Intrigued by the title, not to mention the possibility of hours in the library with Doug, Emily decides to enroll. She figures that she already knows a thing or two about ethics, and if she needs assistance with the agricultural stuff, Doug will be there to help out.

The week before classes are to begin, Doug calls her to say that he is going to miss the first day of classes because his father needs help putting up hay on their dairy farm. Would she take extra-careful notes in Ag Ethics and share them with him? No problem; she's glad to help. On the first day, the instructor passes out a twenty-page syllabus, and Emily takes an extra for Doug. The instructor, Dr. Wright, without taking roll, reads through the syllabus and then discusses its contents with the forty-four students in the class. Sitting in the back row, Emily is surprised by the last two sentences:

> On the first day of class you will read the syllabus, discuss it with the instructor, and have an opportunity to ask any questions; then you will be tested over the contents. You may not make up this quiz if you miss the first day of class.

5

When discussion subsides, Wright puts three quiz questions on the board. Emily is relieved to see immediately that the answers are obvious and she quickly writes them down. While other students are finishing, Emily starts to wonder about Dr. Wright's own ethics. Is it fair to give a test on the first day and not let people make it up? Doug is absent for good reasons, reasons beyond his control, and he is a good student. Is it right for instructors to penalize absent students without first giving those students a chance to explain themselves?

Emily notes that Doug will be forced to start the semester behind the rest of the class simply because he missed the first day. Then an odd thought strikes her. She is proficient at disguising her handwriting by using her left hand and a neat, back-slanting script. Dr. Wright cannot see her behind all these people. Feeling loyal to Doug and skeptical about Wright, she quietly pulls out a second piece of paper, puts Doug's name at the top with her off hand, and writes another set of answers to the three questions. She leans back, wondering whether she ought to hand it in.

Questions flood her mind. She thinks: Is this the right thing to do? Maybe not; maybe so. Doesn't it all, ultimately, come down to this: Who is to say what's right and wrong? Okay, so there is a university rule against cheating. But is that rule a good one? Was it meant to apply to my particular case? Did the people making the rule know Doug's particular circumstances? And isn't it true that other universities, like other cultures, have different rules about cheating? So maybe it is permissible to cheat—in a global, moral, sense—even if it is against the local, legal, rules. Why should I feel compelled to obey a particular, localized version of rules when other people see things differently?

Emily wonders: What's the difference between this university's customs and morality, anyway? Is there any difference between morality and the law? Or religion? Is there anything objective about morality? It seems so subjective, so dependent on people's emotions and feelings. Morality does not seem like science at all, in which there are right and wrong answers and a method for figuring them out.

As she reflects on these perplexing questions, Emily hears Dr. Wright ask for the papers to be handed in.

Turn now to Appendix A and perform Exercises 1.A and 1.B. Then, return to this chapter and continue reading.

DISCUSSION OF ISSUES

Should I cheat? Emily's question is one all students have faced, and nearly all have recognized that it should be answered negatively. Cheating is wrong for a variety of reasons. It is a case of breaking an implicit promise, a promise we have made to our teachers and peers not to take advantage of each other. It is a case of unjustifiable deception; Emily would be lying to her instructor were she to put Doug's name on the piece of paper. And it is an action with potentially deleterious consequences, because in misleading Dr. Wright, Emily may be developing undesirable character traits that will diminish her reputation in the future.

In this particular instance, Emily is contemplating an action that, harsh as the word seems to us, is *immoral*. She ought to resist it. That said, you might think we are moving too fast. We can imagine cases in which the immorality of Emily's act might not be so obvious. What if Emily were Dr. Wright's graduate assistant and the two of them were conducting an experiment to see whether anyone sitting around Emily would turn her in? Or, less plausibly, what if Doug were dying and his last wish was for Emily to turn in her paper as his? You may be able to think of other scenarios in which we would not automatically judge that Emily would be doing something immoral. If we hesitate to embrace the judgment that it is morally wrong for Emily to write someone else's name on work she completes, the reason may be that we fear we do not yet have all of the morally relevant facts. There might well be extenuating circumstances inclining us to approve of Emily's "cheating."

One way to navigate these potentially murky waters is to keep separate cases separate. Try the following exercise. Describe a different set of circumstances in which Emily's "cheating" would not be cheating at all. Let your imagination run here. For example, you might suppose that Emily attends a European university at which professors do not give grades to students, and for a professor even to try to give students a grade on the first day would be unimaginable, an offense against good judgment and convention. Or, you might imagine that Emily comes from a culture in which, no matter which university a student attends, students are *expected* to put others' names on papers under these conditions. Or, you might suppose that the world is very different from the one we live in and an evil god has arranged things as follows. If Emily does not put Doug's name on the paper, the entire city of Wheaton, Illinois will blow up. Emily, knowing about the arrangement, regretfully decides to cheat rather than be responsible for the destruction of an entire town.

Let us call the entire class of imaginative cheating cases "*other* cheating cases." Call Emily's case, and every case like it, "*ordinary* cheating cases." In general, as *other* cheating cases show, we should not rush to moral judgments because we may be in the dark about important facts. That is, we may not be dealing with an ordinary case. Therefore, we must always strive to collect *all* of the information relevant to a case before deciding that some action is immoral. Caution clearly is a virtue when doing ethics.

For present purposes, assume that no hidden agenda or unusual context exists here. Assume that Emily's case is simple, straightforward, and mundane. It is a situation that thousands of students face every day. And thousands of students know, as Emily does, what the right thing is to do. We do not cheat. (Or, we cheat knowing that what we are doing is wrong.) We are making moral decisions and, insofar as we successfully resist temptation, our decisions are *correct* decisions.

The problem is that when people mention *ethics*, they are not ordinarily referring to easy questions like Emily's. They are thinking of the tough questions. Should we eat veal from calves kept in anemic conditions in small confinement crates? Should we allow doctors to help people in untreatable pain end their lives? Should we engage in premarital sex when our religious faith proscribes it? Should we allow the Boy Scouts to bar gays from leadership roles? Should we allow genetic counselors to recommend abortion to patients carrying fetuses with very severe genetic abnormalities? When someone says, "Now there's an ethical issue," the statement almost always means, "Now there's a controversial case."

It is important, however, that we not let contentious issues (about which we disagree) blind us to the broad range of uncontroversial moral judgments (about which we agree). The fact seems to be that regardless of the culture, religion, or time period, people everywhere converge on a vast number of particular values, and honesty is one of them. Let us therefore try to formulate and make more precise some of the key reasons that cheating is wrong.

First, it fails to show respect. When we enroll in a class, we enter an implicit social contract, an unwritten agreement that we will do our own work, assume responsibility for our own grades, and not appropriate the work of others as our own. Not to fulfill these implied promises is to disrespect others in the class.

Second, it leads to bad consequences. It cheapens the value of the grade one earns in the class; it poisons the classroom atmosphere, causing people to guard their work and mistrust others; and it increases the likelihood that the cheater will break promises and plagiarize the work of others in the future. If cheating were to become widespread at an institution, the value of the degrees granted by that institution would also be diminished. Cheating has many negative and few positive effects.

Third, cheating undermines a sense of community. Universities are, on the whole, civil places where diverse people pursue goals of a better life while being exposed to ideas and traditions unlike those with which they were raised. If Emily could guarantee that no one would ever discover her deception, she might not be guilty of threatening this spirit of cooperation and working together to mutual advantage. But Emily cannot make that guarantee; she cannot ensure that others will never find her out. And if they find her out, they may cheat. If they cheat, others may cheat as well. Eventually, the spirit of trust and collegiality essential to university life will be badly frayed.

To reason about the intrinsic nature of the act of cheating and its potential consequences is to do ethics. Here are some frequently asked questions about ethics, and some very brief answers.

Frequently Asked Questions about Ethics

Q: What is ethics?

A: Ethics is a branch of philosophy. Philosophy has other branches, which include the following: Logic (the study of principles of good reasoning); epistemology (the study of how and what we know); metaphysics (the study of reality, e.g., minds, brains, souls); philosophy of religion (the study of supernatural beings). The branch of philosophy called ethics involves the study of arguments and theories about what actions are right (or wrong) and which states of affairs are good (or bad).

Q: What are the major theories within the study of ethics?

A: Utilitarianism, Deontology, Virtue Ethics, Feminism, Ethical Relativism, Natural Law, and Divine Command Theory

Q: Who are the major figures known for their contributions to ethics?

A: Deontology: Immanuel Kant (German, d. 1804); Utilitarianism: Jeremy Bentham (d. 1832) and John Stuart Mill (d. 1873), both English; Virtue ethics: Alasdair MacIntyre, Carol Gilligan (Americans); Relativism: Gilbert Harman (American); Divine Command Theory: Karl Barth (Swiss, d. 1968)

Q: What's the study of ethics good for?

A: Answering questions about what's morally right, wrong, good, and bad.

Q: Does it hold any answers?

A: Yes! However, we don't yet know all the answers (or questions).

Q: Where does the study of ethics fit within history?

A: The religious traditions of various cultures have historically been the primary teachers of virtue and morality. Major religious figures, therefore, are important, including Moses, Jesus, Muhammad, Confucius, and Buddha. In the modern period and especially since the Enlightenment, the secular study of ethics has gained prominence, and ethics in the contemporary university often proceeds with little reference to theological claims.

Two theories are widely discussed in contemporary secular ethics: utilitarianism and rights-based theories. Utilitarianism is a form of consequentialist reasoning in which an action is judged to be right if it leads to the greatest balance of good consequences over bad consequences. Rights-based theories, on the other hand, judge actions to be right if they respect persons, regardless of the consequences.

More recent developments include a movement to downplay the importance of ethical theory (*anti-theory*), and one to merge feminist and ecological thought (*ecofeminism*).

When we "do ethics," we are trying to reach a conclusion about how we ought to act by examining the reasons for and against each of our options. We think about as many of the morally relevant features of the act as possible and then figure out which option has the strongest set of reasons to support it. As we are conducting this procedure, we try to adopt what is called "the moral point of view."

Philosophers are divided about what exactly are the constituents of the moral point of view. But in general they agree that, unusual circumstances aside, we should try to reason impartially, without undue bias or prejudice. We should try to put ourselves in the position of each of the parties who will be affected by our eventual decision; then we ask ourselves whether we would be helped or harmed by each decision. Thomas Nagel calls the moral point of view "the view from nowhere"; Henry Sidgwick called it "the point of view of the universe"; and Charles Taliaferro calls it the view taken by an "ideal observer," or God.[1] As mentioned, it is a matter of controversy just what the moral point of view is and, even more so, to which particular moral judgments it leads. But moral reasoning is distinguished by all comers from narrowly self-interested, purely emotional, or money-grubbing forms of reasoning.

In ethics we inquire into a large range of difficult questions, such as: Is it moral to place conditions on food aid to the starving? Should we develop biotechnologies that will displace large numbers of workers? When, if ever, is it permissible to place an embargo on grain exports in order to keep the price of domestic food low? Which uses of animals in research are acceptable and which are not? Applied ethics is what we do when we try to figure out the correct answers.

As previously noted, ethics often is interpreted to mean hotly disputed matters. And sometimes it does mean that. When it does, we must think carefully about our response. But ethics is not always, not only, an attempt to figure out answers to new and puzzling questions. It is sometimes an endeavor in which we simply try to articulate, and remind ourselves of, deeply shared values.

Sophomores often doubt whether there really are any shared values. Are there? To make some progress on this question, complete Exercise 1.C (see Appendix A) called "Shock Treatment for Naïve Relativism."

If we collect all of our claims, we will have begun a fairly substantial list of particular moral judgments on which we agree. Consider that:

It is morally right, all other things being equal, to:

(1) Rescue your two-year-old cousin who is drowning
(2) Feed your sister's cat while she is gone on vacation
(3) Help a blind person who has asked for assistance in crossing a busy intersection
(4) Give your students the grades they deserve on exams

It is morally wrong, all other things being equal, to:

(5) Drown the two-year-old cousin you have been asked to babysit
(6) Poke needles in your sister's cat's eyes to see whether the cat will squirm
(7) Push blind people into busy intersections
(8) Give students grades far below the grades they have dutifully earned

Few would disagree with these claims, unless they were working hard to surround the claims with very unusual circumstances. In that case, they would no longer be thinking of *our* eight claims at all; they would be thinking of different claims, that is, "other" cases.

Okay, someone objects, so there are a few moral judgments held in common. But there are not many judgments of this sort. You can count them on two hands.

How would you reply to this challenge? One way would be to ask the objector to perform the following thought experiment. Leaving everything else the same in proposition (1), substitute various ages for the two-year-old cousin. Won't we all agree that it is morally right to rescue a drowning cousin irrespective of their age, irrespective of whether the cousin is one year old, two and a half years old, three years old, thirty years old, or sixty years old? By simply substituting various ages, we can generate dozens of new particular moral judgments on which we agree.

Still confining our attention to proposition (1), now substitute various relations for cousin. Won't we agree that it doesn't matter whether the person drowning is our cousin or sister, brother or mother, father, grandparent, friend, or teacher? Perhaps we might agree that we should rescue the person even in the event that she is a total stranger, but I hesitate to suggest this possibility (because it would probably generate debate). Depending on how active our imaginations are, we can quickly generate hundreds of noncontroversial moral judgments.

Imagine, further, doing similar thought experiments with (2) through (8). Substitute various animals for the cat, various physical challenges for the blind person, various social relations for the teacher-student relation. In a matter of minutes we will have thousands of particular moral judgments on which we all agree, using nothing more than the eight judgments I suggested off the top of my head.

It is important to begin ethics with a robust sense of our common moral judgments. If we gathered answers to the shock treatment exercise from everyone and then expanded them in the way just suggested, chances are that the class could easily fill up an entire wall by writing on it "things it is always absolutely, positively ethically wrong to do," and another wall with "things it is always absolutely, positively ethically right to do." The walls of our classrooms, appropriately filled with noncontroversial dos and don'ts, would provide all the evidence we need for the following claim: We share a vast number of uncontroversial, particular moral judgments about right and wrong.

We can now offer a first, provisional definition. *Ethics* is the branch of philosophy that studies morality. Ethics has two tasks. One task is to try to provide reasoned answers to difficult moral dilemmas. We do this, in part, by trying to form an ethical theory, a clear, noncontradictory, comprehensive, and generalizable set of rules intended to govern all human behavior and resolve conflicts among values. We then apply that theory to the question at hand. A second task is equally important, however. In ethical pedagogy, we teach our children, and remind ourselves, of the particular moral judgments we hold in common. And we encourage others to try to form their lives by these judgments, a task easier said than done.

If ethics is the study of morality, what is morality? Sometimes it seems to be indistinguishable from a society's customs. Sometimes it seems to be the same thing as the law. Many people think that it derives from religion. And many think that *ethics*, the branch of philosophy that studies the justification of morality, is not at all objective, not at all like science. Here I take up the relation of morality to custom and then to law. I wait until the next chapter to investigate the relation of morality to religion.

Are *Morality* and *Custom* the Same Thing?

When we use the word *morality*, we mean different things by it. Sometimes we mean the various sets of rules that implicitly guide the conduct of some group or other.[2] With this interpretation, a variety of "moralities" exist because different groups have different sets of rules. Think about the different attitudes toward animals found in various societies. Jewish cultures traditionally encourage animal husbandry: the selective breeding, raising, and slaughtering of animals. Jews see this practice as justified by a divine mandate that instructs humans to try to perfect the world by cooperating in God's creative activity. Although Judaism permits the killing and eating of some animal species, however, it prohibits the killing and eating of others, such as mollusks. And it prohibits the consumption of blood.

On the other hand, Jain cultures in India have markedly different attitudes toward animals. Whereas Jains agree with Jews that killing mollusks and consuming blood are wrong, Jains further believe that killing any animal whatsoever is wrong. By extension, they do not use traditional methods to selectively breed cattle. Holding that all life is sacred, Jains think that animals should be left alone, that we should do no harm to animals. In their view, animals and humans are linked through the cycle of karma and reincarnation, so that all living things are interdependent parts of one another.

Animals, therefore, are entitled to live out their normal life spans without being exploited by humans.

Jews have one "morality," Jains have another "morality," and Christians have a third. Christians historically have encouraged the breeding, raising, and slaughtering of food animals and the eating of shellfish, and have not felt bound by the Hebrew proscription that outlaws the consumption of animal blood. Christians believe that God revealed the goodness of all life and its fitness to be eaten in a vision given to St. Peter recorded in the Book of Acts. This is the most permissive dietary morality of the three religions just introduced.

If ethics meant nothing other than the study of these different customs, the ultimate goal of the study of ethics would be an empirical, descriptive project: to survey and articulate the various moral codes of the world. Such a survey, however, would reveal a striking fact: the moralities conflict. Most Christians think it permissible to kill and eat animals; Jains do not think it permissible. Shouldn't ethics help us answer the tougher question, Who is right?

Indeed it should, and ethicists (people who do ethics) give different answers to that question. Those we might call naïve cultural relativists answer that both the Jains and Christians are right. The moral code of the Jains is true for Jains, and the moral code of the Christians is true for Christians. Such relativists grant that the codes appear to conflict and that the conflict would be a problem. However, the conflict is, for the relativist, merely an appearance and, therefore, not a problem at all. How so? Because moral codes do not make truth claims. Relativists, intent on honoring differences among the world's cultural and religious traditions and fearful of dominant societies aggressively imposing their values on others, resist cultural, political, and moral imperialism. They conclude that moral rules are simply expressions of people's feelings. What's morally right for George is morally right for George. But the same thing may be morally wrong for Jorge. For relativists, *there is no absolute universal moral code, no truth in ethics*. Therefore, one society's moral codes can't conflict with another's.

A relativist might respond to Emily's case in the same way, saying that although it may be wrong for Emily to cheat, it is not necessarily wrong for someone else to cheat. Emily has her morality, you have your morality, and I have my morality. Don't try to impose your values on me and I won't try to force mine on you.

These attitudes sound admirable on the surface because they recommend tolerance and acceptance. But appearances can be misleading. In fact, naïve cultural relativism is indefensible. Notice, first, that the relativist will have trouble securing any kind of respect for tolerance, period. To argue for relativism on the basis that it is more respectful of differences among cultures is already to assume that respect is a universal moral good. But relativism insists that no such nonrelativized moral goods exist, revealing a deep and objectionable internal inconsistency. How can someone be proud of his or her view because it is so tolerant when tolerance itself is not tolerated by some societies?

We can go on to ask whether relativism itself really is so tolerant. Just how tolerant is it to insist that all those people who think that one single objective morality applies universally are wrong? Is a position truly tolerant if it has no room for objectivists and absolutists? Relativism seems inconsistent here a second time because it cannot make good on its desire fully to honor every culture's morality.

Some cultures hold that a single universal moral code exists: Jews, Jains, and Christians all believe this. If, however, Jain morality holds that there is truth in ethics, and that there are, for example, right and wrong ways to treat animals, then it follows that part of Jain morality contradicts part of Christian morality. Jains do not believe that animal life is intrinsically valuable only to Jains; they believe that animal life is intrinsically valuable to everyone. Were naïve relativism true, then Jains would be mistaken in this belief, because relativism holds that Christian morality is true for Christians, yet Christian morality denies the intrinsic value of animal life. (Or, to be more precise, most Christian ethicists in the past denied the intrinsic value of animal life; there are a few, more recent, Christian theologians who defend, for example, vegetarianism.)

If Jains and Christians disagree about this matter, then one of three things must be true. Either relativism must be false because it denies the possibility of disagreement, or the Jains must be confused about Jain values, or the Christians must be confused about Christian values. The latter two options seem difficult to accept, however. Who knows more about Jain values, Jains or naïve relativists? And who knows more about Christian values, Christians or naïve relativists? The best answer, therefore, must be the first one: Naïve relativism is false.

To summarize. Christians think they are right and Jains wrong about killing and eating animals. Jains, in turn, think they are right to place a high value on animal life and Christians are wrong not to. If relativism were true, this situation would be a mirage because there could be no grounds on which Jains and Christians could disagree. Both would have to be right. Yet, both *cannot* be right since they disagree with each other. So, the relativist's interpretation of the disagreement between Jain and Christian morality fails to account for the explanations given by Jains and Christians of their own morality. In short, relativists cannot honor the morality of Jain and Christian cultures because these cultures are absolutist and anti-relativist. The relativist can hardly say, "Yes, there is merit in these cultures' absolutist views, and I accept them both, as I accept all views." The reason is that in agreeing with any one absolutist morality, the relativist is either renouncing the basic relativist principle (there is no truth in ethics) or imposing a relativistic interpretation on a culture that emphatically denies being relativistic. So much the worse for cultural relativism, which cannot make good on its promise fully to honor every culture.

It is tempting to adopt a position of naïve relativism in part because the study of ethics is hard work. How easy is it seriously to engage foreign cultures, examining their arguments and traditions, comparing their theories with theories from other cultures, all the while subjecting every value one finds to critical scrutiny? It is a challenging task, to say the least, to set out to find the correct view. In the face of all this complexity, history, and conflict, articulating a generalizable and impartial set of rules is no easy matter. True, the very thought that rules might exist that apply to anyone at any time in any place is an idea fraught with the danger of imperialism and colonialism. We do not want to force our values down others' throats. And how do we know which of two conflicting moral positions is the right one? These are critical questions. Yet the ethical journey is one we must take, despite its many pitfalls. Committing to set out upon the ethical journey represents our best hope for resolving cultural conflicts in a peaceful, just, and impartial manner.

Here then is a second, more precise, definition of ethics. *Ethics is the intellectual attempt to decide which action one has the best moral reasons to undertake, irrespective of one's inherited traditions.* On this definition, a difference exists between the "moralities" we have been discussing—moralities that reflect the customs of various societies—and what we might call (admittedly grandiosely) "true" morality, which is universal, impartial, and applies to everyone. On this definition, "true" morality does not instruct George that it is permissible to kill a cow in circumstances q, r, and s while telling Jorge that it is impermissible to kill the same cow in the same set of circumstances.

Distinguishing between moralities, which can vary, and true morality, which cannot vary, provides us with one conceptual tool with which to try to explain what Jains are doing when they criticize Christians for undervaluing animal life. Jains are disagreeing *ethically* with Christians; they are claiming that Christians have the best reasons to undertake actions that highly value animal life *even though* this attitude is not part of the morality the Christians inherited.

Distinguishing between "moralities" and "true morality" helps to bring into focus the difference between ethics and custom. We sometimes perform actions out of habit that are not ethically justifiable. Some taxi drivers customarily give blank receipts to their fares on the understanding that the person will inflate the price paid, receive a higher amount in reimbursement from his or her company, and pass a bit along to the taxi driver. The mere fact that drivers and customers act this way does not make it right for them to do so, just as the fact that some people beat their horses does not make it right to torture animals. Then again, separating moralities from ethics (true morality) allows us to observe that some actions that are not customary are not necessarily unethical. We do not usually explain the sordid details of a recent divorce proceeding to strangers who casually ask us how we are doing, but it is not immoral to do so. There appears to be a real difference between customs, on the one hand, and ethically justifiable customs on the other. True morality and custom, therefore, are not the same thing.

If real conflicts exist among the world's moralities, they cannot be settled by turning to one tradition's inherited customs. How then can they be settled? Perhaps by turning to the law.

ARE MORALITY AND THE LAW THE SAME THING?

Some hold that a society's morality is reducible to whatever laws a society adopts for itself. Law is indeed not only a body of rules governing how people ought to behave but also a tutor, helping to instruct and encourage good behavior. But here again, a problem arises in conflating morality with a close neighbor. Societies can adopt laws that are clearly unethical (for example, requiring African-Americans in Alabama to sit in the back of a bus), and societies can fail to classify as illegal actions that are clearly immoral (for example, allowing cruel psychological abuse of a child or spouse). Therefore, some things are legal but clearly unethical, and some things are illegal but not necessarily immoral. Although communities should strive to form laws in accordance with ethical standards, we should not automatically assume that the one can be collapsed into the other. As with ethics and customs, a gap exists between ethics and the law. Morality and the law, therefore, are not the same thing.

If we do not find out what ethics requires by consulting customs or laws, how do we find out? In the United States, a very common response is: God's will. Let us now turn our attention to that answer.

NOTES

1. Thomas Nagel, *The View from Nowhere* (New York: Oxford University Press, 1986); Henry Sidgwick, *The Methods of Ethics*, 7th ed. (New York: Macmillan, 1907); Charles Taliaferro, *Contemporary Philosophy of Religion* (London: Blackwell, 1997).
2. A very useful introduction to these matters is James Rachels, *The Elements of Moral Philosophy* 2nd ed. (NY: McGraw-Hill, 1993). I also must acknowledge the patient advice and instruction of Margaret Holmgren, whom I am lucky to count as a colleague at Iowa State University.

Chapter 2

RELIGION

Gary L. Comstock

CASE STUDY: RICH THE ATHEIST

Students in Emily's Ag Ethics course are still discussing cheating during a class several days later. Emily hesitates to get drawn in, but eventually cannot stop herself from raising her hand to say that one of the reasons that she decided not to cheat is that she is a Christian. Honesty, honor, love, and respect are central virtues of the Christian faith, she explains, and cheating seems distinctly un-Christian to her.

Rich, who sits in the front row and has already distinguished himself as an active participant in discussions, loses no time.

"Dr. Wright, I mean no disrespect to those with religious beliefs, but we aren't going to get involved in this class with questions about what the Bible says, and what God wants, or what the Pope thinks, are we?"

"Well," the professor replies, "you raise some good questions. But why do you ask?"

"Because I don't think religious discussions ever get anywhere when it comes to talking about morality. First, not everyone in the discussion believes in God, so why should atheists be forced to adhere to standards that they don't agree with? Second, even those people who do believe in God don't agree about morality. Liberal Protestants say abortion is okay under virtually any circumstances; traditional Catholics say it isn't permissible under any conditions; and you have the full spectrum of views in between. Third, how can anyone know what God commands people to do? The Bible is full of contradictions, isn't it?"

Emily squirms a bit in her seat.

"Wait a minute," she replies. "There are methods in my religious community for determining better and worse interpretations of Scripture; there isn't that much disagreement among Christians on abortion—well, at least not in *my* church; and everyone, sometime, has to adhere to standards he or she doesn't agree with. So I don't see why religious arguments should automatically be excluded from the conversation. Religious traditions are important in teaching values, and they can help us to form our children and our communities in the right way."

Dr. Wright responds by saying that the class will not be able to spend a great deal of time on the subject of religion, but it must consider one ethical theory that depends heavily on religious belief. That theory is the Divine Command Theory, in which moral standards are thought to be necessarily related to God's will.

"But isn't the Divine Command Theory simply false?" Rich persists.

"Let's withhold judgment on the matter until we have at least had time to get the theory out on the table," says Dr. Wright. He looks around the class. "Any other questions?"

The classroom is very quiet. No hands are raised, and everyone seems to be avoiding eye contact with everyone else. As the bell rings, Emily rises from her seat feeling alone. Outside the building another student, Dawn, approaches her.

"Hey, I just wanted you to know that I'm a Christian, too," says Dawn. "And I support you a hundred percent. But we have a problem; we don't know how to talk about our religious convictions in this class. It doesn't seem that the instructor, or this campus, is very open to honest discussion of beliefs in anything supernatural."

"Oh, thanks so much for telling me," says Emily.

"If it's any comfort, I would guess that the majority of the students in the class feel the same way that you and I do. We're all kinda religious, but we're also kinda intimidated by the secular atmosphere of the university. We want to learn how to talk about our religious beliefs—we want to learn what we believe!—but it's pretty clear that our instructors are not very friendly to belief."

See Exercise 2.A in Appendix A for discussion questions to consider.

DISCUSSION OF ISSUES

Rich's comments direct attention to the source of ethical values. In the previous chapter we observed that ethics is not derived from custom or law. From whence does it come? Perhaps the right answer is Emily's answer: God's will. Because religion is so powerful and its relationship to ethics so complex, the relation between the two subjects deserves thorough investigation.[1]

ARE MORALITY AND RELIGION THE SAME THING?

Some hold that moral systems may be reduced to the values of a society's religion. This is an important point because those of us in the United States live in a very religious culture. Harris polls show that more than 75 percent of all U. S. college students believe in God.[2] In 1994, more than eighty percent of Americans said that they believed that Jesus was resurrected from the dead and that Heaven exists.[3] The United States has some 900,000 religious fellowship groups; on average, that amounts to twenty thousand religious groups in each of the fifty states.[4]

Where we find religion we typically find instruction in morality. Although the aberrant, hateful religious organizations are the ones that make headlines, the truth about religion is more mundane and hopeful. It is a rare religious community that does not teach honesty, integrity, love, reciprocity, caring for others, and civility. According to Nancy Rosenblum, the influence of religion permeates our entire culture, creating the general "expectation that our pain and indignation at day-to-day unfairness and abuse will not be met with indifference, and thus [religious belief] may cultivate the iota of trust necessary for democratic citizens to speak out about ordinary injustice."[5]

Historically, the ethical values of cultures have resided within religious traditions.[6] The faith traditions have been the primary incubators and champions of virtue and character, whether you think of Jews, Muslims, the Nuer in Africa, the ancient Greeks and Romans, or the Lakota Sioux. In these traditions, rules about permissible and impermissible behaviors are closely aligned with religious beliefs. Morality is intimately tied up with religious beliefs about the power of deceased kin, the whims of capricious gods and goddesses, the will of a single omnipotent deity, or the power of the karma of one's past volitions.

Because religion both teaches moral rules and provides motivation for adhering to them, it cannot help but be a close neighbor to ethics. So close a neighbor that we sometimes fail to distinguish between them. As James Rachels points out, when New York Governor Mario Cuomo appointed a special panel to advise him on medical ethics, he did not select professors of ethical theory or trained applied ethicists.[7] He chose Christian clergy and a Jewish rabbi. We commonly think of spiritual people as moral experts, and we commonly resort to our religious traditions when trying to decide about contentious moral issues.

Religion is not only a close neighbor but also a powerful one. The price of sin and moral transgression is not only the sanction of God but also the disapproval of one's religious community. The power of religion and its proximity to ethics is especially critical today, when most Americans are concerned that the nation is going down the tubes morally. In a 1996 poll, more than 85 percent of Americans believed that "something is fundamentally wrong with America's moral condition," citing as proof the prevalence of "teen-age pregnancy, unwed childbearing, extramarital affairs, easy sex as a normal part of life."[8] (It is worth noting that Americans, ironically, do not seem to think that racism, sexism, speciesism, environmental degradation, and the growing income gap between rich and poor are further evidence of this moral decay. Indeed, one might interpret the following fact as underscoring the possibility that the typical American's worries about "moral decay" are *not* connected to issues of race, equality, and distributive justice: Twice as many Americans believe that "'lack of morality' is a greater problem in the United States than 'lack of economic opportunity.'"[9])

Americans appear to be very interested in spirituality, and concerned with the moral state of their country. Curiously, however, we seem not to be particularly skilled at analyzing our problems in religious language. Consider the behavior of various U.S. leaders. Almost every recent president—George W. Bush, Clinton, Bush Sr., Reagan, Carter—all claimed to be devout Christians and most of them went regularly to church. Each one consulted with the evangelist Billy Graham. But, in public, the most sophisticated theological pronouncements they seem capable of making is the puzzling phrase they repeat over and over: "God bless America." A masterful expression, but one not particularly well suited to subtle theological analysis of complex public policy.

Because morality and religion are proximate, powerful neighbors, those of us who are religious as well as those of us who are not need to think carefully about their relation. I begin with a definition of religion.

Defining Religion

It helps to have some paradigmatic cases before us when we try to define a term. Representative religions include Judaism, Christianity, Islam, Hinduism, Buddhism, Taoism, Confucianism, Sikhism, the Ojibwa and Sioux of North America, and the Yoruba and Ibo of Nigeria. Religions are complex and consist of many different components. They contain narratives, such as the Yoruba creation story, the synoptic narratives of Jesus's suffering, death, and resurrection in the New Testament, and the autobiographies of individual believers. They feature rituals, such as the Christian Eucharist, baptism, and last rites, the Jewish bar mitzvah, and the Lakota Sun Dance. They include institutions, such as the universal Roman Catholic Church, the local Foursquare Gospel prayer meeting, a neighborhood ladies missionary circle, and Jewish synagogues. And there are beliefs, I argue, about the supernatural, immaterial places, states, or beings whose effects, powers or actions are not explicable in terms of material causes and effects. The supernatural is anything to which people refer when they use other-worldly terms such as God, Krishna, Yahweh, Allah, Creator, karma, ancestral spirits, the All, the One, the Divine, miracles, heaven, hell, nirvana, damnation, salvation.[10] I summarize this discussion by offering a definition.

> Religion is that complex dimension of human activity involving beliefs about the supernatural, beliefs that are expressed in propositions and narratives and enacted in rituals and institutions. These beliefs authorize the group's moral code and answer the question, What is the best way of life overall?

Note that this is a substantive rather than a functional definition of religion. It is a substantive definition because it insists that a religion must contain beliefs about the supernatural. Social theorists such as Emil Durkheim and Clifford Geertz proceed differently, using a functional definition. They note that social order is required in order for any people to live together, and they call whatever glue that ultimately binds a group together that group's religion. Functional definitions therefore don't require a religion to include supernatural beliefs. A religion is anything that *functions* in a certain way to bind a culture together. For a functionalist, Confucianism in China counts as a religion, even though Confucius himself did not believe in supernatural phenomena and explicitly denied the reality of ancestral spirits. For a functionalist, certain atheistic forms of Buddhism in China and India count as religions, as do communism and secular humanism in the West.

But we may ask: *Should* these traditions, which deny the existence of the supernatural, count as religions? Are they not instead cultural traditions? Perhaps we should reserve the term *religion* for those forms of Confucianism, Hinduism, Buddhism, Christianity, and Judaism that contain not only a cultural binding force but also a belief in the extraworldly. I have argued elsewhere that functional definitions of religion are not particularly helpful because they exclude nothing.[11]

A substantive definition of religion, by contrast, provides a good tool to think through the relationship of religion and ethics. Every religion has certain moral rules, such as "Treat others in the way you would like to be treated," and "Do no harm to any living creature." These rules are sometimes implicit and unarticulated, but they are sometimes explicit, worked out in treatises such as the Catholic Church's encyclical "Culture of Death," the Pope's attack on the permissibility of abortion, capital punishment, and euthanasia in modern Western culture.

Clearly, moral rules and ideals are found in religious traditions. But if we assume that not every tradition or person is necessarily religious, then moral rules and ideals can exist apart from religion as well. Many people do not qualify as adherents of religion, and yet they have moral principles and lead lives of moral integrity. I think of atheist colleagues I admire who teach philosophy or religious studies, of the members of the society of secular humanists, of the liberal Jews and Protestants who do not believe in a transcendent being and yet live lives of courage, decency, tolerance, and love. It appears impossible to insist that true morality, thought of as good behavior, is the exclusive property of religious people.

Assuming that religion refers to human activities involving beliefs about the supernatural and that people can be virtuous even if they do not believe in the supernatural, then morality can be independent of religion. To help us keep this fact in mind, I use the phrase *rational morality* for the next few pages to refer to any institution of morality that exists separately from religion. I use the phrase *rational applied ethics* to refer to all non-theologically based attempts to develop general public policies, that is, public policies meant to apply to everyone, whatever their religious tradition. When we do ethics with the intent of influencing public policy, one of our most important jobs is to study arguments: premises, conclusions, and the validity of moving from premises *a, b,* and *c* to conclusion *d.* You will learn how to evaluate moral arguments in the next chapter.

As noted in Chapter 1, applied ethics has two tasks. One is to try to answer difficult moral dilemmas. The other is to remind ourselves of the astonishing number of particular moral judgments we hold in common. Religions typically help to teach these common values by offering their members moral instruction. Now, some religions teach values not found in rational applied ethics. For example, Jain morality teaches that one should not kill insects, and Christian morality teaches that one should love one's enemy. It is difficult to find justification for these judgments on rational grounds. But these values are the exception rather than the rule. More commonly, the world's religions teach their youngsters what I have called the moral truisms, the lists of rights and wrongs we have previously generated in our thought experiments: Do good, avoid evil, seek justice, honor your mother and father, help the needy.

Religion, in sum, is one vehicle through which children learn right and wrong. To put it another, perhaps more controversial, way: Religion teaches rational morality. But, of course, religion is not necessary in order to teach moral truisms or to explore ethics. Consider one anecdotal piece of evidence for this claim. Religion plays at best a marginal role in ethics courses offered at U.S. state universities and virtually no role at all in ethics discussions in Europe. Typically, philosophy instructors spend at most one or two days on the Divine Command Theory (discussed later), and that is the extent of the treatment of religious approaches to ethics. Moreover, philosophy instructors typically conclude discussion of the Divine Command Theory with the claim that the theory is false. Indeed, it is not unusual for ethics professors to issue explicit disclaimers that appeals to religion will not be allowed to settle matters in the class. As a result, religion appears very little, either in classroom discussions or in the papers submitted by students. In my experience, nuanced and careful talk about religion is about as prevalent in university ethics courses as it is in public political discussion in France and Sweden, where it is virtually nonexistent. So, ethics is being taught without religion.

A religious person might think this an objectionable state of affairs. But is it? Consider three points.

First, it may be that at least some basic moral values can be justified rationally, without drawing on religious premises. This discussion explores this point in more detail soon with the Divine Command Theory.

Second, religious people have several basic values, often including religious freedom: the right of each individual to behave and believe religiously in the way dictated by his or her conscience. The beliefs and rituals of one religion should not be imposed on those who do not share those beliefs, and no one should be forced to worship one way or another. In a democratic setting that contains a plurality of religions, all people, and especially the very devout, have good reasons not to impose their beliefs on others.[12]

Third, we can reason *impartially* about our values, developing policies that apply not only to the members of our own religious community but also across the board. University classrooms often include students who do not accept the beliefs of any religious community. Which moral principles will they consent to? To answer this question is to begin to reason impartially.[13]

One feature of morality distinguishes it from economic calculations of costs and benefits, from prudential calculations of what will best serve one's own interests, and even from religious considerations about revealed truths. Morality has the quality of *overriding* these other considerations. The overridingness of morality is the feature of ethics that insists that the right thing to do is not determined by the polls, our preferences, economic utility, or the results of democratic votes. The right thing to do is determined by the actions we have the best reasons to perform. Whatever is the ethical thing to do is the thing we *ought* to do; the right thing trumps all other choices.

Even religious choices? We should do the right thing rather than what God tells us to do? This is a sensitive and controversial issue because God apparently sometimes has told individuals (Abraham) to do the wrong thing (kill his innocent son Isaac). But such instances are extraordinarily rare. In the ordinary case, and in the public secular arena, we place higher value on the dictates of morality than we place on the freedom of religious thought. Consider one example. Rational morality tells us that difficult cases in which young children with treatable leukemia whose parents refuse medical treatment for them on religious grounds should be settled in favor of saving the child's life rather than sacrificing the child to respect the parent's religious beliefs. When it comes to life and death issues, courts in Western culture insist on doing the right thing. When in such cases the dictates of rational applied ethics override fundamental spiritual convictions, we see—for better or worse—that religion is marginalized in secular courts.

The marginal character of religion is underscored when people review the particular moral codes specific to their professions. A *professional ethical code* is a summary of the rules regarding what is considered to be right and wrong in a profession, such as the National Cattlemen's Association's code of ethics and the Veterinarian's Oath. Such codes typically articulate noncontroversial and widely held beliefs about the responsibilities that attach to one's role. Veterinary scientific and cattle associations all disavow dishonesty, fraud, and disrespect for the law. All commend the use of professional skills for the benefit of society. Religious leaders make up a profession, and there are ethical standards that apply to them. In the Evangelical Covenant denomination, for example, male pastors are strongly discouraged from meeting alone in counseling sessions with women parishioners.

How is religion related to professional ethical codes? To my knowledge, and apart from the codes of the clergy, no twentieth century professional ethical code makes reference to a supernatural power. The Hippocratic Oath (c. 370 BCE) invokes Apollo and Panacea and "all the gods and goddesses" as witnesses, but the American Medical Association's code has discreetly dropped such references. The Boy Scout's oath refers to God, but Boy Scouts are not professionals. Religion, then, is nearly nonexistent when it comes to the official ethical statements of today's professional associations. Of course, religious beliefs and traditions may be the basis of moral thinking for many individuals within the professions, even though these beliefs and traditions do not appear in their official codes.

I have noted that morality seemingly can be taught without religion. But is this correct? Does morality not need religion in order to be justified? To answer this question we must do some work in *ethical theory*, the philosophical study of what makes things good or bad and actions right or wrong. Theorists inquire into questions such as: What is the standard for judging things to be moral? Is it God's will? Individual rights? Pursuit of the greatest good? What is the relationship between moral and nonmoral explanations? Can moral language be reduced to naturalistic language? How should ethical theories be constructed and justified? On certain rational or religious foundations? Or by a process of comparative reasoning that considers our intuitions, scientific knowledge, and moral principles?

How is religion related to ethical theory? Two possible answers exist: necessarily and not necessarily.

Necessarily

The idea here is that moral laws logically must derive from divine commands. This idea is found in the Divine Command Theory, which holds that *an action is right if and only if God commands it*. A classic exposition of this theory is given by C. F. H. Henry, who writes that Biblical ethics *discredits* rational morality. Biblical ethics is superior because it

> . . . gives theonomous ethics its classic form—the identification of the moral law with the Divine will. In Hebrew-Christian revelation, distinctions in ethics reduce to what is good or what is pleasing, and to what is wicked or displeasing to the Creator God alone. . . . The good is what the Creator-Lord does and commands. He is the creator of the moral law, and defines its very nature.[14]

The virtue of this theory is that it renders morality objective, absolutist, and enforceable. Ethics is not a matter of etiquette, feelings, evolutionary adaptation, or do-what-you-will. Things are not right or wrong based on what you happen to think about them; they are *objectively* right or wrong, and there are moral facts about whether it is right to rape and steal. A standard exists by which we can tell what is good and bad. The Ten Commandments, for example, is one statement of the standard. Notice, too, that this theory carries with it a police force and judge as well as sanctions for disobedience. We ought to be moral on pain of punishment on Judgment Day. The theory also has the theological virtue of respecting God's omnipotence and sovereignty. God is the creator of rational morality, and God's actions are not constrained by a law higher than God. The slogan here might be that no ethical theory exists without religion.

Two of the most prominent German theologians of the twentieth century, Karl Barth and Emil Brunner, both argued for this theory. It has at least three interpretations:

1. "Morally right" means "commanded by God."
2. No moral reasons exist for acting one way or the other that may be known independently of God's will.
3. Morality logically must originate with God.

Each of these premises has problems.

I begin with the first interpretation. Whenever anyone says that "x is morally right," what the person really means is that "x is commanded by God." But it does not seem correct to say that this is what people mean who do not believe in God. If proposed as an explanation of what people everywhere mean when they use moral terms, then the Divine Command Theory seems obviously false. Now, someone could argue that we should just stipulate that this is what morally right means, and that whenever we use the term this is what we mean. But this strategy would beg the question, rendering our inquiry pointless. Why try to find out *whether* rational morality requires religion if we are simply going to assert from the very start that it does? This move certainly will not settle the question of whether morality requires religious justification. So the first interpretation is defective.

Now consider the second interpretation of the Divine Command Theory. If no moral reasons exist for acting one way or the other that may be known independently of God's will, then the claim, "God is good," becomes meaningless. On the Divine Command Theory, to say that "God is good" is redundant; it is to say the equivalent of "God is God." The reason is that the statement "God only does what is good" comes to mean "God does whatever God wants to do," and the statement "God commands us to do what is good" is reduced to the tautology "God commands us to do what God commands us to do." But when we say, "God is good," we do not generally think that we are uttering an empty tautology; we think instead that we are ascribing a property to God. Furthermore, it seems that even in the absence of divine revelation, people can and do know that it is wrong to poke pins in cats' eyes and right to assist the needy. (The Catholic theologian St. Thomas Aquinas argued as much.) Therefore, the second interpretation seems unsatisfactory.

Finally, regarding the third interpretation of the Divine Command Theory, if morality originates with God, then what is right is reducible to what God says is right. But if whatever God says is right, then moral norms become arbitrary and unreliable. This is the problem we know from the ancient Greek philosopher Plato (d. 347) who, in a dialogue called The Euthyphro, asked whether something is good because God wills it or whether God wills something because it is good.[15] God commands us, for example, not to starve our children to death not because God is capricious and happens to decide at the moment that murdering children is distasteful. Rather, murdering children is wrong, and God, being omniscient, knows that it is wrong. Being omnibenevolent as well, God is good and commands us not to do what is wrong. *God is a good God.* That's an informative sentence, not a tautology. Indeed, we can imagine good gods and bad gods; bad gods are those who command us to do evil. We would not be able to imagine evil gods were it the case that whatever the gods command is necessarily what ought to be done.[16]

To see the concern that philosophers have come to call "the Euthyphro problem," we must use our theological imaginations and be willing to entertain different possibilities in our idea of God. The traditional God of Western religions, of course, is omnipotent, omniscient, and omnibenevolent. But it is not a logical fact that God must have these characteristics, and other cultures have had, and continue to have, very different pictures of the deity. For example, the ancient Greeks believed that before the Olympian gods came to power, the Titans ruled the heavens. What if God were not the loving God of Western religions but rather Cronus, the giant Titan god who castrated his father, married his sister Rhea, and killed and ate his children.' If the universe is ruled by Cronus and if the Divine Command Theory is true, then castrating your father, having sex with your sister, and killing babies are good things. Why? Because whatever God wills is good, and Cronus—who, we are imagining, is God—wills these things. Consequently, having sex with our sisters is not only permissible but also something we ought to do. But that seems wildly counterintuitive and offensive.

Obviously, what is right or good is not necessarily the same as what any particular religion teaches. A religion that taught obedience to Cronus would teach prejudice, rape, discrimination, and murder. This fact would does not make prejudice and rape right.

There is another problem with the third interpretation of the Divine Command Theory. If God can make morally good what seems morally heinous, then the right theory of ethics seems to be that might makes right: whoever is at the moment the most powerful gets to declare what is right. In other words, if morality originates with God and there is no independent standard by which we can judge God to be a good or a bad God, then our moral standards are completely at the mercy of divine whims and we may think that abhorrent actions are good actions.

Consider three defenses of the Divine Command Theory.

First, some writers, such as G. E. M. Anscombe and Fyodor Dostoevsky, believe that people will not behave morally unless they believe that bad behavior will be sanctioned—punished—by a divine lawgiver. With regard to civil laws, people must believe that an authority will punish them if they break the law or else they will not obey it. Without sanctions, laws lack teeth. Indeed, without sanctions, laws may not even count as laws; they may function only as suggestions or requests.

So it is with moral laws. If no divine authority enforces it, agents will not experience the law as binding. Just as civil laws demand police forces and judges, so moral laws demand a divine police force and lawgiver. Kant held that in order for morality to inspire adequate motivation for compliance, a God must exist who enforces the law and who rewards and punishes us in the afterlife. Anscombe, a twentieth century British philosopher, basically argued that rational ethics makes no sense. And in the *Brothers Karamazov*, the Russian novelist Dostoevsky had his character Ivan Karamazov assert that "If God doesn't exist, everything is permissible." If morality has reason alone as a basis, then morality fails to account for the overridingness of moral values, is uninspiring, and fails to tell us why we should be moral.

All the writers just mentioned were theists who sought to underwrite rational morality by giving it a religious foundation. Another philosopher, who held that God is dead, agreed with part of what these theists believed. That philosopher, Nietzsche, thought, however, that rational morality, like God, ought to be dismissed, and he sought to undermine morality, which he viewed as prophylactic principles invited by the huge numbers of society's weakest members to protect themselves from willful and strong individuals. Ironically, atheistic nihilists such as Nietzsche share this belief with Divine Command theorists: that religion is essentially related to ethics. If religion disappears, so does morality.

Problems are identifiable here. Are there really no sanctions other than the deity for our actions? The following, if they exist, might all exercise a powerful influence dissuading us from bad behavior: conscience, moral facts, cultural taboos, and the evolutionary advantageousness of altruistic behavior. In ethical theory, God is not the only possible psychological enforcement mechanism for morality. So it seems that this first line of defense of the Divine Command Theory fails.

A second line of defense argues that rational ethical theory ignores the twin facts of sin and forgiveness. Selfishly egoistic actions and attitudes offend God, but a nontheologically-based ethical theory has nothing to say about those people on whom God has mercy, even though they commit moral transgressions.

Here is a response: In order to believe in sin and divine forgiveness, one must believe in God because sin is not just any moral transgression; it is, rather, a moral transgression against a supernatural power. However, can we believe in *sin* or *divine forgiveness* unless we first believe in the existence of God? It would not seem possible. And yet the point of our inquiry here was to figure out whether ethics needs God in the first place.

So to object that rational morality ignores sin is to beg the question of whether there is a God.

A third line of defense proposed by Robert Merrihew Adams responds to the charge that the Divine Command Theory makes morality arbitrary. Adams argues that the nonarbitrariness of divine commands is insured by God's character. God's character is not that of a mercurial, evil-minded arbitrary being; God is a constant loving Parent who wants the best for us.

My response is that Adams' argument seems only to push the problem back a level. What does it mean to do something that is "loving"? On Adams' Divine Command Theory, it must mean "to do whatever God commands," because no independent standard exists of what is loving or hateful. Therefore, to say that "God commands what is loving" is to say that "God commands what God commands." Are we not stuck in the same quandary noted previously in response to interpretation (3) ? In Adams' account, the problem seemingly has only been transferred from the term "good" to the term "loving."

We might conclude, therefore, that religion is not essentially or necessarily related to ethics. Fortunately, there is another way to construe the relationship.

Not Necessarily

Having considered the ways in which religion might be necessarily related to ethical theory, I turn to the other alternative: not necessarily. The idea here is the following. If divine commands exist, they are always issued in accord with moral laws so that when God commands something, God commands it because it is good. Humans, therefore, can discover what God wills in the moral realm by consulting our conscience, reason, intuitions, and sense of justice. The theory of natural law holds that moral principles are rational and that our faculty of reason is the divine image within us. Morality is given by God but it is discoverable within the bounds of reason alone. Even on this Thomistic view (that is, a view inspired by the medieval Catholic theologian Thomas Aquinas), however, agents can discover what is morally right or wrong without special revelation so that Natural Law theory does not require a divine command giver.

Now, some will object that if moral standards exist that are independent of God's being, then monotheism is compromised because something exists that God did not create. Even worse: If moral standards exist independently of God's will, then God is not the author of morality; something exists that God did not create and God is not free to make God's own laws. Rather, God must obey the laws of morality.

The answer to this worry is that even God seems to be bound by certain laws, such as the laws of logic and morality. God cannot make a married bachelor or a color that is simultaneously red and green. There appear to be some things that God cannot do: God cannot make it the case that God does not exist. God cannot both love us and hate us simultaneously, or call an action that is clearly evil a good action.

To conclude, then, it seems that what is right or good is not necessarily identical to what a particular religion teaches. There is the Cronus problem, that some religions teach prejudice and discrimination, and there is the Euthyphro problem, that God commands something because it is right. To put it another way, morality is independent of God's will. Therefore, we should not conflate the spheres of piety and morality.

Good reasons exist to separate public policy decisions and the revelations of particular faiths, and not only because religious people disagree among themselves about what is right. Countries that try to separate church matters from matters of state attempt to make regulations and laws not on the basis of sacred truths revealed to a few but rather on the basis of broader principles upon which people from diverse religious backgrounds—and no religious background—can agree. Reaching a consensus about moral issues is possible without invoking religious authorities. Consider one example: In the United States, many people once believed that allowing women to vote was morally wrong. Some traditions thought it imperative on biblical and theological grounds to keep women out of the public sphere, whereas other traditions supported the suffrage movement on grounds that were equally theological and biblical. However, after the culture removed the issue from the sphere of religion and looked at the facts about women, it could not justify its view that women should not vote. The general population came to a consensus that the policy should be changed because *justice* demanded it. There was no need to settle the vexing theological questions; the question was settled, and in the right way, on nonreligious grounds. Strictly put, then, morality is not the same thing as religion.

Before ending this discussion, please notice three implications that do not follow from my argument:

- It does not follow that God does not exist. Nothing I have said should raise any doubts in your mind about the existence of God. Other things may be able to raise these doubts, but I have not said them here.
- It does not follow that the moral teachings of Christianity, Judaism, Buddhism, or any other religion are incorrect or faulty. To the contrary, I think it is clear that our religious traditions have through time been the repositories and incubators of some of our highest ideals.
- It does not follow that people do not need religion nor that secular philosophy can tell you all you need to know about how to lead your life. Morality is only part of human life. It does not do everything. It does not, for example, reward us if we try to worship it.[17] Nor does it seem to touch upon all aspects of our life. Many dimensions of life do not necessarily have anything to do with morality: the beauty of a cello concerto, the drama of an NCAA basketball game, the complex history of the Lewis and Clark expedition, the meditative quality of a Cormac McCarthy novel, the silence of prayer, the difficulty of spiritual repentance, the sculpture of an unplowed tall-grass prairie.

We are multifaceted beings. If an omnipotent and benevolent God created us, then it may well be our primary end in life to worship and enjoy that being. In that case, religious activity is a vehicle by which the various dimensions of our lives are given coherence, our discordant activities harmonized. If our chief purpose is to glorify God, then religion is unlike morality in important ways. Religion's primary role is not to answer questions about what is morally right and wrong but rather to answer questions about how in general we ought to live. Which activities should be subordinate to others? What is the relative importance of parenting, prayer, aesthetic experience, professional obligation, and worship?

Returning to the ideas raised in the case study at the start of this chapter, Rich may justifiably believe that religion *is not* a necessary part of ethical theory. Emily may justifiably believe that religion *may be* necessary for full human flourishing. In other words, anyone may without contradicting themselves believe both of the following propositions:

We can know what is morally right or wrong independently of religion.

We cannot live a complete human life independently of religion's beatific vision.

NOTES

1. I presented versions of this chapter between 1994 and 1998 at Bioethics Institutes at the University of Illinois, Michigan State University, Purdue University, Iowa State University, North Carolina State University, and Oregon State University. Many thanks to the participants of those institutes whose questions and criticisms helped me to refine the presentation.

2. Note that the number of U.S. college students who said that they believed in God in March 1965 was more than 97 percent. University of North Carolina Institute for Research in Social Science Study Number S1522. On the web at: http://www.irss.unc.edu/tempdocs/12:58:39:2.htm

3. Survey Collection: Harris/941104. IRSS Study Number: S941104. At: http://www.irss.unc.edu/tempdocs/13:06:55:2.htm, and at: http://www.irss.unc.edu/tempdocs/13:11:21:4.htm

4. Nancy Rosenblum, "The Moral Effects of Associational Life," *Report from the Institute for Philosophy and Public Policy* 18 (Summer 1998), 11. Rosenblum refers to the work of Robert Wuthnow. See Wuthnow, *Sharing the Journey* (Princeton, NJ: Princeton University Press, 1994).

5. Rosenblum, 11.

6. Apart from the modern Western period in which the morality called secular humanism has developed in explicit opposition to religion, the only historical exception to the rule that morality develops within religion is probably Confucianism in China. According to many interpreters, Confucius (d. 479 BCE) did not believe in supernatural phenomena and denied the reality of one's dead ancestors, yet Confucius developed a very clear moral system based on the principle of *ren*, or benevolence. Ren is "the attitude and habit of reciprocity in moral thinking." Confucius once

summarized ren as "Do not do to others what you would not like yourself. " In the ethic of self-discipline and justice that characterized the Chou political court, we have an example, if my interpretation is correct, of a morality that did not rely on the sanction of transcendental beliefs or religious authorities. In our culture, secular humanism is a twentieth century manifestation of a similar phenomenon.

7. James Rachels, *The Elements of Moral Philosophy*, 45.

8. "A Call to Civil Society: Why Democracy Needs Moral Truths," Institute for American Values, New York, 1998, 5. The report cites Daniel Yankelovich, "Trends in American Cultural Values," *Criterion* 35 (Autumn 1996), 2–9.

9. "A Call to Civil Society," p. 5. The report cites "Real National Unease—Especially on 'the Moral Dimension,'" *The Public Perspective* (October/November 1996), 24.

10. By "transcendent," I mean supernatural, not simply a mental realm that exists outside the body. One may be an atheistic mind-body dualist, such as Descartes would have been had he not been a theist, and not believe in the transcendent in the sense I am using it here. Atheists may believe that human identity consists of something more than the material transactions happening in our brains, but that does not make them believers in "transcendence," at least as I am using the term here.

11. See Gary L. Comstock, *Religious Autobiographies* (Belmont, CA: Wadsworth, 1995).

12. Unfortunately, philosophy instructors often presume that helping students learn to reason for themselves requires that one talk dismissively about religion. The best kind of reasoning includes reasoning about matters near and dear. Perhaps professors need to worry less about stopping religious students' illegitimate appeals to authority and worry more about enabling religious students' attempts to draw legitimately on religious traditions as moral sources.

13. University instructors may need to be reminded of the possibility that some rationally justifiable ethical principle or other may best be disseminated, as a practical matter, through the resources of some religious community or other. To imply that students should cut themselves off from their theological resources unnecessarily constrains not only moral development but ethical reasoning.

14. Nielsen, 1.

15. Rachels, 48.

16. In his book, *Contemporary Philosophy of Religion* (Blackwell, 1997), Charles Taliaferro articulates an ethical theory in which normative judgments are hooked into the concept of an ideal observer. Morality, in his view, may depend metaphysically on such an ideal observer and, because such an observer bears many similarities to standard Western conceptions of God, Taliaferro's proposal might be construed as a defense of a (modified) divine command theory.

17. Cf. Susan Wolf, "Moral Saints," *The Journal of Philosophy*, August 1982, 419–439, and Robert M. Adams, "Saints," *The Journal of Philosophy*, July 1984, 392–401.

REASONING

LILLY-MARLENE RUSSOW

CASE STUDY: KAREN THE ETHICIST

Rich, Dennis, Ken, and Karen are heading for the cafeteria after class. Rich says, "I hope Wright doesn't intend to waste any more time talking about religion."

Ken responds, "Yeah, I want to talk about real issues, like how we can protect the environment."

"Is there anything in the syllabus about that?" asks Karen.

"Yes, ma'am," answers Dennis brightly. "Looks like we'll be going beyond the ordinary stuff we talked about in high school, about preserving endangered species and wilderness."

Ken says, "People say the environmental movement has gone too far. But they don't realize that the ozone layer has not stopped disappearing, that the earth has not stopped getting warmer, that people have not stopped killing whales and seal pups, and that rainforests continue to be cut down. We're part of the problem. I see guys empty soda cans and leave them sitting in classrooms every day. They could easily drop them in recycling bins; they're all over campus now."

Karen: "I'm a forestry major and an environmentalist. But sometimes I wonder if I know more about *what* I believe than about *why* I believe it."

Ken: "Huh?"

Karen: "And, more important, why I think others should accept my goals and practices as their goals and practices."

Ken looks at Dennis. Dennis grimaces. "I don't see your point, Karen. And I certainly don't agree with green Ken."

Ken: "What's your problem?"

Dennis: "My problem is that everybody agrees we ought to use the earth's resources wisely. But the reason is that humans need it! We need it for food, for recreation, for oil production, for timber for our homes. And who are we in the rich countries to tell poor people in developing nations to protect endangered species when they're worried about how to get enough beans on the table to sustain their kids?"

Karen: "That's my point. There are conflicting views, so we need to figure out not just what we believe, but why."

Rich: "Say some more about your distinction between *what* and *why*."

Karen: "Just that I'm not as interested right now in the so-called 'right answers.' I'm more interested in how anyone would arrive at them. I want help figuring out what method to use in assessing how concerns for the environment can be balanced against concerns for humans. I don't feel very skilled at defending my views and, as a forestry major, I am going to have to learn how to do it."

Dennis: "Eh, why bother?"

Karen: "In the end, because I want to learn how to convince people to adopt environmentally responsible practices. But, before that, to learn *why* I'm so certain about my values."

Rich has been quietly putting away his barbecued beef sandwich. He says, "I understand the conclusions you both *want* to reach. However, neither of you has given me any reasons whatsoever to support one position or the other, and neither of you has cited important factual or empirical data that would be relevant. You also haven't shown how that information would convince me that your position is right, that your conclusion is true."

Karen: "Right. Those are the kinds of things I want to investigate. Isn't the question really one of cost-benefit analysis? If people want to pay enough to preserve a wilderness area or a species, let them do so. Some environmental groups, such as the Nature Conservancy, recognize that economic reality. However, if a community decides that a new more urban development is what they want, and can pay for it, that's what should be done. The wise use movement is trying to do just that: Encourage development and use, but do so wisely."

"Very nicely put, Karen." It's Dr. Wright. He has just come through the cafeteria line and asks whether he can join them.

"Only if you promise to answer Karen's question," Rich says, laughing.

"Which is?"

Karen repeats what she has just said.

Wright responds: "Well, let's start with the meaning of words. What does the wise use movement mean by 'wise'? And why does it focus on 'use'? Is it possible that the environmentalists and the developers may have different meanings of 'wise' in mind?"

Karen: "So, is that one of the first steps in ethical analysis? Getting clear about the meaning of the words we use?"

Wright: "Yes. And from there we begin to use those words to figure out what arguments someone is using to support his or her view. For example, Emily might believe that the Bible is the source of morality and that it tells us to have dominion over the earth. Meaning, we should use natural resources for our benefit. But Ken might think that this argument is not a sound one because the Bible says many things about nature, and one should look at the whole text rather than just pick out specific passages. He might accept the Bible but think that it leads to environmentalist values. Emily and Ken would now have to analyze not only the meanings of their words but also the reasons and arguments each of them have for their conclusions."

Karen: "I see. Get clear about the words we're using, reconstruct the arguments and reasons that we string together, and then try to decide which reasons take priority, which arguments are better than others."

Wright: "You've got it."

Rich: "But how do we do that? Is there some ethical method for telling good arguments from bad ones?"

Wright: "Well, yes, actually. We can first distinguish factual reasons from philosophical reasons and then test the factual reasons scientifically to see whether they're true."

Dennis: "I think I know how to test scientific claims, but how do you test philosophical claims?"

Wright: "Various ways, depending upon what the claim is. If it's a moral principle, then we can try to imagine all of its various implications. If some of its practical implications are simply unacceptable to everyone, then we have a principle with 'counterintuitive implications.' This result gives us good reasons for doubting the validity of the principle. If the claim is an argument, on the other hand, then you examine the argument to see whether its conclusion follows from the various premises. If the reasoner has cheated in stating the conclusion, then we have found an invalid argument. If the argument is invalid, we have a good reason to reject it."

Karen: "And if it's valid?"

Wright: "We ask whether it's sound."

Karen: "A sound argument is different from a valid argument?"

Wright: "That's correct. And it's a very important difference in ethics."

Karen: "How do you tell?"

Wright: "If it's valid, you simply ask whether all of the factual premises are true and all of the normative premises are justifiable."

Karen: "Sounds complicated."

Wright: "It is. But these are the basic skills involved in rigorous moral thinking, and we will spend the next few days working to develop them."

Karen: "I said it sounds complicated, but I didn't mean to imply that I wasn't interested. It's actually very exciting. It's exactly what I took this class for."

Claims about what we should or should not do, moral precepts, and general claims about what is right or wrong need not be arbitrary.[1] They can and should be supported by reasons. It is therefore important to understand how we can evaluate those reasons and to distinguish good and bad arguments. This chapter is intended to provide some insight into the process of evaluating arguments and developing good arguments for claims about ethics. Since moral reasoning is a special case of reasoning in general, we begin with a general look at critical thinking and then consider some special features of moral reasoning. Although many of the examples involve the analysis of someone else's reasoning, it is essential to realize that the same principles apply to one's own moral reasoning.

IDENTIFYING REASONS AND CONCLUSIONS

Philosophers and logicians apply the term *argument* to any group of statements, some of which (the premises, or reasons) are intended to help convince us that one or more of the statements (the conclusion[s]) are true. Thus, an argument need not involve a dispute or disagreement. In many cases, the argument structure is clear; we can easily pick out the premises and conclusions and see how they are supposed to fit together.

Thus, consider the case of Sam, who has volunteered to work for a private environmental group working to preserve natural parks and other important habitats in Hawaii. Upon his arrival, he learns that one of his duties will be to set snare traps to catch feral pigs living in these delicate ecosystems. His supervisor explains that it is necessary to get rid of the pigs because they are a non-native species, descended from domestic pigs escaped from European settlers who introduced them as farm animals. They also have no natural predators and are rapidly destroying endangered plants and the habitat and nests of native birds. Although snare traps cause more suffering than most other forms of trapping, the supervisor tells Sam that it is important for the health of the ecosystem to get rid of the pigs and that all other forms of capturing the pigs have proven to be ineffective. Sam tells his supervisor that he will accept the assignment but in fact (a) sets his traps in ways that are designed to fail, (b) deactivates other snare traps, and (c) notifies local and national animal rights organizations in the hope that they will organize protests against the policy. Sam thinks to himself, *I should not use traps to kill pigs because it will hurt them.*

Sam has provided us with an example of moral reasoning because he has been thinking of an argument. What is an argument? It would not be correct to say that he has been "arguing with himself." When we say that someone is "having an argument" (with themselves or others) we mean that a heated debate is going on, or an expression of disagreement, whether or not it involves any reasoning. This is *not* the sense of "argument" that interests us. Throughout this book, the term *argument* refers to something else, something produced through reasoning.

An argument is simply a collection of statements in which someone reaches a conclusion by relying on a reason, or reasons. How did Sam reach his conclusion? We do not know for certain, but let us suppose that he reasoned that the trapped pigs would suffer. We can articulate that argument by placing Sam's premise in (1) and his conclusion in (2).

(1) Using snare traps to kill feral pigs will cause suffering to the pigs.

(2) Causing suffering to the pigs is morally wrong.

Sam has wrapped (1) and (2) into a single sentence, demonstrating that moral arguments are sometimes found in very simple expressions: *I should not use traps to kill pigs because it will hurt them.* Arguments are often complex, however, and single sentences cannot express all of their components. Arguments usually consist of several premises, some for and some against a particular idea. They often contain chains of premises, with some arguments leading to ideas that serve as parts of another argument. For example, a statement might be the conclusion of one argument and then be used as a premise in another.

The conclusion of an argument and each of its premises must be a *statement* or a group of statements. Statements are either true or false; they assert, either truly or falsely, that something is the case. Statements are expressed either by uttering a string of words or by writing them down in *sentences*. For example, the sentence "Pigs can feel pain" expresses a statement because it makes a claim that is either true or false.

Not all sentences express statements. For example, genuine questions are not statements, nor are commands such as "Do not feed the elephants!" Genuine questions must, however, be distinguished from *rhetorical questions*, which are not questions at all but rather are statements. For example, "Who can deny that torturing little kittens is morally repugnant?" is not a request for information but a forceful way of asserting that torturing animals is morally repugnant. Similarly, in certain cases, commands and exhortations should be interpreted as statements when they appear as the conclusions of arguments. For example, the exhortation "Vote for Brown because she is the best candidate for mayor!" is really an argument in which the conclusion is the statement "You should vote for Brown." The premise in this argument, of course, is that "Brown is the best candidate for mayor."

Of particular importance in arguments are sentences that use "either . . . or . . . " and "if . . . then . . . " The sentence "Either Tom is out on the trail or he is at the beach," expresses a single statement. It does not assert that Tom is on the trail, nor does it assert that he is at the beach; it states only that one or the other of these alternatives is the case. Similarly, "If Brown receives the backing of the labor unions, she will win the election," asserts only a single conditional statement about what will happen if Brown is supported by the labor unions. Finally, an argument can be expressed by a single sentence if the sentence contains at least two appropriately linked statements, for example, "This law must be struck down since it discriminates against the handicapped."

Arguments are meant to support their conclusions and thus *rationally* motivate us to accept their conclusions *as true*—to believe them. They purport to represent good reasoning that is a reliable decision-making process. Giving an argument is thus distinct from following hunches or intuition, trying to persuade through emotional appeals or trickery, simply stating one's opinion, however forcefully or eloquently, or merely describing the position one wishes others to adopt without providing any supporting reasons.

A hunch or intuition stands alone. To call something a hunch implies that one has no evidence for it and it is not the result of reasoning. Trickery or emotional appeals may lead someone to accept a statement, but these appeals work by short-circuiting the reasoning process. Mere descriptions or statements of opinion simply put forward a point of view without any reasons or evidence. Since these procedures are irrelevant to the truth of statements, they are unreliable. By contrast, arguments aim to give reasons in the premises that *are* relevant to the truth of their conclusions. Thus, if an argument is good, it can rationally motivate us to accept its conclusion on the basis of its premises.

Deciding whether a piece of writing constitutes an argument, or contains one, is sometimes quite difficult. Practice your skills in identifying arguments by working through Exercise 3.A in Appendix A.

Philosophers typically express arguments in a particular form. We list the premises, then draw a line, and then list the conclusion. Obviously, some arguments are good arguments and some arguments are bad arguments. Here is a bad argument, written in standard philosophical form:

(1) There are not many family farms any more.
(2) I would like to see Congress save the family farm.

---

(3) (Therefore) living on a family farm is the best way to live.

This is a bad argument because premises (1) and (2) do not give us good reasons to believe (3). While the first two premises mention family farms, they do not mention reasons to believe that living on family farms is the best way to live. They mention interesting claims about family farms, but these claims are irrelevant to the specific claim made about family farms in the conclusion.

Here is an even worse argument:

(4) There are not many family farms any more.

(5) Madonna is not married to the artist formerly known as Prince.

(6) (Therefore) we should not buy clothes made in China.

Obviously, (4) and (5) have nothing to do with each other, much less with the alleged conclusion (6). Later, we will discuss the elements of a good argument. For beginning purposes, however, it is useful simply to recognize the parts of an argument, apart from the argument's *validity*, a technical term that will be explained presently. Turn to Appendix A and perform Exercise 3.B; then, continue with this chapter.

Notice now that some arguments are *moral* arguments and some are not. Moral arguments typically support conclusions that claim that someone *ought* or *ought not* do something, or that a certain sort of action is either *right* or *wrong*, or that a certain sort of thing has *positive* (goodness), or *negative* (badness) moral value. One rough way to tell whether an argument is a moral argument is to figure out whether the conclusion tells us or implies what, if we are trying to act in a morally good way, we should do; for example, situations in which someone may be *harmed* are almost always moral situations because they involve something that we should not do: produce harm. So if an argument concerns potential harms to pigs, farmers, young people who wish to drive, or even, perhaps, island ecosystems, then the argument is a moral argument. Other indications that you are dealing with a moral argument would be phrases such as "*x* is wrong," "from an ethical perspective," and "conscience would tell us not to do *x*." Moral claims can be expressed in many other ways, but these examples should give you a good starting point. Turn to Appendix A and perform Exercise 3.C before continuing with this chapter.

GETTING TO THE POINT: THE CONCLUSION

The purpose of an argument is to give reasons for thinking its conclusion is true. Thus, to evaluate how good an argument is, we must begin by identifying its conclusion, that is, what is being argued for. The more lengthy or unclear the argument, the harder this first step becomes. This is especially true when one is dealing with a *complex argument* or *argument chain* consisting of several *intermediate steps*. Each of these steps might be an argument with a conclusion of its own. These *intermediate conclusions* then combine to support the *final conclusion*.

To look for the *final conclusion*, follow these three general rules:

1. Ask yourself what the main idea is. What is the author trying to establish or work toward?

2. In a complex argument or argument chain, determine what the intermediate steps point to. Do the intermediate conclusions contribute to the support of one overall idea? More generally, which statements lead to or support other ideas?
3. Look for *clue words* that indicate the author's organizational scheme.

When presenting an argument of your own, you can use *clue words* to direct attention to your conclusion. The following words are often used to signal that a conclusion follows:

Clue Word	Meaning
Consequently	It follows that
Therefore	Suggests that
This proves that	Points to the fact that
So	Entails
Since this is so	Implies
Hence	Thus

In arguments that use no clue words, you must rely on the first two rules. In complex arguments or argument chains that do contain clue words, the clue words might signal the conclusion of an intermediate step, thus directing your attention away from the final or main conclusion. For this reason, even when clue words are present, it is advisable to test the use of the third rule with the other two rules. To illustrate this point, consider the following argument:

> Age discrimination is often fostered by economic motives since younger workers generally have less experience, hence can be hired more cheaply.

The clue word *hence* directs our attention to the claim "[Younger workers] can be hired more cheaply," but this is not the final conclusion. If we look for the main idea, we see that it is the first statement, "Age discrimination is often fostered by economic motives." The claim "[Younger workers] can be hired more cheaply" is the conclusion of an intermediate step in the chain of reasoning.

To see how the three rules operate in a more complex argument, consider the next argument, in which each sentence has been numbered for ease of reference.

> (1) Should you repeal the present 55 mph speed limit? (2) This question cannot be decided on economic grounds alone. (3) Raising the speed limit to, say, 70 mph will save time in transporting goods and hence tend to reduce costs. (4) But it is unlikely that this will result in a significant economic benefit, since driving at higher speeds consumes more fuel. (5) Even critics of the present speed limit concede that it has helped to reduce the number of deaths and injuries in automobile accidents. (6) The vast

amount of money we spend on health care shows that saving lives is more important to us than saving dollars. (7) The 55 mph speed limit saves lives, and the economic advantages of changing it are uncertain. (8) So, it should be retained.

The first rule of looking for the final conclusion directs us to look for the main idea. At first we might think that the main idea is expressed in the second sentence, but, reading further, we see that the overall point the author is trying to establish comes right at the end of sentence 8.

The second rule serves as a check on the first. Having picked out "It [the 55 mph speed limit] should be retained" as the final conclusion, we now go back to see which of the other statements point toward the idea and support it. The statements that support the final conclusion most directly are "Saving lives is more important to us than saving dollars" (from sentence 6), "The 55 mph speed limit saves lives" (from sentence 7), and "The economic advantages of changing it are uncertain" (from sentence 7). The first two of these are intermediate conclusions that work together to support the final conclusion. Each of these, in turn, is supported by further statements in sentences 5 and 6. The main support for regarding the economic advantages of a change as uncertain comes in sentence 4.

The third rule is the simplest but needs to be applied thoughtfully. The argument contains three different clue words or phrases signaling that a conclusion follows: *hence* (in sentence 3), *shows that* (in sentence 6), and *so* (in sentence 8). Only the last of these indicates the final conclusion and might conceivably have been omitted. *Shows that* signals an intermediate conclusion in the overall argument. *Hence* points to the conclusion of an entirely separate argument.

Sometimes a conclusion is signaled not by using clue words but rather by *juxtaposition*. It is a common practice to make a claim (the conclusion) and then follow it with a statement or a set of statements of the evidence that is supposed to support it (the premises).

Finally, some arguments have final conclusions that are not explicitly stated at all. The arguments have *implicit conclusions*, since letting readers "draw their own conclusion" is often rhetorically effective. In these cases, the premises are usually presented in such a way that only one "obvious" conclusion can be drawn from them. Thus, readers are not really drawing their own conclusions but merely making explicit the implicit conclusion the author intended.

Turn to Appendix A to perform Exercise 3.D; then, continue with this chapter.

GIVING REASONS: THE PREMISES

The premises of an argument are those statements that lend support to the conclusion. From our earlier discussion of arguments, it is clear that the support we are looking for is of a special kind. We want reasons that point to the truth of the conclusion. Thus, reasons or premises must be distinguished from the following sorts of statements that often occur in the course of a discussion:

1. Introductory remarks that merely mark out the topic, set the context, or explain why someone might be interested in the issue.
2. Comments that merely restate or elucidate a position without giving reasons that support it.
3. Mere persuasion, such as use of emotional language or seductive appeals that are not evidence for the truth of the conclusion.
4. Disclaimers, that is, remarks that discount a statement or possible criticism without actually providing an argument, such as "One might think that taxes should be increased, but I oppose any such measure."

Consider the following pair of examples

Many of the biologists in the environmental movement support left-wing causes.

Since none of those who declare that nuclear power plants are safe is willing to live within a mile of one, we should be skeptical of such claims.

Unlike the first of this pair, the second is clearly an argument. The second has a conclusion and a statement, which is a reason, however weak, for thinking that the conclusion is true. On the face of it, the first example of the preceding pair is just a single statement and hence not an argument. But if we encountered the first example in the context of a debate over which scientists we should trust on environmental issues, we might be justified in regarding it as an argument with the implicit conclusion that we should ignore the views of many biologists in the environmental movement. But the only reason given for this implicit conclusion is that these biologists are supporters of left-wing causes, which is irrelevant to its truth. Thus, if the first example is an argument, it is a very bad one.

Examples such as the first one of the preceding two raise difficult issues about when something should be considered an argument. If it is clear from the context that someone has reasoned, however poorly, from one statement or statements to another, then we should treat those assertions as an argument, adding the implicit conclusion if necessary.

Just as there are words that signal a conclusion, there are also terms that are often used to identify premises. The clue words on the following list may be introducing a premise. If you are presenting an argument, you can use these signals to help your audience identify your premises more readily:

Clue Word	Meaning
Since	For the reasons that
Because	May be deduced from
As shown by	Follows from
Seeing that	May be inferred from
Is proved by	Is suggested by

As with clue words for conclusions, you should not rely blindly on these signals. Check to make sure that what you have identified is actually a premise. For example, the word *since* does not always indicate a premise, since it can be used in a temporal sense (for example, "Personal computers have become much more powerful since they were first introduced in the 1970s"). *Because* is sometimes used in stating a claim about the cause of something rather than in stating a reason for thinking a statement is true (for example, "The car stopped because it ran out of gas").

To sum up, the first step is to ask whether we have an argument at all. If no conclusion exists, we must locate the premises. The premises are those statements that provide evidence that the conclusion is true; other statements might "color," explain and clarify, or set the stage for the conclusion without giving reasons, but these are not premises. All of the considerations considered so far apply to both the constructive and analytical enterprises, so you should keep them in mind when you develop or present your own arguments.

When trying to create or evaluate an argument, we must direct our attention to the premises or reasons and refuse to be distracted by the other sorts of statements. Clue words often help us to identify the premises.

Turn to Appendix A and perform Exercise 3.E before continuing with this chapter.

Notice that many of the arguments we have been discussing, including Sam's argument about the trapping pigs in Hawaii, are far too brief, as they stand, to work as complete arguments. The reason they are incomplete is because there are other, unstated (*implicit*) premises that must be identified and stated before we can understand the argument.

It is quite common, and not necessarily a flaw, to leave some premises of an argument unstated. However, when we want to evaluate an argument, we need to state those premises explicitly. The guiding idea here is the *principle of charity*: try to find the most reasonable statement, or the statement that is most reasonable from the arguer's perspective, that will complete the argument. Consider the following two possible additions to Sam's argument:

(4) Causing suffering to animals is always wrong.
(5) Causing avoidable suffering to animals without overriding justification is wrong.

Unless strong evidence exists to the contrary, the principle of charity suggests that we should choose (5) rather than (4) because it is the most plausible and lends the most support to Sam's case.

The principle of charity applies to attempts to reconstruct any argument, but this example illustrates two features that are distinctive of moral reasoning. The first is that a complete moral argument will almost always involve at least one premise that makes a factual or empirical claim, and at least one that appeals to a general moral principle. We can refer to factual claims as *empirical premises* and premises that talk about what is right or wrong, what we should or should not do, what is good or bad, as *normative premises*. When we are analyzing, developing, or evaluating a piece of moral reasoning, we should look for both these parts and make them explicit or more precise if necessary.

Here is another example, drawn from an article by Peter Singer. If we were to summarize Peter Singer's argument about famine relief, we would get something like the following:

(1) Death by starvation is a very bad thing.
(2) By sending substantial amounts of money for famine relief, we can prevent death by starvation without sacrificing anything of comparable moral worth.
(3) If we can prevent something very bad from happening by doing X, and if we can do X without sacrificing something of comparable moral worth, then we have a moral duty to do X.

---

(4) (Therefore) we have a moral duty to send substantial amounts of money for famine relief.

Statements (1) and (2) are empirical premises; although (1) seems obviously true, there is some debate about (2). Statement (3) is a normative premise, and that, too, would need closer scrutiny. Some statements in an argument seem to combine elements of empirical and normative premises. In those cases, it is helpful to rephrase the argument to separate and identify the premises, since important differences exist in the ways in which empirical and normative premises are evaluated.

The second important feature of moral reasoning, whether we are evaluating an argument such as Sam's or Singer's or constructing arguments to support our own conclusion, is to consider whether the empirical premises are complete enough, that is, whether all the relevant known facts have been included, and also whether there are other normative premises that would either strengthen or weaken the argument. What other facts might be relevant to Sam's decision and do other moral principles exist that should be weighed? Is (2) in Singer's argument true, and again, should other relevant moral principles be factored in?

In a subsequent section, I discuss ways of determining whether this is a good argument. The point to keep in mind when developing or evaluating an example of moral reasoning is that it must contain both factual claims—the first two premises—and a general moral principle, such as the one stated in (3) of Singer's argument. When considering an example of moral reasoning, we need to look for both of these and make them explicit if they are not already stated.

Turn to Appendix A and perform Exercise 3.F before continuing with this chapter.

Another basic question to consider when reconstructing, developing, and identifying arguments is that one must decide whether the argument is intended fully to establish the truth of the conclusion or merely to show that it is more probable that the conclusion is true. More precisely, we need to consider whether the argument is intended to be understood such that if the premises were true, the conclusion *must* also be true, or instead, merely that if the premises were true, the conclusion is *more likely* to be true. This marks the difference between deductive and inductive arguments. Inductive arguments come in many forms; generalizations, predictions about the future, and inferences from the best explanation are familiar types. Although many moral arguments are inductive, both of the arguments reconstructed so far are deductive, which therefore determines how we evaluate them. I discuss the difference between inductive and deductive reasoning in detail later in the chapter, but it is useful to be aware of the distinction at the outset, as we try to determine the precise content of the argument and how it should be interpreted.

OUTLINING THE STRUCTURE OF ARGUMENTS

Whether we are evaluating an argument or developing one of our own, one technique that is sometimes helpful is outlining an argument. Its usefulness is limited by the following considerations:

- Sometimes arguments are so short or simple that outlining is unnecessary: the structure of the argument is already clear.
- Some presentations are so dense that paraphrasing the main points is easier than trying to sort through all the author's statements.
- Some arguments are so incompletely stated that an outline does not give much sense of the fully reconstructed argument.

Even with those caveats, outlining is a good discipline for helping to ensure that you have correctly identified and understood an argument. It is also often useful in constructing your own arguments. The basic technique involves three steps:

1. Find the final conclusion, underline it, and put brackets around it. If the final conclusion is implicit or if it has not been appropriately stated, write out the conclusion in your own words.
2. Enclose each separate premise in brackets and assign each a number. If you are not sure whether something is a premise, go ahead and give it a number but be prepared to leave it out later, if it should turn out not to be a premise. Take care to separate each distinct thought, but do not break up a single idea. If the conclusion has also been bracketed, give it a number.
3. Draw an outline of the argument by writing down the number assigned to the conclusion and draw arrows pointing from the numbers of statements that directly support the conclusion. Continue adding arrows and plus signs where appropriate, as explained in the following example.

Here is an example of a very simple argument: Animals feel pain just as people do; therefore, it is wrong to torture them.

And here is how you would apply the outlining technique to it:

[(1) Animals feel pain just as people do];
{therefore,}
[(2) it is wrong to torture them].

Figure 3.1 shows how to draw a picture of it.

1   **FIGURE 3.1**

2

The conclusion is as follows: "It is wrong to torture them (animals);" the clue word "therefore" introduces this conclusion. The only other statement is a premise that points to the conclusion, and this premise is indicated by the arrow in the outline that points from 1 to 2. Before going on to consider more complicated arguments and their outlines, note a few potential difficulties.

First, the conclusion is not always stated, or it may be stated in the form of a rhetorical question or in some other oblique way. In these cases, formulate a statement of the conclusion in your own words and make a note of it. Thus, if the argument had as its second sentence; "Why, then should we feel we can torture animals without justification?" instead of the original version, you could rewrite the conclusion as: "We should not feel that we can torture animals without justification."

The second difficulty actually includes two things, both connected with the problem of bracketing individual statements correctly. Statements will not always coincide with sentences, so if a sentence contains two distinct statements connected by *but, and*, or a similar conjunction, you should distinguish the statements and give each its own number. Thus, the sentence "All citizens of a country have an obligation to obey the laws of that country, but this obligation does not override the greater duty to do no wrong" should be broken up into two distinct statements, separated by the word *but*:

[(1) All citizens of a country have an obligation to obey the laws of that country]
{but}
[(2) this obligation does not override the greater duty to do no wrong].

The second consideration when bracketing statements is that you must be careful not to break up statements that are unified. This temptation is especially strong in cases in which you are dealing with a complex sentence that has the form "if . . . , then . . . ", or the form "either . . . or . . . ", or some form equivalent to either of these two. The sentence "If animals feel pain the same way humans do, then it is wrong to torture them" does not contain two separate statements, one to the effect that animals feel pain and the other claiming that it is wrong to torture them; the sentence remains neutral as to whether they do feel pain and as to whether it is wrong to torture them. Rather than contain two separate statements, the sentence expresses a relation between two factors; it presents the idea that feeling pain and the wrongness of torture are linked in some way. So, it would be wrong to break these two factors apart. You enclose the whole statement in a single pair of brackets and give it one number. The same holds true of the following examples:

• Either you donate some money to Oxfam or you spend it in other ways.
• You cannot save endangered species unless you protect their habitat.
• When it snows, elk migrate to lower elevations.
• He who hesitates is lost.
• If family farms are to survive, they will need government support.
• If family farms are to survive and large corporate farms will not profit unfairly, regulations controlling government support will have to be rewritten.
• A species can survive only if it has a sufficiently heterogeneous gene pool.

The general principle is to watch for statements that express a relation between two or more factors and *not* break them up. These relations are most commonly found in "if . . . , then ... " statements, "either ... or ... " statements, or variations of these forms.

More complex arguments require more complex outlines. The argument might, for example, have an argument chain, in which case one of the statements may be the conclusion of a preliminary argument and serve as a reason for the final conclusion. An example of this is shown in Figure 3.2.

[(1) Decisions about water rights are often fostered by economic rather than environmental motives,]
{since}
[(2) larger corporations have more money to hire lobbyists,] and
{hence}
[(3) corporations can exert more influence on legislators].

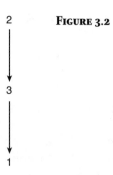

2

3

1

FIGURE 3.2

Many arguments will give more than one premise for a conclusion. Frequently, the conclusion is dependent on both premises taken together. Sometimes, but not always, this connection is signaled by a conjunction such as *and*. The fact that two premises are dependent on one another is indicated in the outline by a + between the numbers of the two statements, as shown in the following example:

[(1) Animal research needs to be continued]
{because}
[(2) there are many serious diseases that still need to be understood] and
[(3) animal models are the only way of conducting rigorously controlled studies].

The point to note is that neither (2) by itself nor (3) by itself gives any reason for the conclusion. It is necessary to combine them before they are relevant to (1). Thus, the outline should look like that shown in Figure 3.3.

It makes sense to have an arrow leading from a plus sign to a number because the arrow indicates that combining two statements produces something that supports the

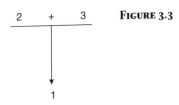

FIGURE 3.3

conclusion. You should never have an arrow pointing to a plus sign. An arrow must always point to a conclusion, either an intermediate conclusion or the final one.

Alternatively, you may have arguments in which two premises work independently to support the conclusion. Each one, taken in isolation, lends at least some support to the conclusion. The example in Figure 3.4 and its outline show how this is handled:

[(1) Family farms promote traditional virtues] and
[(2) are generally more environmentally friendly than large corporate farms]
{so}
[(3) Federal policies ought to promote family farms.]

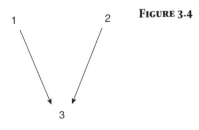

FIGURE 3.4

In complex arguments, several of these things may be going on simultaneously but the outlining method remains basically the same. In an argument with more than two or three statements, starting with a small chunk here and there may be easier to do before trying to combine all the various pieces into one outline. This step-by-step process is illustrated in the following example:

The widespread alarm about use and abuse of drugs in sports probably arises from some genuine, and perhaps rational, concern; but [(1) it is difficult to discern the basis for that concern in present policies and discussions.] [(2) If it is based on unfairness, it is irrational.]
{For}
[(3) there are far greater sources of unfairness.] And [(4) whatever is due to drugs can be neutralized by a system that allows all athletes equal access to drugs.] [(5) If it is based on paternalism, it is disingenuous and misplaced.]

{For}

[(6) the risks of sport itself far exceed the demonstrated risks of those drugs that arouse the greatest concern.] [(7) If it is based on some notion of naturalness, we need more conceptual work to tell us why synthetic vitamins are considered natural and naturally occurring hormones are considered unnatural.] [(8) We are not even clear on the moral difference, if any, between a food and drug,] [(9) nor is there a clear understanding of those terms.] (From Norman Fost, "Banning Drugs in Sports: A Skeptical View," Hastings Center Report, August, 1986.)

After picking out and underlining the conclusion as well as bracketing and numbering the various statements, you might notice a few of the closer connections.[2] Thus, Stage 1 of the outline might be rather fragmented, representing those connections that are easier to spot, as shown in Figure 3.5.

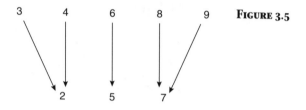

Figure 3.5

After you have done this, it becomes easier to see how these various pieces can be combined to make up the full outline (Fig. 3.6).

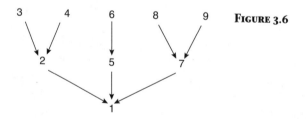

Figure 3.6

The outlining technique is useful in another way. It can help you plan and organize your own arguments by making you think about the connections between ideas before trying to construct a full argument. It also helps to identify and correct *pincushion* arguments. These are arguments that contain many unrelated and undeveloped reasons for a conclusion: the result is a diagram that looks like a pincushion (Fig 3.7).

The outlining technique is the same as the one used earlier to analyze someone else's argument: arrows run from reasons to the ideas that they support, reasons can work independently or dependently and you might have argument chains. The only difference is that you will be supplying the claims that are organized by the outline.

FIGURE 3.7

Often we argue for a conclusion simply by giving a single reason. For example, if a person is asked whether she thinks a university ought to adopt a policy of not buying athletic equipment from firms that rely on child labor, she might say, "Yes, because that would encourage companies to treat employees more fairly." If we number the two ideas, we get:

(1) Universities should not buy equipment from companies that rely on child labor.
(2) Boycotting a company can force it to change its labor policies.

Since only two statements are contained in Figure 3.8, you don't have much to organize.

FIGURE 3.8

People who disagree with the conclusion stated in (1) might list their reasons as follows:

(3) Child labor is often a necessary source of income for families living in poverty.
(4) Companies who employ children are often helping entire families to raise their standard of living.

After these reasons have been articulated, you can step back and ask how they are best related to the conclusion (unstated: that we ought to continue buying from companies that use child labor) and to each other. Upon examination, the first claim, (3), seems to support the second, (4), which in turn leads to this conclusion:

(5) We ought not to concern ourselves with labor practices in companies from which we purchase goods.

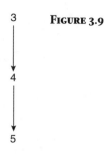

FIGURE 3.9

So, your outline would look like that shown in Figure 3.9.

The conclusion in (5) would lend support to those who oppose (1). If further reasons are thought of and incorporated into the argument of (3) and (4), you may want to refine the conclusion. Eventually, you might reach a conclusion that is the exact opposite of (1), such as:

(6) Universities *should* buy equipment from companies that rely on child labor.

So far, however, no explicitly stated normative premise is stated in (3) through (6), so you would need to use the principle of charity to determine how the normative premise should be formulated.

This procedure for creating an argument can be summarized as follows:

1. Try to develop a preliminary statement of the conclusion—of what you are arguing for. As you think about the subject more, do not hesitate to go back and change this statement to make it clearer, more precise, or a more accurate representation of the position you want to defend.
2. Make a list of the ideas that you think are relevant to that conclusion and assign each a number. At this stage, do not worry about connections or development; that will come later. In the case of moral reasoning, remember that you will need at least one empirical premise and at least one normative premise.
3. Try to find an outline that reflects the natural or intuitive connections between these ideas; in doing so, you may find yourself adding ideas to the list in order to fill out the outline.

This procedure, like outlining a very short argument of someone else's, is not really needed if you have come up with only one or two ideas. (Even here, though, it gives you time to stop and think whether your numbered statements really point to or support your conclusion, and whether they work together or independently). In cases in which you have come up with a longer list, the outlining technique allows you to break down the task of organizing your thoughts into more manageable parts. It's also a good idea to try to develop the strongest argument you can think of *against* the conclusion you are defending; doing so will help you spot gaps or weaknesses in your original argument. Consider the following list of reasons for advocating special support for family farms:

(1) Children on farms will learn the importance of caring for animals in a humane way.
(2) Family farmers are sensitive to environmental issues.
(3) Family farms are not under the control of large corporations that lack understanding of local conditions.
(4) Family farms involve more close contact with crops, water, and livestock.
(5) Large corporate farms are more likely to use chemical controls such as pesticides and antibiotics on a routine basis.
(6) Families are more likely to care about preserving land for future generations.
(7) Large industries tend to emphasize short-term profits.

Since this list moves from one strand of thinking to another, you should organize these ideas into an argument that is more focused and easier to follow. To do this, you need to organize some of the subsections and then tie the subsections together. Noting that some of the statements have to do with reducing the negative effects of farming owned or controlled by large companies, whereas others emphasize the positive value of the family farm, you might begin with one of those areas.

If you look for the positive side, you note that (1), (2), (6), and possibly (4) emphasize the positive value of the family farm rather than dwell on defects in alternative farming methods. Since the statements are just meant as starting points, many other equally good ways of grouping them exist. Remember that you are trying to develop an organizational pattern, not discover one that is already determined. Thus, the reason for saying that (4) *possibly* falls into this group is to indicate that, by itself, (4) is rather cryptic and this line of reasoning might be developed in several ways. In the complete argument, you will probably want to add statements to link it more clearly with the other parts of the argument within which you choose to locate it.

You can begin the outline, then, by focusing on the benefits of family farms, looking for links between the points you have already formulated and introducing other statements that might help to clarify the points you are trying to convey. You can mark these additions with an asterisk if you find that doing so makes keeping track of them easier. Thus, a preliminary attempt to outline the positive part of the argument might look like the argument in Figure 3.10.

(8) People who are in close contact with the land are better able to perceive the impact of different farming practices.

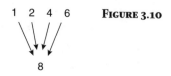

FIGURE 3.10

As noted previously, this is a creative process, so other organizational plans would also yield effective arguments. If you choose a different plan, keep in mind the potential need for additional premises that will help clarify the argument, as I did by adding statement (8).

Going back to the list, note that several of the statements, namely (3), (5), and (7), have to do with the bad effects of large corporate farms but that you have no general statement that conveys the broader objection.[3] So, you add to the list:

(9)   Corporate farms are more likely to adopt practices that degrade the environment.
(10)   State and federal policies should encourage family farms rather than large corporate farms.

These additions allows you to organize a second part of the argument, with an outline like Figure 3.11.

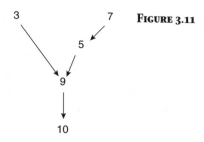

FIGURE 3.11

The bad effects mentioned in (9) are regarded as a reason for thinking that conclusion (10) is true.

Although it has not happened in this particular example, you may often find that one or more of the statements on your initial list have not yet been used in this stage of the outlining process. Depending on your goal and audience, you may decide to pursue this missing topic by adding further ideas to your list and formulating a new section of the outline, or you may decide to abandon it as unhelpful. Whichever you do, you must eventually try for a final formulation of your position and bring together all the sections of the argument. In doing so, you may notice other connections between statements in different subsections; for example, a statement in favor of family farms might also be used as a reason against promoting large corporate farms. These can also be indicated on the outline as shown in Figure 3.12.

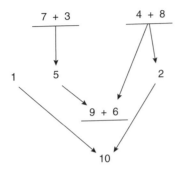

FIGURE 3.12

If you now wanted to present your argument in words, you could simply begin at the top of one branch of the outline and follow it down step-by-step, reformulating your numbered statements to make their place in the development clearer. When one branch is complete, you move to another, until the entire argument has been followed through to its final conclusion.

This sort of outline can help you see how the arguments that are implicit in the outline can be constructed in the strongest possible way. The outline helps you recognize what sorts of evidence might be relevant, how the premises must be formulated, what needs to be added or deleted, and so on. It helps you group related ideas together and gives you time to think about the relations between these ideas—which ones lead to what and how they fit together. They help you locate gaps and provide an opportunity to supply missing pieces. By working out the pattern of the argument before you state it in final form, you can offer your ideas (which started off as numbered statements formulated in no particular order) in a way that shows your audience how they fit together. This ability to develop a clear and well-organized argument provides a firm foundation for learning how to reason correctly. When outlining an argument that has been presented to you, you also complete an essential first step in evaluating the effectiveness of that argument.

VALIDITY AND SOUNDNESS

The main goal of critical thinking is to evaluate arguments, either your own or those offered by others seeking to persuade you. The first steps toward this goal are locating the conclusion, finding the premises, and outlining the structure of the argument, which are the points I covered in the previous section. I now discuss the question of evaluation.

Two main factors make an argument good or bad: (1) the relationship between the premises and the conclusion; and (2) the status of the premises. I first concentrate on the relationship between the premises and the conclusion.

I have already noted that *deductive arguments* are intended to show that the conclusion *must* be true, so I begin by introducing two concepts for evaluating deductive arguments. The first, *validity,* has to do solely with the relationship between the premises and the conclusion. The second, *soundness,* concerns both the relationship between the premises and the conclusion as well as the status of the premises.

To understand the concept of validity, remember that the purpose of an argument is to present a reliable form of reasoning, and that reliability has to do not merely with usefulness, practicality, or appeal but also with truth. Ideally, you would like for arguments to have the following feature: If you start with true premises, those premises must lead to a true conclusion. In other words, you want your arguments to be truth preserving. An argument that is truth preserving in this way is called a valid, or deductively valid, argument.[4] With this feature in mind, you can define a valid argument in any one of three equivalent ways:

1. An argument is valid if and only if it is not logically possible for its conclusion to be false when all its premises are true.
2. An argument is valid if and only if its conclusion follows logically from (or is logically implied by) its premises.
3. A valid argument is one in which its premises are related to its conclusion in such a way that if all its premises were to be true, then it would have to be the case that its conclusion is true also.

Though all three of these definitions are equivalent, the first is sometimes more useful, especially when it is not clear whether the premises of an argument logically imply its conclusion.

It follows from these definitions of validity that any argument that is not valid is invalid and vice versa: both these terms are "all or nothing." If an argument cannot *guarantee* the truth of its conclusion on the basis of the truth of its premises, it is simply invalid. There is no such thing as an argument that is somewhat valid or mostly invalid. You will see later, however, that some invalid arguments are inductively strong. Remember, the terms "valid" and "invalid" describe arguments, not isolated statements. Similarly, the terms "true" and "false" should be used to describe only statements, not arguments.

How do you tell when an argument is valid? Any argument that exemplifies a valid *form* of argument is valid. So, the next question is: How do you recognize a valid form of argument? By a valid form of argument, I mean a pattern such that any argument that has that form or follows that pattern exactly will automatically be valid.[5]

To examine the form of arguments more easily, with fewer distractions, I will frequently substitute symbols or letters for actual words, phrases, or statements. Doing so allows you to look at the form completely apart from the specific claims made by the premises. This method shows, for example, that any argument of the form "All *A*s are *B*s, all *B*s are *C*s; thus, all *A*s are *C*s" is valid, which tells you that whenever you substitute terms for A, B, and C that make the premises true, the conclusion will be true, too. Of course, if your substitutions make one or both of the premises false, then anything can happen—the conclusion might be true or it might be false, even though the argument is still valid.

Try the following exercise: Which of the following examples are *valid* arguments?

(1) All whales are fish, and all fish are cold-blooded; therefore, all whales are cold-blooded.
(2) All whales are fish, and all fish live in water; so, all whales live in water.
(3) All whales are fish. All fish suckle their young. Therefore, all whales suckle their young.
(4) All whales are mammals. All mammals suckle their young. So, all whales suckle their young.

You have, no doubt, correctly seen that each of the arguments in the previous exercise is valid, even though three of them have at least one false premise and one has a false conclusion. Each argument is valid because if all its premises were true, its conclusion would also have to be true; all four arguments exemplify the same valid form of argument. If a valid argument does have all true premises, the conclusion must also be true. On the other hand, if one or more of the premises of a valid argument is false, the conclusion might be true or it might be false; you have no guarantee either way.

How do you show that an argument is invalid? Doing so can be difficult since many invalid arguments have true conclusions. The crucial point is to prove that even if all the premises were true, the conclusion is *possibly* still false. For that reason, paying attention to the form of the argument can help again. So, you begin by learning how to identify an invalid form of argument.

Since an invalid form of argument leaves open the possibility that true premises can lead to a false conclusion, you can show that a form is invalid by constructing an example of an argument following that form in which all the premises are true and the conclusion is false. Consider the following very simple example:

If we destroy the habitat of a species, then the species will become extinct. We did not destroy the habitat, so the species will not become extinct.

Although this argument may appear to be valid, it is actually invalid, which you can discover by examining its form. The form of argument is: "If p, then q; not p; therefore, not q." You can show that this form is invalid by substituting statements for "p" and "q" that make all the premises true and the conclusion false, as in the following example:

> If Margaret is a Purdue graduate, then Margaret is a human being.
> Margaret is not a Purdue graduate, so she is not a human being.

Given that Margaret is a human being but not a Purdue graduate, this example conclusively shows that in all arguments of this form, the truth of the premises does not guarantee the truth of the conclusion; therefore, this is an invalid form of argument. You can use the same technique with more complicated arguments.

The second concept, soundness, builds on the idea of validity. A sound argument must satisfy two criteria: it must be a valid argument and all its premises must be true. If either one of these criteria is not met, the argument is unsound. This definition tells you two things. As with validity, soundness does not admit of degrees—an argument is either sound or it is unsound. Second, a sound argument will always have a true conclusion. To determine whether an argument is sound (in contrast to determining its validity), you must evaluate both its form and the truth of its premises. This is particularly difficult with normative premises, or premises that make claims about moral principles. In general, your judgments of soundness will be qualified, based on how well the premises can be justified in comparison to alternative principles. However, the fact that a sound argument must have a true conclusion allows you to use a variation of the technique that you used in the previous example. If you want, for example, to criticize Peter Singer's argument described earlier, we might try to show that the argument is unsound by offering something like the following parallel argument:

(1) The illness and death resulting from kidney failure is a very bad thing.
(2) By giving one of our healthy kidneys to someone who needs it, we can prevent the illness and death resulting from kidney failure without sacrificing anything of comparable moral significance.
(3) If we can prevent something very bad from happening by doing X, and if we can do X without sacrificing something of comparable moral worth, then we have a moral duty to do X.

---

(4) (Therefore) we have a duty to give one of our kidneys to someone who needs it.

Unlike the previous example, this does not provide a conclusive refutation, since we cannot demonstrate with certainty that the conclusion is false. However, since the first two premises are true, the form of the argument is valid and mirrors Singer's, and the conclusion is highly dubious, we are justified in questioning the soundness of the argument and focusing our attention on the truth of (3). This technique—constructing an argument that (a) is valid, (b) some of whose premises are clearly true, but (c) leads to a conclusion that is false or highly dubious—is known as a *reductio ad absurdum*, or simply a *reductio argument*. (The term *reductio ad absurdum* is Latin for *reduce to absurdity*. That is exactly what we do when we show that a statement, when it is used in the way just described in (a)–(c), must be false or highly dubious.)

The second broad category of arguments to consider is that of *inductive arguments*; they play a particularly important role in moral reasoning. Inductive arguments are, by definition, always invalid but that does not mean that they are unimportant or bad arguments. True premises make it more likely that the conclusion is true but do not guarantee the truth of the conclusion. However, there are criteria for inductive arguments that are similar to the standards of deductive validity and soundness. Inductive *strength*, like deductive validity, looks at the relation between premises and conclusion. An inductively *reliable* argument is an inductively strong argument with true premises. Unlike validity and soundness, however, strength and reliability are always a matter of degree. Moral reasoning often involves inductive rather than deductive arguments, which leads to the fact that although you can conclude that a decision about the right or wrong thing to do is probably true, you are less likely to reach a conclusion that you can assert with absolute certainty.

The two most common forms of inductive argument that appear in moral reasoning are *generalization* and *inference to the best explanation*. Generalizations attempt to identify a common thread running through specific cases and fit in with goal of "completeness," which I discuss more fully in the next section. The goal is to find a broad or general rule that explains why specific actions are right, wrong, or acceptable. Here is an example:

(1) Killing off "capstone" predators degrades the environment.
(2) Polluting streams harms the ecosystem.
(3) Introducing non-native species can have serious negative impact on native species.
(4) All the previously mentioned effects are undesirable.

---

(5) Whenever possible, we should avoid changing a natural ecosystem.

The first three premises are all empirical premises and should be evaluated in terms of their factual accuracy. Statement (4) is a normative premise. The most important thing to note, though, is that even if all the premises are true, they do not guarantee that the conclusion is true. They make it more probable that the conclusion is true, but either a more narrow conclusion (such as "artificial manipulations of an ecosystem are wrong") or a broader one (such as "we should actively prevent any perturbation of an ecosystem") might be better, in the sense that it is more likely to be true, more complete, or both.

When evaluating generalizations, you need to consider two main points. First, the specific examples should cover an appropriate range of cases; the broader the conclusion is, the more different types of examples should be considered. Second, you need to be scrupulously honest about looking for counterexamples: cases that count against the generalizations. Such examples make the argument inductively weak and may also suggest ways in which the conclusion should be reformulated. Thus, in the preceding argument, if you find cases in which introducing a new species has actually benefited the ecosystem, (for example, Canadian wolves in Yellowstone or ring-necked pheasants all across the U.S., although both cases are controversial) the entire argument would be weaker than it first appears.

A second common type of inductive argument frequently used in moral reasoning is *inference to the best explanation*. It shares certain similarities with generalizations in that both try to identify a common factor among cases mentioned in the premises. In fact, the line between these two sorts of arguments can get quite fuzzy. However, generalizations just suggest a broader category into which the cases mentioned in the premises might all be located, whereas an inference to the best explanation offers, as the name implies, an explanation about what makes the premises true. Thus, it offers a way of identifying, defending, or evaluating moral principles that takes you beyond mere generalizations. Here is an example:

(1) Subjects in experiments should be given enough information about the nature of the project so that they are capable of giving informed consent.

(2) Government policies should allow farmers to decide what sort of crop, and how much, they plant in any given year.

(3) Although we might encourage people to donate money to famine relief efforts, we should not require them to do so.

---

(4) Legitimate moral actions should respect the autonomy of moral agents.

The distinguishing feature of inferences to the best explanation is that they try to identify a common theme that runs through all the situations described in the premises. As with any inductive argument, even if the premises are all true, they do not guarantee that the conclusion is true. What is distinctive about this form of argument is that it attempts to explain why the premises are true. An inductively strong inference to the best explanation provides a plausible hypothesis, theory, or explanation. It is interesting to note, since many people think that science and ethics employ entirely different forms of reasoning, that inferences to the best explanation are at least as common in science as they are in ethical reasoning.

EVALUATING MORAL PRINCIPLES AND THEORIES

As noted in the previous section, although we cannot demonstrate conclusively the truth of a moral principle or normative premise, we can try to judge whether it can be better justified than its competitors. In this section, I examine some of the criteria by which to evaluate the justification of a moral claim. Collectively, these criteria can be called "the Four Cs": clarity, coherence, consistency, and completeness. They do not provide a mechanical algorithm (that is, a foolproof method that guarantees the right answer if you follow the rules) for evaluating moral claims—as far as we know, no such algorithm exists—but they do provide an informal checklist that we can use when trying to formulate or evaluate moral claims.

The first step in evaluating a moral claim is to make sure that you understand what it means and what it applies to. This is the criterion of *clarity*. For example, most of us would quickly agree with the statement "murder is wrong," but the criterion of clarity asks us to take a deeper look. We should try to be clear about what *murder* means: standard definitions agree that killing in self-defense is not murder, but what about engaging in some action that has death as a predictable side-effect? The same criterion of clarity is relevant to the example of Sam with which we started: Is undermining the trapping of feral pigs, when one's primary goal is to avoid causing suffering to and killing sentient creatures, have the inevitable side-effect of degrading a fragile ecosystem, a case of an irresponsible treatment of the environment? When evaluating the clarity of a moral claim, we should also consider what it applies to. Staying with claims about killing, do they apply to a fetus? A dog? A species? Plants as well as animals? It may seem as though questions about clarity are not really criteria for the evaluation of moral claim but rather raise issues that must be settled before you can decide whether a claim is justified. To some extent, this is true, which is why it is the first criterion. But some claims are simply more carefully formulated and clearer than others. All other things being equal, a clear premise is better than a vague or ambiguous one.

The second criterion, *coherence*, asks whether our moral principles fit together in a reasonable way. A classic example of incoherence is the pairing of the claims that (a) it is always wrong to kill a person and (b) convicted murderers deserve to be executed.

If we encounter an argument that includes the claims (a) we ought to respect each person's autonomy in all matters that do not involve harming others and (b) homosexuality is wrong, a more subtle coherence problem arises. It is more subtle because ways might exist of making the two fit together in a more general moral theory, but that theory would need to be spelled out and evaluated.

We are lucky because we do not have to start from scratch when we try to spell out and evaluate a general moral theory. Philosophers have been working at that for several thousand years. That does not mean that "they have finally gotten it right" if, by that phrase, you mean "have the final, correct, or definitive answer" to questions such as "What is the best moral theory?" Philosophy can give us, however, a sort of "roadmap" by identifying the most promising candidates for a reasonable moral theory, refining the early or simpler versions of a particular theory, and, above all, supporting their claims with good reasons.

The fact that nobody has found "*the* right theory" does not mean that anything goes, it's just a matter of opinion, or that all claims about theories are equally good or equally bad. We still need to evaluate the arguments. Entire moral theories can be evaluated according to the Four Cs, just as individual moral claims can.

The roadmap supplied by a formal study of ethical theories, sometimes called *metaethics*, starts by pointing out that there are two major sorts of moral theory. Together, they capture a very large portion of the ethical theories that have been proposed. The two sorts are utilitarian theories (or just utilitarianism) and deontological theories. By learning to recognize these two types, you can get at least a rough idea of what a specific theory looks like, or even use clues in a single moral claim to tell you on which theory the claim probably is based. After you have identified the theory type, you can draw from a rich philosophical background to focus on questions about or typical problems with, say, utilitarianism.

Utilitarianism, then, is a theory that argues that whether a particular action is right or wrong depends solely on the consequences of that action. The basic idea of utilitarianism is that an action is morally right if and only if it results in a better balance of benefits and harms than any alternative action that the agent could have chosen. The agent must be careful to be impartial and to give equal consideration to everyone and everything that is benefited or harmed. Sometimes this may involve identifying "the lesser of two evils," but more generally involves looking for an action that produces more good than harm, and does that better than any of its alternatives. You might be able to identify a specific moral claim as relying on utilitarian principles if you see a phrase such as "look at how much good we can do if..." or "this choice will cause less harm than...." When you know that, you should also think of the typical strengths and weaknesses of utilitarianism. For example, utilitarianism is notorious for having trouble with considerations of justice and fairness.

When we apply the Four Cs to utilitarian theories, we are likely to find that such theories get high marks for consistency and completeness. However, they may well have a problem with clarity, because it has to be clear what constitutes a benefit or harm and how those things can be measured. Economists tend to define benefits and harms in terms of dollars, whereas some others might focus on happiness and unhappiness.[6] The biggest problem for most utilitarian theories, however, is coherence, or fitting well with some very basically and carefully considered intuitions about whether an action is right.

One fairly typical way of pointing out the problem with coherence that a utilitarian theory might well face is to look at a specific case. A standard example would be something like this: Suppose that you find a man who is homeless, unemployed, and is either unwilling or unable to do anything to make society, or even the lives of another person, better. He is unhappy about his own life and he has no friends or relatives who would be saddened by his death. Utilitarianism theories often entail that if we can save several lives by killing him in order to get many organs and tissues that can be transplanted, that would be the right thing to do. This conflicts with a basic, considered moral intuition that it is simply wrong to kill a person for those reasons.

The second sort of theory, deontological, is often specifically developed in response to utilitarianism's problems, with examples such as the one just described. Immanuel Kant is the philosopher who is most closely associated with the foundation of deontology, and he created his arguments specifically as a counterbalance to flaws he identified in early forms of utilitarianism—flaws that are revealed by cases such as the one just discussed.

Deontological theories are much harder to characterize than utilitarianism is. The first step in defining these theories is to say what they are not; deontology is a *nonconsequentialist* approach to ethics, which means that actions might be wrong or right regardless of their consequences. Deontologists would probably say, in response to the example of killing a person to harvest his organs, that any instance of cold-blooded murder is morally wrong, no matter how many benefits we might gain. More generally, they would argue that there are moral principles that must be considered and that such principles would make an action right or wrong, regardless of consequences. Although Kant did not use this terminology, these principles are often expressed in the form of trying to identify some basic (human) rights.[7] When you say that "every human has a right to freedom of speech," you are basically setting some boundaries, or limits, and saying that restricting someone's freedom of speech cannot be justified by pointing to the benefits that might be gained. For this reason, another common name for deontological theories is "rights-based theories." Clue words or phrases that indicate that a claim probably falls within the deontological theories might be "we have a basic right to..." or "no matter what, doing *such and such* is just wrong," or, more subtly, "unless this action falls under one basic principle but conflicts with another basic principle, it is morally...."

The most common problem for deontologists is how to *justify* claims that "x is a basic right" or "y is a basic principle that ought to be followed." Kant argued that all such principles would be entailed by certain "maxims" that could be justified by reason alone. Although he gave several formulations (which he claimed were equivalent), the maxim that "defines" deontology in many people's minds is the principle that no person should be treated *solely* as a means to an end, but must be recognized as an end in itself. This maxim might be roughly translated as "you cannot treat other people merely as instruments to accomplish some goal or consequences; you must recognize that they have a good of their own which must be respected." Other people might defend their position by appealing to some other foundation (for example, the Bible or the U.N. treaty on human rights) but, as you have already seen, some of these defenses need to be looked at more closely.

You should think through for yourself the question of how well deontological theories are likely to fare when judged by the Four Cs. For example, if you hold "Thou shall not kill" as a basic principle or maxim, are you applying it consistently? And although deontological theories seem to be good generally in terms of coherence, there

are several standard counterexamples that show that at best, some of these theories cohere with many but not all of our carefully considered moral intuitions.

These definitions and examples are not intended to give you a full picture of ethical theories. There are some theories that do not fit into either category, for example, virtue ethics, or some versions of so-called feminist ethics. By trying to improve their arguments and claims or to respond to criticism, both main theories have evolved and often been divided into several subtypes. However, going into those subtleties would take at least one full course in ethical theory.

Coherence is basically a question of how well our moral claims fit together; coherence goes beyond questions of logical contradiction. In some cases, questions of coherence involve factual matters but more often they direct our attention to the more general moral theory in which a specific claim is embedded. Thus a series of arguments that switches back and forth between utilitarianism and deontological claims would also suffer from a lack of coherence. Generally. then, considerations about coherence require us to move beyond one specific moral claim and try to formulate the moral theory that supports the claim.

The third criterion, *consistency*, must be applied very carefully because it has some built-in pitfalls. The criterion asks whether a moral principle conflicts with our basic, deeply held moral intuitions. It is often our most important standard: No matter how clear and coherent a theory or principle might be, if it leads to the conclusion that it's morally acceptable to torture a two-month old infant because "I wanted to see what it would feel like to do that," we ought to reject it. The pitfall is equally clear: What we think of as basic, deeply held moral intuitions may in fact turn out to be prejudices or ungrounded values. People who happily eat pork chops but identify "It's wrong to eat dogs" as a basic moral intuition will have to dig a bit deeper.

Two important tools can help in testing the consistency of a principle or theory. The first draws on cultural relativism as an *empirical* fact. Even if we reject cultural relativism as a moral theory—the idea that what is right or wrong is simply a matter of one's society and its standards—the fact that different societies do, in fact, disagree about some matters is potentially enlightening. Just being aware that some people eat dogs frequently and others are revolted by the idea of eating a pig provides a starting point for examining our own intuitions. The second tool is often called *reflective equilibrium*. It is the idea that although we are aiming at consistency and we want our intuitions and moral theories and principles to be in equilibrium, it is not always the intuitions that remain fixed. Rather, we may have to make adjustments on both sides to find the proper balance or equilibrium.

The last of the "Four Cs" is *completeness*. Completeness is a matter of how much of our moral life, moral problems, and moral decisions is covered by the principle or theory in question. Moral theories usually aim at completeness: utilitarianism offers a standard for determining whether any action is right, wrong, or neutral. Most of the

moral principles we use and encounter in arguments are less grandiose, but a principle that applies only in very limited cases should be examined carefully. Someone who relies on very different standards for the treatment of research dogs and pets should think carefully about the criterion of completeness and ask whether a more general (that is, more complete) principle exists that covers both cases.

The criteria of completeness, coherence, and consistency together figure in a useful strategy for analyzing theories and principles. This strategy is sometimes called *the argument from morally relevant difference*. The basic idea is that if we think we are justified in assigning different moral evaluations to two different cases, we must be prepared to identify what the difference is between the two cases, and why the difference is morally relevant.

Obviously, then, both general principles of logic and critical thinking, and specific features of moral reasoning, should be employed to fullest extent possible, either when we are trying to understand and evaluate someone's claim about what is morally right or wrong, or when we attempt to formulate our decisions about ethical issues. Practicing these skills should help us avoid succumbing to the myth that ethical decisions are 'just a matter of opinion'. In an academic setting, they can help us formulate more thoughtful responses to case studies. In real life, they may help us reach better decisions about how to act.[8]

Turn to Appendix A and perform Exercise 3.G.

NOTES

1. Materials for the sections "Identifying Reasons and Conclusions" and "Getting to the Point: The Conclusion" have been adapted from Lilly-Marlene Russow and Martin Curd, *Principles of Reasoning* (New York: St. Martin's Press, 1989) and from material prepared by Martin Curd for the 1996 Iowa State University Model Bioethics Workshop at Purdue University.

2. You may wonder why the last sentence has been divided into the two statements 8 and 9. Does this not violate our rule that we should never break up an "either . . . or . . . " statement? The answer is "no." In this example, "nor" means "and it is not the case that . . . ." In general, "neither A nor B" means "A is not the case, and B is not the case."

3. When we do this, we revise the emphasis of some of our original statements. The basic idea is the same; we can worry about precise formulations after we have decided how the argument should be organized.

4. The terms *valid* and *deductively valid* are equivalent and can be used interchangeably. The only reason for adding the term *deductively* is to emphasize the difference between deductively valid arguments and inductively strong arguments.

5. It may be noted that there are many valid arguments according to the fundamental

(semantic) definition of validity (definition 1) that do not possess a valid argument form in either categorical or propositional logic. For example, "John runs quickly" validly entails "John runs." Definition 1 guides us in what we recognize as a valid form of argument. Unfortunately, the construction of such systems of logic is still incomplete. So, there remain valid arguments that, as yet, are not recognized as instantiating a valid (syntactical) form of argument.

6. The two philosophers who gave utilitarianism its basic formulation are Jeremy Bentham and John Stuart Mill. Both of them argued that pleasure is the most basic standard for comparison. This way of defining benefits versus harm is known as *hedonistic utilitarianism*.

7. The term *human* is in parentheses because many debates in ethics hinge on the question of whether only humans can have rights.

8. Earlier versions of this material were developed with the support of NSF Grant # SBR–9601759.

Chapter 4

METHOD

Gary L. Comstock

CASE STUDY: DENNIS THE RELATIVIST

"All this discussion of cheating being wrong seems utterly obvious to me," says Nancy, a graduate student acting as Dr. Wright's teaching assistant. She's having a strawberry daiquiri in a local bar. She continues, "I think it is wrong to cheat just as it is wrong to spread lies about a colleague to get a grant for which we are both competing. Pass me the pretzels, would you please?"

Dennis, a graduate student in molecular biology, hands her the snack. "What do you mean by 'wrong'? That no one should do it? That's not what I mean by 'wrong.' I mean an action that someone does not want others to perform. You don't *want* people spreading lies in that situation. But I see things differently. If spreading lies were the only way for me to keep my job and feed my family, or avert widespread ecological catastrophe, then spreading lies in that situation is something I would like them to do."

Nancy replies, "Well, perhaps I should have added to my initial statement the qualifier 'all other things being equal.' But there's a deep problem here. You think the term 'wrong' means 'something I don't like.' But I think it means 'objectively impermissible.' Wow. Those are two very different concepts."

"You're absolutely right about that. But there's a deeper problem: People have widely different values. Even if people agree about the meaning of an ethical term, they will find a way to disagree about the situations to which it applies. You claim that it is wrong to spread lies to get grants, but I don't think so. Not always. Suppose that the competing colleague had first spread lies about you. I wouldn't think it would be wrong to even the score; you would just be making the playing field level."

Nancy sighs deeply. "I don't agree with that analysis at all."

"Well, let me add one more complaint while I'm at it," says Dennis.

"Fire away."

"You hold the belief that cheating on colleagues is wrong only because you fear that it might happen to you, and you would not like that."

Nancy can't sit still. "Now, wait a minute. I have good grounds for holding to my original belief about the wrongness of lying. I believe it because I think human beings should be respected, and lying fails to show persons the respect they are owed."

Dennis is not persuaded. "You only believe *that*," he retorts, "because you heard it in that moral theory course you took last year."

"No," Nancy replies, "I believe it because I have good reasons. Persons are rational creatures and are able to make free choices. To lie about them is to interfere with their distinctive capabilities and therefore is wrong."

Dennis will not give up. He thinks he has caught Nancy in a circularity. "Look," he retorts, "you think it is wrong to interfere with a person's distinctive capacities only because you do not want others to do it to you. And you don't want others to do it to you because it would hurt you. And that's what I said about your dislike of lying a few moments ago; you don't approve of lying simply because you fear lies and liars. But you are going around and around, not getting anywhere. Why don't you just say, 'I don't like lies,' and be done with it? Why try to dress up your feelings in fancy philosophical language about human *rights* as the foundation of *respect* that *rational persons* are owed? Everyone knows that's just gibberish that some philosopher made up."

"Hey, don't get me wrong; I don't think this is your fault. Anyone who tries to 'do ethics' is caught in the same trap. It's all completely circular and based on emotion. Not at all like what we do in biology and science, where we have well-established methods for getting objective and verifiable results."

Turn to Appendix for A Exercise 4.A, "Discussion Questions."

DISCUSSION OF ISSUES

The sciences are descriptive disciplines in which we try to discover and articulate natural laws and regularities that in fact govern the behavior and relationships of objects in the natural world. We commonly believe that scientific laws are *discovered* in the world and that science provides real knowledge about the actual workings of nature. Ethics, on the other hand, is a normative discipline aimed at prescribing conduct; in ethics, we try to discover and articulate moral laws that ought to govern human behavior. Insofar as ethics is unlike science in this fundamental way, might it be that ethical rules are *socially constructed*, that is, simply invented by individuals and groups that cook up these rules? Is ethics therefore completely unscientific?

In his debate with Nancy, Dennis has focused attention on three reasons, discussed in the following sections, that are commonly offered for thinking that ethics bears no relation to science.

PEOPLE HAVE CONFLICTING DEFINITIONS OF ETHICAL TERMS

Dennis and Nancy are surely right about this one. There *is* great ambiguity in our vocabulary when we use basic ethical terms such as *wrong*. The word *wrong* can be used as Dennis uses it to mean *something I don't like*. Or, it can be used as Nancy uses it to mean *objectively impermissible*. These two uses point to two very different, and probably irreconcilable, meanings. Many people would agree with Dennis that the ambiguity of ethical terms is a good reason to believe that ethical words are *always* open to whatever interpretation people want to give them. But perhaps Nancy is right to protest that this relativistic conclusion does not necessarily follow?

PEOPLE MAKE CONFLICTING ETHICAL JUDGMENTS

Dennis and Nancy agree about this claim too and, again, they are both right. It does initially appear that a wide variety exists in our moral assessments. We often assume, therefore, that this is a reason to believe that ethical judgments are relative to the values of an individual or group, that no commonly shared ethical judgments are possible. But, given our experience in working through the shock treatment for naïve relativism (see Exercise 1.C in Appendix A), must we accept this conclusion?

PEOPLE CANNOT ESTABLISH THE RELIABILITY OF ETHICAL JUDGMENTS WITHOUT USING CIRCULAR ARGUMENTS

Even if two people agree on their terms and on some of their judgments, they are bothered by the fact that their judgments seem to rely on a host of unargued assumptions. The judgment that it is wrong to spread lies, Dennis complains, assumes not only that we know what *lying* and *wrong* mean but also that we can tell an instance of lying when we see one; that we are not commonly deceived about the difference between right and wrong; that people deserve respect; that lying shows disrespect; and on and on. We simply must assume the truth of all these claims in order to justify any particular moral judgment. Doesn't that mean that ethics lacks foundation?

Well, maybe. It is true that Nancy does not want others to spread lies about her because it would hurt her, and her desire not to be hurt is one of her reasons for thinking it wrong for anyone to tell lies to get grants. She is indeed drawing on her feelings, somewhat surreptitiously, in order to justify her Kantian argument about the need to respect persons as rational creatures. Furthermore, she has not provided any reasons for the legitimacy of her desire that others not spread lies about her. She thinks she probably could provide such reasons, but she has not done so yet, and she wonders, indeed, whether she would not have to draw on Kantian claims in order to do so.

It seems, at this point, that Dennis is right. Nancy is apparently caught in several illegitimate moves. First, she is trying to justify her original assertion by relying on premises that are only implicit. Logically, one should not make use of premises in an argument when one has not spelled out those premises. Second, the truth of the suppressed premises has not been established. Third, and perhaps worst of all, it is not at all clear that the truth of the suppressed premises could be established without invoking some form of the Kantian premises that are currently at issue. In other words, in order to establish any grounds at all for believing the hidden premises that she is using to support her conclusion, Nancy would have to assume the truth of something like her conclusion. The reliability of Nancy's original judgment can be secured only by invoking other beliefs, beliefs that in turn can be rendered reliable only if we assume the credibility of the first assertion. This circularity seems to be vicious, as if we are always begging the question whenever we try to justify ethical judgments.

Does ethics contain no objectivity, no truth of the matter? Dennis's question is one of the most difficult in moral philosophy because it raises the issue of whether moral judgments can be justified. To address it, I first examine how we justify scientific judgments. Getting a clear picture of that procedure will provide us with a foundation from which to explore the question of whether any analogous method exists in ethics.[1]

Many of us typically think that scientific judgments are justified on the basis of what we have learned to call *the* scientific method. But, as Ernan McMullin and others have pointed out, science uses many different kinds of methods.[2] The Babylonians, interested primarily in astronomy, were interested in prediction. Aristotle, on the other hand, was primarily interested in explanation. Evolutionary biologists and geologists typically are less interested in predictive power than in natural history. Scientific endeavor consists, in sum, of different models, aims, and, most significantly, methods.

Nonetheless, many students at least seem to think that one method exists, a "value-free" form of inquiry, that begins with pure observations, leads through experiments to facts, and ultimately leads to true theories. We begin with observations, and on the basis of reliable observations, we construct hypotheses, which we then test using controlled experiments. When we have proven a hypothesis, we have the basis for constructing a scientific theory. This method, we are told, provides objective knowledge that cannot be doubted, is infallible, and does not need support from other sources. *The* scientific method provides knowledge that is indubitable, incorrigible, and independent. It is knowledge that is fundamental, the standard against which all other kinds of knowledge claims are to be judged.

Examining the "Scientific" Method

Following is a discussion to see whether this is how modern biology actually works. Take the case of the discovery of Pfiesteria Piscicida, a toxin that has been killing fish in eastern coastal waters of the United States and that killed half a million fish in the lower Neuse River in North Carolina during five days in July 1998.[3] Pfiesteria is a dinoflagellate, a microscopic, mostly one-celled organism that lives in tidal estuaries. It is an intriguing organism. Botanists seem to think that dinoflagellates are plants because some of them thrive through photosynthesis. But other dinoflagellates eat protozoans, so these organisms probably ought to be classified as animals. The growth of Pfiesteria appears to be spurred by agricultural fertilizers, urban runoff, and animal wastes from hog confinement operations that leach into waterways. The microbe causes lethal lesions to grow on fish and biologists fear that it may affect the health of humans as well. People who have come into contact with Pfiesteria have complained of disorientation, temporary memory loss, and skin infections. The prospect that Pfiesteria might attack people if they so much as wade in North Carolina streams is not welcome news and is especially troublesome for North Carolina agribusiness and tourist industries.

Intriguing ethical twists in the Pfiesteria story revolve around the aquatic botanist Dr. Joanne Burkholder of North Carolina State University, perhaps the world's leading authority on the organism.[4] Following is a review of the rather mundane steps by which Pfiesteria was discovered.

In 1988, a Ph.D. student named Stephen Smith was working in the lab of a fish pathologist, Dr. Edward Noga, at the Veterinary College of North Carolina State University. Smith was trying to figure out how the immune systems of tilapia, a small and common African fish, would handle exposure to parasitic organisms. Smith believed that the dinoflagellates killed fish by attaching themselves permanently to gill tissue and mucus membranes. But before Smith could even begin his experiment, all the fish in his 300-gallon holding tank died as he was raising the salinity level to approximate that of the ocean.

What was wrong? He had a hunch that something had failed with his equipment, that the biological filter was defective or the air supply hose had been unintentionally disconnected. I quote from Barker's book:

> Using a kit designed for just these situations, he checked for the usual environmental toxicants and was surprised to find that the ammonia levels were just fine, as were the nitrite and the pH. Since he was unable to figure out what was wrong with the water by the obvious methods, and as there was obviously something clouding the water, he took a sample up to the laboratory and looked at it under a light microscope, where he found the specimen swarming with weird little organisms (29–30).[5]

Smith had just discovered Pfiesteria. How did he do it?

Intuitions

Smith began with a sort of nondescript feeling: the fish in the holding tank were killed because of a *failure of equipment*. It seems appropriate to call it a feeling because it was completely unsubstantiated. Smith had no evidence for it; it came as much from his training as anything and he landed upon it more or less by default. Years of working in fish toxicology labs made it second nature for him to suppose that when something goes seriously wrong, the first thing to do is check the equipment. Notice that this feeling was part of his personal agenda. Smith *wanted* to figure out the problem, he was *interested* in the solution, and, for all we know, he *desired* to get the project back on track quickly so that he could go on a brief vacation with his wife and daughter. The feeling was not disassociated with his personal values or from the interests of his scientific community. He cared about the integrity of his work, and his community cared about the accuracy of his experimental results. So, having been acculturated in the scientific community's usual ways of proceeding, Smith easily hit upon the idea that he ought to check the equipment. The feeling, in other words, was not unmotivated; it was rather, to use the current jargon, "theory laden." It came from what we might call "the theory of fish toxicology lab science."

All this points to one conclusion: Although Smith's feeling was a good one, he was nonetheless in a rather insecure epistemological state. He had no evidence that his hunch was true and he had not articulated to himself or anyone else any reasons that he ought to believe it. He was simply acting on an inherited tradition, on a belief he had acquired from his community. Were it not so distasteful a word to the scientific community, one could almost say that he was acting on intuition.

Checking Procedures

Hunches and intuitions can be made well or poorly, and they can be based on good training or self-interested bias. Did Smith get the feeling that the equipment malfunctioned because he wanted to get out of the lab as soon as possible and this seemed to be the easiest solution? Or was his feeling based on his best scientific lab instincts? Notice that I am not (yet) asking whether the intuition was true or false but only whether it occurred in good faith. The intuition may have been motivated by defensiveness: Smith's fish die; he lays the blame on someone else. We can imagine Smith saying, "Ken, that graduate student idiot, must have sabotaged my experiment by yanking the hose off the tank."

Smith probably did not know at first whether his initial feeling was biased or not. So he ran a few checks: Did he have reason to suspect Ken? Should he have discounted his hunch because of his poor relationship with Ken? Or because of other defects in Smith's personality? Was Smith prone to believing conspiracy theories—did he think his neighbors were out to get him? Was he sick of working with Ken, fed up with tilapia, frustrated by North Carolina State University, resentful of Dr. Noga? Did he have some secret reason to sabotage the experiment? Was he simply absent-minded, turning off the air supply while flicking off the lights as he left for the evening?

Suppose that as Smith was examining his motives, he found that the answer to each of these questions was negative; he was as honest and diligent as the next postdoc and he has only the best of relations with everyone in the lab. He might then check his assessment of himself against the assessments of others in the lab, with his wife, with others in the department, with his friends in his neighborhood.

Suppose that all these tests proved satisfactory. Everything was in order; Smith found no reason to think that his initial hunch was a defensive strategy for avoiding a real problem. Having completed a checking procedure of examining motives for his belief, Smith now had a sufficient reason to think that his intuition might be sound.

Considered Judgments

We might call Smith's epistemological state at this point one of "considered judgment," in which he possessed an intuition that he had carefully scrutinized and then reaffirmed. On the basis of this and other "screened intuitions," Smith was now entitled to perform some quick inductions and so to produce an hypothesis.

Hypothesis Formation

Smith's hypothesis needed to explain the following two facts. First, in prior experiments using the same tank, water source, and species of fish, no massive die-off occurred. Second, during the most recent experiment, under the same conditions, all the fish died rapidly. We can imagine Smith forming the following hypothesis: "The fish do not die if I insure the proper functioning of all equipment. They do die if crucial pieces of equipment malfunction." Smith could then test the hypothesis and, in the real-life case, he did. The result surprised him. The equipment functioned properly, but all the fish died. Replicating his experiment enabled him to falsify his hypothesis.

Notice how many assumptions Smith had to make to test his hypothesis. He had to assume that:

- The fish shipped to him by the scientific supply company really were tilapia and not a near relative.
- The hose supplying air to the tank was not infected with a toxic substance after it passed the quality assurance test of its manufacturer.
- The hose was not infected after it reached his lab.
- The chemicals he used to disinfect the hose before installing it were not contaminated with foreign compounds.

- The glasses he was wearing to read the labels on the chemical jars were not distorting his vision causing him to think he was reading something other than what the manufacturer printed on the jar label.
- His eyes were trustworthy.
- He was not suffering hallucinations from nerve damage.

Were we to pause for a few minutes, we could quickly fill up several pages, and eventually volumes of books, with entries detailing everything that Smith had to take on faith in order to complete the most mundane of experimental procedures. If we had the time, we could compose long lists of propositions stating assumptions of Smith's experiment. And, as the last two items on the list suggest, these assumptions would reach all the way down to the reliability of Smith's own cognitive and perceptual capacities. For he was taking it on faith—he was, in other words, not testing the assumption at the moment—that even his eyes and neural system were reliable. In trusting our own senses, our own observations and memories, we have no independent deductive argument to justify us in accepting the claims of the apparatus in question. What we have are circular arguments—arguments based on many assumptions. Again, there is no shame in this condition, because science apparently works perfectly well in spite of the fact that scientists cannot independently prove their every assumption.[6]

Of course, none of the assumptions need remain an assumption forever. We can decide to hold any one of them up to the light of critical scrutiny. We just cannot hold all of them up for scrutiny *at the same time*.

Smith took his intuition and turned it into an empirically testable hypothesis. How did he find out that it was false?

Hypothesis Testing

He tested it, and proved that it was not faulty equipment that was causing his fish to die. He did not know at first know what the reason was, but he and his mentor, Dr. Noga, suspected that it might be the tiny dinoflagellates clouding up the water. They knew little about these organisms, so they contacted Dr. Burkholder. What was the first thing she did? She repeated Smith's experiment and her test results corroborated his findings. To test a hypothesis, we hold all things constant except for one or perhaps two key variables. We make a prediction based on our expectations about what ought to happen and then we see whether we are right. We then replicate the results.

After Smith's original intuition proved false, Burkholder performed a variety of novel experiments that proved another of Smith's original beliefs false. At the beginning of his work, you may recall, Smith believed that parasitic dinoflagellates attacked tilapia by permanently attaching themselves to fish tissue. Burkholder showed that some Pfiesteria do lethal damage to fish in one stage of their life cycle; then they detach themselves from the fish, transform themselves into another stage, and drop to the bottom of the tank. Repeated experiments by other labs later confirmed Burkholder's hypothesis.

Scientific Principles

Burkholder produced what was, apparently, a new scientific explanation, or principle, in the history of aquatic ecology: "Pfiesteria produce toxins that kill fish without permanently attaching themselves to the fish." Notice that this principle does not purport to state merely Joanne Burkholder's own personal opinion, nor an opinion she happens to share with Smith. Nor is it a statement summarizing the results of a vote among Pfiesteria specialists. If her principle is true, it is true whether or not she believes it, whether or not Smith wants it to be true, and whether or not the state of North Carolina Chamber of Commerce has a favorable attitude toward it. And, if the principle is false, it is false whether or not she believes it and whether or not Smith and the Chamber of Commerce believe it. It would be very strange were it the case that Burkholder's principle was true for her but false for the Chamber of Commerce, true for blue-eyed Methodists but false for brown-eyed Catholics. The reason is that true scientific explanations state some fact about the universe and this fact remains whether any humans know it or not.

Of course, our degree of certainty in the truth of the principle may be very low, in which case we will want to be diligent about reviewing it. If a higher-powered scanning electron microscope comes on the market, we may want to take better pictures of the guts of the Pfiesteria to confirm prior results. If a new form of an even smaller dinoflagellate is discovered and we suddenly have a reason to suspect that it is doing the damage formerly attributed to the larger Pfiesteria, then we ought to reexamine the principle. On the basis of new observations and tests, we are justified in revisiting, and are even required to revisit, scientific principles we previously accepted. We test again and again. For that is the way science proceeds: begin with intuitions, check them in an initial screening procedure, form hypotheses, test them, reach scientific judgments about their truth, assign to them appropriate degrees of confidence, retest them when they are called into question, and so on.

But this is not the end of the story.

Scientific Theories

Scientific principles are, as Ernan McMullin puts it, questions rather than answers. Individual explanations are not satisfying on their own and they seem to invite attempts at systematization. We see groups of individual scientific principles and wonder, what is the whole explanation? Higher-order general explanations are called scientific theories.[7]

Smith and Burkholder were alone in their inquiries. Scientists have performed a wide range of experiments over the course of several years, and their conclusions all point in a single direction; Dinoflagellates kill fish by interfering with certain biological pathways. Now, if we wanted to turn this explanation into a theory, we would have to enlarge its scope in order to explain a large body of perceived regularities. A very general scientific theory about the way in which toxic dinoflagellates kill fish and cause sickness in humans would have to include a huge range of claims from not only the disciplines of cellular and molecular biology but also genetics, marine ecology, epidemiology, and even sociology. It would have to include such laws as:

- Basic environmental interactions can be explained in terms of natural processes such as aerosol-based dispersal of contaminants and blood-based dispersal of toxins.
- Human interactions with the environment can be understood as a set of interacting subsystems of the larger earth science system.
- One ecological subsystem consists of relationships that can be characterized in part in terms of potentially harmful chemicals produced by aquatic microorganisms.
- There is an epidemiological subsystem at the level of the individual organism (such as a person), the terms of which allow us to explain causal relations between the presence of symptoms such as disorientation and pneumonia in a patient and the presence in that patient of toxins produced by Pfiesteria.

The overarching background theory that lends coherence to these various laws will be comprehensive and detailed, and will include many other statements, such as:

- Humans with high levels of exposure to environmental pathogens are more likely to experience dizziness, disorientation, and hastened mortality than humans with low levels of exposure.[8]
- Each of the two genes at a locus has a 50 percent probability of being the single gene at that locus carried by a particular gamete (Mendel's law).[9]
- Basic biological elements can be characterized by atomic weight and chemical composition.

And so on. We come to accept overarching scientific theories not on the basis of observations alone but also on the basis of their coherence, simplicity, and elegance, along with their capacity to synthesize, unify, and explain, as William Alston puts it, "a vast body of otherwise heterogeneous and unrelated empirical generalizations."[10]

The best background theories are also fertile. On the basis of the theory and a host of additional empirical assumptions, we can make predictions about the outcome of new experiments. For example, on the basis of the theory, we might now predict that the outcome of a new experiment will lead to the following scientific judgment: The incidence of newly reported cases of pneumonia will be higher when streams are infested with Pfiesteria than when no stream is infested with Pfiesteria. Just as background theories are built up out of screened intuitions and tested hypotheses, so theories in turn serve to generate new hypotheses and intuitions. In an appropriately roundabout way, this feature of scientific inquiry helps to confirm the suspicion with which we began: that theories influence observations. The reason, in part, is that theories are themselves fecund, giving rise to new ideas.

A good theory is robust and we are justified in clinging to it even in the face of a handful of experiments that render anomalous conclusions. Good scientists do not give up on a robust theory on the strength of one contrary observation. We are justified in holding to theories, even in the face of initially contradictory evidence, until an accumulated weight of evidence from a variety of sources begins to suggest that the theory needs revision. Part of the reason is that the laws of the theory are stated at such a level of generality that a single low-level observation is unlikely to call the theory into question. However, such an event is not outside the realm of possibility, and during a time of scientific paradigm change, an accumulating number of low-level observations may in the end point to an anomaly that will make us decide to change the theory.

One hundred years ago, physical theorists believed that matter was indestructible, but an accumulation of observations has led them to reverse themselves. The fact that they reversed themselves is not a good reason, however, to think that physics is entirely subjective and naively relativistic. Not so long ago, molecular biologists held that information could flow only from DNA to RNA, but the accumulation of observations has destroyed this theory as well.[11] So, although scientific theories appear to be inductively constructed purely on the basis of value-free observations, the actual relationship between theories and observations is dialectical. Observations do not provide scientists with an indubitable and incorrigible foundation, nor do theories, hunches, or hypotheses. These various sources of scientific beliefs provide us with a web of beliefs that may forever be in need of mutual correlation, revision, and adjustment.

Scientific theories sometimes conflict and we must figure out how to evaluate them. Very complex methods for theory assessment exist, and through these methods we try to assess which theory is most adequate in explaining the phenomena; that is, which method proves to be the most coherent, simple, and fertile. The mere fact that scientific theories may conflict, however, is not a sufficient reason to suspect that we do not properly understand the phenomena that the theories are designed to explain. Conflict between theories may signify simply that we have not yet reached a level of understanding sufficient to decide which is the best theory.

The best theory will also be one with predictive power. Some sciences lend themselves more readily to predictions than others because it is easier in some sciences to deduce testable consequences from the theory. Making predictions in some branches of chemistry is reasonably easy; making predictions in some forms of ecology is notably difficult. The relationship between theories and predictions is straightforward. If the theory entails a prediction and the prediction is true, then the prediction confirms the theory. If the theory entails a prediction and the prediction is false, then the theory must also be false.[12]

I have avoided saying anything about whether nonobservable entities postulated by scientific theories actually exist. This issue calls attention to the complex debate between realists and anti-realists in the philosophy of science.[13] I think the account I have given of the way scientists justify their judgments is neutral on the issue of whether scientific judgments disclose real structures in the world.

SCIENTIFIC INQUIRY AND HUMAN INTERESTS

The way contemporary biology actually proceeds is very different from the way my undergraduate students think it proceeds. They think the scientific method is unilateral, foundational, and value free. They think that observations have no connection to the personal motives, values, and theories of the investigator, or to the social contexts of the discipline, or to the political machinations of the scientific grant award process. In fact, however, scientific inquiry is inextricably bound up with human interests. Students also typically think that the scientific method gives them facts that cannot be doubted and are free of other assumptions with which the students are working. In fact, however, it is impossible to generate any hypothesis, much less submit it to empirical verification, without making many assumptions. Students think, too, that scientific theories provide an incorrigible foundation upon which all other knowledge can be constructed. In fact, however, scientific theories have been, can be, and will be overturned.

Students also think that science is independent and self supporting. In fact, however, no way exists to provide absolute foundations for science. Any such attempt must appeal to premises derived from human observations, and human observations are themselves part of the perceptual practice of science. To appeal to the truth of observations when one is trying to establish the reliability of the cognitive method that itself relies on observations is to beg the question. Science has no sufficient noncircular argument to secure the truth of the scientific method of acquiring knowledge. This fact does not mean that scientific knowledge is subjective or untrustworthy; it is simply the way the world is. I hope I will not be misunderstood; I am not arguing for anti-realism or that science is unobjective. The description of scientific knowledge offered here does not lead to skepticism. It leads only to appropriate epistemic humility about science and healthy doses of circumspection when passing along its findings to others.

How Do We Justify Ethical Judgments?

I want to suggest that in ethics we are in approximately the same position as we are in science when it comes to finding warrants for our judgments. In ethics, we often begin our inquiry with little more than an intuition, and we make thousands of assumptions in trying to defend moral judgments. We cannot question all our assumptions simultaneously, but neither is any assumption above individual scrutiny. Neither intuition nor theory provides an indubitable foundation for our values, and there are competing, mutually contradictory ethical theories. These features of ethics might incline us to agree with Dennis, that ethics is entirely unscientific; however, given the previous account of scientific justification, you may already see that such a conclusion would be exactly the wrong one. Ethicists seem to be in no worse epistemic shape than are scientists, and scientists seem to be subject to as many assumptions as are ethicists. Indeed, all the features just mentioned are the ones that render ethics most like science.

There is, for example, at least one method for checking the reliability of moral intuitions and justifying moral judgments. The method is called *coherentism* and has been developed during the last three decades as a method for theory construction and decision making in ethics. John Rawls, a Harvard philosopher, outlined it originally, and it has been developed by prominent philosophers convinced that theory acceptance in ethics is analogous to theory acceptance in science. The underlying idea is that ethics involves bringing together a variety of moral and nonmoral beliefs, considered intuitions, and background scientific theories so that all our values can be rigorously examined and, through mutual adjustment, formed into a coherent system. The goal of ethical inquiry is, in Rawls's phrase, to attain "reflective equilibrium" between these various inputs.

The following discussion examines how this method might work by applying it to the Pfeisteria case.

Intuitions

Joanne Burkholder has been a lightning rod in the scientific community in part because she represents ethical values that are widely accepted. Now, to my knowledge, Dr. Burkholder has not publicly revealed what her ethical conclusions are or how the argument might go for those conclusions. But suppose that a fictional character called Jean Burmeister, who is in a position similar to Burkholder's, expressed the following moral judgment:

> The state of North Carolina should fund my scientific research program
> because it will protect the people of North Carolina from Pfiesteria.

Notice that this is a normative claim; the word *should* gives us a clue that Burmeister is making an ethical assertion about what the state of North Carolina morally ought to do. Normative claims cannot be assessed using only empirical techniques; we must use philosophical techniques to determine whether a normative claim is justified. In response to Burmeister, I can already hear Dennis objecting, "Well, that's only her opinion." Would he be right?

I think so. At this stage of the inquiry, the normative value stated previously appears to be a kind of feeling, a hunch on Burmeister's part about the obligations of state governments to citizens and about the role of state-funded scientific researchers in protecting public health. Burmeister, we may assume, has no evidence for the feeling and has landed upon it by default. She has no other explanation of her views at present, and this one is familiar to her from her days of thinking about her role as a tax-supported scientific researcher. She has worked in the role of public servant for years and has been acculturated in our secular democratic political system. Her feeling, as Dennis might point out, comes from her environment and is motivated by her own interests in securing funding.

All this is true. Our initial moral hunches are not free of our personal values or communal upbringing. Burmeister cares about the health of North Carolina residents and wants very much to do the right thing in her professional life. That is why she articulates her initial feeling in the way that she does. Her feeling is theory laden; it fairly drips with the ideal of the modern liberal state.

I have stipulated that our fictional Dr. Burmeister does not have the conceptual tools or knock-down arguments at the beginning of her ethical inquiry to justify calling her feeling anything more than a feeling. Because I have set up the thought experiment this way, we may say that she is in the same epistemological state that Dr. Smith was in when he had the feeling that he ought to check his equipment. Neither one can articulate sophisticated reasons for his or her starting point, but neither needs to do so. We start with intuitions in ethics and in science. No problem, for one might well ask: Where else *could* we start?

Checking Procedure

Hunches, as Dr. Smith found out, can be wrong. Burmeister's moral intuition might be wrong, too. Does the state have the obligation she identifies? To answer that question will require some work in ethical theory. But there is a prior set of questions that she must address. Her moral feeling is not that some Pfiesteria scientist or other has the right to receive the North Carolina taxpayers hard-earned dollars. It is, rather, that *she* has that right. Does she? Or is she espousing this value only to support her contention that she ought to get a grant? Perhaps Burmeister is flirting with duplicity here, not endorsing the feeling stated previously at all, but mouthing it only to give the appearance of moral respectability to her greed for funds.

The second step in ethical inquiry is to check our intuitions to make sure that we are not acting merely out of self-interest. Is Burmeister deceiving herself and us, espousing a moral value only because it serves other, darker, motives of hers? There are widely accepted ways to proceed here. Burmeister can ask herself whether she has a secret agenda. Am I prejudiced? Overly self-interested? Do I have a habit of saying things I don't believe? Suppose that she carefully considers each question and honestly answers no. She might then check her judgment against the views of others. Suppose that everyone says, "Jean, you are scrupulously honest and fair minded, a citizen of great integrity, and you have nothing personally to gain from your moral intuition." If everyone agrees, then she has some reason for believing that her intuition is not distorted by personal preference. Someone may even point out to her that the intuition might endanger some of her own self-interests, because the intuition might be taken to imply that scientific research should be peer reviewed, meaning that she should compete with other scientists for scarce research tax dollars. When we personally have something to lose as a result of one of our moral intuitions, we can usually assume that we are not biased in espousing it.

At the end of her review, Burmeister finds no good reasons to think that she is lying to herself. She has done what she could to check her moral intuition for bias. She has every epistemic right to proceed.

Considered Judgments

Passing a test for distortion does not prove that an intuition is true. Burmeister now has sufficient reason to justify calling her belief a "considered judgment," a moral conviction in which she can have confidence, because she has ascertained that the intuition has a low probability of representing merely her own individual prejudice. On the basis of such judgments she may now perform some quick inductions and produce the ethical equivalent of a scientific hypothesis: a particular moral judgment.

Particular Moral Judgment (PMJ) Formation

The judgment here needs to link the factual conditions of the decision Burmeister faces with the normative dimensions of her intuition. The result will be a particular moral judgment (PMJ), a judgment about what morally ought to be done by a specific person or group of persons in a particular context. Here is one formulation she might come up with:

PMJ #1: It is wrong for the state of North Carolina knowingly to expose its residents to unacceptable risks of disease from exposure to Pfiesteria toxins by failing adequately to fund Pfiesteria research.

Burmeister has converted her initial hunch into a particular moral judgment. How does she find out whether it is justified?

Particular Moral Judgment Testing

We test a scientific hypothesis by devising experiments to test its factual claims. We test PMJs by determining whether good arguments exist to support them. Moral arguments consist of at least one factual claim, at least one general moral principle (GMP), and the conclusion, which is the particular moral judgment. Here is a plausible, valid argument to support PMJ #1:

Fact #1: By failing adequately to fund Pfiesteria research, the state of North Carolina will knowingly expose its residents to unacceptable health risks.
GMP #1: It is wrong for any state knowingly to expose its residents to unacceptable health risks by failing to fund Pfiesteria research.

PMJ #1: It is wrong for the state of North Carolina knowingly to expose its residents to unacceptable risks of disease from exposure to Pfiesteria toxins by failing adequately to fund Pfiesteria research.

We know how to test scientific judgments. How do we test moral judgments? At least three ways are available. First, we test the factual premises using scientific means. Is Fact #1 actually true? Should it turn out to be false, then this argument cannot support the PMJ #1. Of course, PMJ #1 would not thereby be proven false, because other arguments, still to be considered, might justify it. Second, we ascertain whether the argument is valid by asking whether we have made any logical mistakes in drawing the conclusion from the premises. In this case, the conclusion could not be false if Fact #1 were true and GMP #1 were justified, so the argument is valid. Valid arguments can be unsound, however, so the third test is to assess the general moral principle. Is it morally wrong for a state to expose its residents to unacceptable health risks? How do we assess such a claim?

General Moral Principles

One way to test a GMP is to examine its implications. In the case of GMP #1, it seems that at least one counterintuitive implication exists, as follows: Residents of the state of North Carolina may face exposure to organisms other than Pfiesteria that pose much greater health risks than the risk posed by Pfiesteria. Residents of all states face all manner of disease risks, including the risk of widespread chronic diarrhea, malnutrition, and death from waters polluted with human wastes. The state of North Carolina, therefore, regularly spends a large portion of its budget supporting the construction and maintenance of wastewater treatment plants. The state budget is not unlimited, and bureaucratic officials face hard choices.

Suppose that the only way adequately to fund Pfiesteria research in North Carolina is to take money out of longstanding programs designed to protect public health. In that case, the state might well be subjecting its residents to even greater health risks by funding Pfiesteria research. If we accept GMP #1, however, we would be led to the particular moral judgment that the state of North Carolina is *obligated* to fund Pfiesteria research even if it means taking money away from other projects and thereby placing its citizens in harm's way. This implication of GMP #1 is, however, deeply counterintuitive. GMP #1, we may conclude, is not justifiable. So we throw it out, or at least look for ways to qualify it. Here is one idea:

> GMP # 2: It is wrong for any state to expose its residents to unacceptable health risks by failing to fund Pfiesteria research *unless failing to fund Pfiesteria research is the only way to prevent even greater health risks.*

In the course of ethical inquiry, we would then test GMP #2, repeating the procedure by asking whether it leads to PMJs that are counterintuitive. If we find that it has no counterintuitive consequences, and if we find that the principle has many plausible implications, then we have gone a long way toward justifying the principle.[14]

It bears noting that if we substitute GMP #2 into the original argument, we get a new PMJ:

> PMJ #2: It is wrong for the state of North Carolina knowingly to expose its residents to unacceptable health risks from exposure to Pfiesteria by failing adequately to fund Pfiesteria research unless doing so is the only way to prevent even greater health risks.

The aim in ethics is to construct an argument in which all factual claims are true and the GMPs lead to many plausible PMJs and no counterintuitive ones. If we make no mistakes in reasoning from the minor premises to the conclusion, then we have done all that we can to test our PMJ and we are justified in holding to it.

When we arrive at moral judgments that have withstood years of scrutiny of this kind, we add them to our list of moral truisms. Notice that these truisms (it's wrong to drown babies, it's right to do your job, it's right for state governments to protect their people from dangers) do not state mere personal opinions, nor are they the result of votes among moral specialists. If PMJ #2 is justified (I'm not asserting that it is justified but only asking you to suppose that it is), then it is justified whether Burmeister believes it or not, whether you or I believe it, whether the state legislators of North Carolina believe it. In such a case, PMJ #2 would (remember that we are still assuming that it is justified) come as close to stating a moral fact about the universe as a similarly well-justified claim in the life sciences would come to stating a biological fact about the universe. Of course, we have not established that PMJ #2 is justified, and should we discover another widely accepted PMJ that contradicts it, then we would have reason to believe that it may not be justifiable. Or if we came to accept a different moral theory than the one we currently accept, we would also have to see whether the new theory entails the contradiction of PMJ #2. On the basis of new arguments and theories, then, we can be required to go back to values we have accepted as truisms and re-test them. Perhaps they will be overturned. This may seem like a house of cards, but that is the way ethics proceeds. It is not different in science. We begin with intuitions, check them in an initial screening procedure, form a judgment about their truthfulness, test it by reasoning about it, and then assign to it an appropriate degree of fallibility.

Ethical Theories

Now on to the most difficult and complex step. As Ernan McMullin has suggested, scientific laws are not answers but questions demanding a theoretical explanation postulating an underlying causal structure of some sort. General moral principles in turn are not answers but questions demanding a theoretical explanation postulating an underlying rational structure of some sort. As we acquire GMPs in which we have confidence, we begin to wonder whether some meta-principle exists that ties them all together. So we see whether we can raise the level of generality of the GMPs. For example, is there any reason not to revise GMP #2 to apply it to every *nation*, as well as every state? As we accumulate more and more considered justifiable moral judgments, and as we move them to higher and higher levels of generality, a moral theory may emerge. We may find, for example, that a single, simple, overarching principle exists that summarizes many of the GMPs that we accept. For example, we might decide that the following statement sums up most of our GMPs: We should never perform an action that has the consequence of leading to a lower ratio of significant-preferences-being-satisfied over significant-preferences-being-left-unsatisfied.

Or, alternately, we might decide that the following principle forms our theoretical base: We should always perform that action that best respects individuals as ends in themselves.

Here we have statements of two major ethical theories, preference utilitarianism and deontology. In science, theories can be used to make predictions. Is that possible in ethics? Well, yes, although here the predictions will be normative predictions about what we ought to do, not empirical predictions about what in fact will happen. Martin Curd explains how moral theories can lead to practical predictions:

> A philosopher, such as Peter Singer, will take a normative theory (such as utilitarianism) or some general moral principles that appear to be plausible and well-confirmed, and deduce from them consequences concerning our duty to relieve world hunger and to stop raising animals for food. These consequences may be surprising and unwelcome, but if they really do follow logically from a theory that we accept as true, then, on pain of inconsistency, we have to accept them and act accordingly.[15]

Following is an example of a practical prediction (PP) formed on the basis of an ethical theory (ET), and a moral hypothesis (MH). The ethical theory is preference utilitarianism, defined previously.

ET: Preference utilitarianism is true.
MH: If preference utilitarianism is true, then humans ought not to raise and kill mammals for food. (Because: mammals have significant preferences; to kill a mammal is to deprive it of the ability to satisfy significant preferences, and; eating meat from mammals is not a significant preference for humans to try to satisfy. Therefore, killing mammals for food lowers the ratio of significant-preferences-being-satisfied over significant-preferences-being-left-unsatisfied.)
PP: Humans ought not to raise and kill mammals for food.

Notice that, in ethics as in science, we come to accept a background ethical theory not on the basis of considered judgments alone. We also examine the coherence, simplicity, and elegance of the theory; its capacity to synthesize, unify, and explain "a vast body of otherwise heterogeneous and unrelated" (Alston) normative generalizations. To the extent that our best systems of ethical beliefs have been tested in this rigorous way, they provide us with a sufficient reason to assume that any one of our considered intuitions taken individually is justified, unless and until we have a good reason to question it.[16]

But how do we decide which theory is correct? This is as difficult a task in ethics as it is in science. Challenges to each theory will arise from unacceptable implications of the theory. For example, the first principle, the principle of utility, would sanction doing medical research on people we do not like (such as drug pushers). And the second principle would sanction the sacrifice of thousands of innocent people in order to protect one potentially guilty saboteur. But both of these particular moral judgments seem counterintuitive. It seems wrong, for example, to do medical research on people against their will just because they are unsavory to us. And it seems wrong to allow the death of many people just because we do not want to obtain information through torturing an imprisoned informant. When the implications of an ethical theory give rise to action-guides that conflict with our considered judgments, we have a reason to consider readjusting, or giving up on, the theory.

But defenders of utilitarianism and rights theories are not left without a response. Utilitarians, for example, might respond that the counterexamples are unrealistic. Rights theorists, in turn, might respond that the counterintuitive conclusions simply must be accepted.[17] And this is the way theory construction goes in ethics. We work back and forth, revising our particular moral judgments so that they match the premises of our theory, and revising our theory so that it fits with our strongest considered convictions. In sum, we start with paradigm judgments of moral rightness and wrongness and then try to construct a more general theory that is consistent with these paradigm judgments, working to close loopholes and fight incoherence. Then, because we can never assume a completely stable equilibrium, we renew the process, just as in science.

As the moral legal theorist Joel Feinberg notes, this procedure is similar to the reasoning that occurs in courts of law. On the one hand, if a principle commits one to an antecedently unacceptable judgment in a particular case, then one should modify or supplement the principle to render it coherent with one's particular and general beliefs taken as a whole. On the other hand, when a well-founded principle indicates the need to change a particular judgment, the overriding claims of coherence require that the judgment be adjusted.[18] Ethicists, like scientists, reject theories that are inadequate, inconsistent, and fail to account for a wide range of considered judgments.

In conclusion, with this understanding of ethics in mind, I revisit Dennis's original objections.

The first objection was that people have conflicting definitions of ethical terms. Yes, people disagree with each other, and they sometimes react to agreement in ethics by *trying* to redefine ethical terms to produce disagreement. But the mere fact that people disagree about ethical terms is not a good reason to think that ethics is subjective. Consider another case of disagreement. I read that some fundamentalist Christians in the state legislature of Alabama want to enact legislation to redefine the mathematical value of pi as 3.00 instead of 3.1415 (and so on). Their reason is that the Bible says that the ratio of the diameter of the holy altar in Jerusalem to its circumference was 3. Now the mere fact that some people believe that pi has a different value is no reason to think that the value of pi is subjective. People can be wrong.

As in math, so in ethics; people can be wrong in their values. Suppose that a lawyer wanted to sue the state of North Carolina for harms caused by Pfiesteria in the state's drinking water. Suppose that he knows that no hard evidence of Pfiesteria in the drinking water exists but that there is evidence of a harmless microorganism, call it Q. To strengthen his case, therefore, he decides to redefine the class of microorganisms called Pfiesteria so that it includes Q. If he is allowed to have his way, then there will suddenly be ample evidence of "Pfiesteria" in the drinking water. But what would such an absurd claim mean? Our courts would lack all appearance of justice were we to allow willy-nilly changes in the meaning of key terms, and no self-respecting judge would tolerate our imagined attorney's procedure.

As in the law, so in science and ethics. A presupposition of reasoned discourse and inquiry is agreement about definitions and a commitment to hold them stable. Therefore, if in the middle of testing a particular moral judgment, a student suddenly wants to redefine the term *wrong*, we simply must refuse. We would get no further in ethics than we would in science if we allowed wanton obfuscation.

Dennis's next objection was that people make conflicting ethical judgments. Several comments are in order here. First, we have already noted that there are a vast number of PMJs on which we agree. So the extent of disagreement may be overestimated. Second, we can disagree only with claims we understand, and we can understand claims only if we understand all their key terms. Much disagreement on ethical issues may be more rhetoric than reality because the partners to the controversy are using different definitions.

Third, ethics is hard work. It is easy to bail out of an ethical argument by declaring disagreement when one has not done the necessary work of understanding, explaining, justifying, and theorizing. Before we declare that we disagree with someone's moral judgments, we ought to be able to give an account of those judgments that will satisfy our partner. If the disputants committed themselves to even this minimal level of mutual understanding, they might find that they disagree about less than they like to imagine.

Yes, it sometimes seems that we make no progress in ethics but, again, we might be wrong here. In the United States two hundred years ago, few people thought that African-Americans should be free; that women should be allowed to vote; that horses that kick should not be beaten. Today, it would be difficult to find many United States citizens who think blacks should be enslaved, women should be disenfranchised, and animals should be abused. The reason students think of ethics as an area where no progress is made may be that they focus on recent, very difficult questions, such as abortion and euthanasia. A little historical perspective provides an effective antidote to such constricted vision.

Finally, Dennis protested that we cannot establish the reliability of our ethical judgments without using circular arguments and a host of unargued assumptions. True. In ethics we simply assume the truth of a large body of considered judgments (for example, the truisms we collectively produced in Chapter 1), and of an elaborate background normative theory, if we are effectively to test any one particular moral judgment. However, this fact need not undermine our confidence in the reliability of any of our values. In ethics, every judgment is potentially open to revision, no judgment is ever beyond question, and we make thousands of assumptions every time we try to argue about ethics. But there is no reason to be concerned; we proceed in exactly the same way in science.

Further, in ethics no noncircular sufficient argument exists to establish the reliability of any one of our values. In ethics, epistemic circularity is inevitable and, as William Alston explains, to establish the reliability of any claim we must always "make use of premises derived from the practice under consideration . . ."[19] But there is no reason to run and hide here, either. As we have seen, biologists are caught in the same circularity. The circularity exists, but it is not vicious.

Dennis, in sum, is right. In ethics we do not have indubitable, infallible foundations. We have intuitions that emerge from the cultures in which we live. We have a web of beliefs that are motivated by human interests, deriving support from a multitude of sources. We cannot question all these sources simultaneously. But we can work dialectically, back and forth, mutually adjusting considered moral intuitions and general moral principles, examining arguments and testing theories, trying to construct a system of beliefs in which all our sources of information are in equilibrium. Subjective and unreliable? Not at all. It is the way we ought to proceed if we are interested in getting at the truth.

We have focused on one of the most difficult questions in moral philosophy, the question of whether particular moral judgments can be justified. We have seen that there is at least one method. Truths of ethics are truths, as James Rachels puts it, of reason. "The 'correct' answer to a moral question is simply the answer that has the weight of reason on its side" (1993, 40.)[20] In trying to find where the weight of reason lies, ethicists make truth claims, test them according to widely accepted methods, and offer practical predictions and explanations. If this account of ethics is correct, then more similarities exist between ethics and science than we typically realize. Students probably need to raise their opinion of their epistemic position in ethics while lowering their assessment of their epistemic position in science.

In science, students are probably in worse shape than they like to imagine, whereas in ethics they are probably in better shape than they allow themselves to think.

NOTES

1. I presented versions of this chapter at a symposium, "Ethics in the Practice of Science," at the Luso-American Development Foundation, Lisbon, Portugal, May 4–5, 1998, and at the Bioethics Institutes at Illinois and Oregon State. On this subject, I have learned much from Martin Curd, who presented two lectures on this subject at the Purdue Bioethics Institutes, and from Ernan McMullin, who lectured on the philosophy of science at the Lisbon conference.
2. Cf. Ernan McMullin, "A Case for Scientific Realism," in J. Leplin, ed., *Scientific Realism* (Berkeley, CA: University of California Press, 1984), 8–40; reprinted in Daniel Rothbart, *Science, Reason, and Reality: Issues in the Philosophy of Science* (Forth Worth: Harcourt Brace, 1998), 440–464.
3. "Pfiesteria Outbreak," *USA Today* (30 July 1998), 3A.
4. I recommend Rodney Barker, *And the Waters Turned to Blood* (New York: Simon & Schuster, 1997). In that work, Dr. Burkholder is described as complaining about ethical violations in her pursuit of her research. She expresses concerns that funding agencies intentionally ignored her work; that the state of North Carolina was negligent in failing to underwrite her work; and that colleagues competing for funds harassed her. My purpose in describing Burkholder's work is not to weigh the merits of her ethical charges but rather simply to describe the scientific method that was pursued in discovering the dinoflagellates.
5. Cf. "Toxic-algae crusader famous, but still furious," *The News & Observer* (Raleigh, North Carolina), 18 May 1997. Thanks to Ken Tenore for bringing these Pfiesteria resources to my attention.
6. Cf. William Alston, *Perceiving God: The Epistemology of Religious Experience* (Ithaca: Cornell University Press, 1991).
7. In cell biology, therefore, we observe that something in the cell directs the growth

of organisms; we decide to call it a gene. We further observe that biochemical structures in the cell direct the production of proteins; we hypothesize the existence of chromosomes. By inference and explanation, we construct a model designed to account for the phenomena, and we derive a theory of molecular biology. Other sciences proceed similarly. In soil science, we observe that different soils have different filtering capacities, and we theorize that chemical leaching tends to increase with increasing soil permeability and decreasing soil dissipation capacity. Slowly, we build a model of the transport of liquids through soils. Cf. W.D. Reynolds, C.A. Campbell, C. Chang, C.M. Cho, J.H. Ewanek, R.G. Kachanoski, J.A. MacLeod, P.H. Milburn, R.R. Simard, G.R.B. Webster, and B.J. Zebarth, "Agrochemical Entry into Groundwater," in D.F. Acton and L.J. Gregorich, eds., *Toward sustainable agriculture in Canada*, Centre for Land and Biological Resources Research, Research Branch, Agriculture and Agri-Food Canada Publication 1906/E (1995); on the Web at http://res.agr.ca/canis/publications/health/_overview.html.

8. "Career radiation doses for 8,961 male workers at the Calvert Cliffs Nuclear Power Plant (CCNPP) were determined. ... On average the workers experienced mortality from all causes that was 15% less than that of the general population of the U.S., probably due to healthier members of the population being selected for employment," R. Goldsmith; , J.D. Boice, Jr.; Z Hrubec; P.E. Hurwitz,: T.E. Goff; and J. Wilson, "Mortality and Career Radiation Doses for Workers at a Commercial Nuclear Power Plant: Feasibility Study," *Health Physics*; 56(2):139–50, Feb. 1989.

9. A.W.F. Edwards, *The Foundations of Mathematical Genetics* (Cambridge: Cambridge University Press, 1977), 3. Cited in John Beatty, "The Insights and Oversights of Molecular Genetics: The Place of the Evolutionary Perspective," *PSA 1982*, P. Asquith and T. Nickles, eds. (East Lansing, MI: Philosophy of Science Association, 1982). Vol 1, 341–355, reprinted in Michael Ruse, ed. *Philosophy of Biology* (New York: Macmillan, 1989), 210.

10. Alston, 127.

11. Craig Nelson, "Tools for Tampering with Teaching's Taboos," in Wm. Campbell and Karl Smith, eds., *New Paradigms for College Teaching* (Edina, MN: Interaction Book Company, 1997), 64.

12. I owe this point to Martin Curd, from a presentation he made to the 1997 Bioethics Institute at Purdue.

13. See McMullin reference in note 2.

14. Thanks to Fred Gifford for help in formulating this point. I learned much about science and ethics from his lecture, "The Relation Between Science and Ethics," at the 1996 Michigan State University Bioethics Institute.

15. Martin Curd, "Ethics and Science," unpublished lecture given at the Purdue University Bioethics Workshop, May 1997. Quoted with permission.

16. Alston, 306.

17. Cf. J.J.C. Smart, *Utilitarianism: For and Against* (Cambridge: Cambridge University

Press, 1973), 68; cited in Rachels, 113 (and see footnote on 204).

18. Beauchamp and Childress, 23, citing Joel Feinberg, *Social Philosophy* (Englewood-Cliffs, NJ: Prentice-Hall, 1973), 34–35.

19. Alston, 146.

20. James Rachels, *The Elements of Moral Philosophy,* 2nd ed. (NY: McGraw-Hill, 1993).

Part

LIFE SCIENCE ETHICS

Chapter 5

ENVIRONMENT

Lilly–Marlene Russow

CASE STUDY: MARIE THE ENVIRONMENTALIST

As Emily and Doug are settling into their chairs on Friday of the third week of Ag Ethics, they notice a woman standing at the front of the class next to Dr. Wright.

"We begin today," announces Dr. Wright, "to think about our duties to the environment. I'm sure you will all agree that there are obvious reasons to try to preserve nature. Farmers want to preserve the fertility of their soil so that their farms will be profitable years into the future. Eco-tourists want to preserve pristine wilderness areas so that they can get away from the hustle and noise of city life. City-dwellers want clean water and air so that their children can grow up in a healthy environment."

"Notice that each of these reasons," he continues, "is an 'instrumentalist' reason. The farmer, eco-tourist, and city dweller all want to protect nature because nature is a useful instrument and it can be used as a tool as they pursue their various goals. There is an altogether different kind of argument often given for environmental protection, however. This is a 'noninstrumentalist' argument, and we must consider it carefully."

"So what is it?" interrupts Rich.

Dr. Wright looks at him. "The argument is that nature itself has value. And this value exists in nature even if humans do not recognize it. The environment is significant even if it is not, has not, and never will serve as an instrument to some person's goals."

"Who believes *that*?" asks Rich, incredulously.

"Well, many people believe it, including today's guest speaker."

At this point Dr. Wright turns to the woman standing beside him. "This is Marie," he says. "She has explicitly asked me not to say anything more about her by way of introduction, except to add that she is a Friend of the Chatham River."

Marie laughs. "Yes, I am a Friend of the Chatham. As you know, the Chatham is a river that runs not far from here. I am the president of an organization, the Friends of the Chatham, that is dedicated to preserving the river. As you may know, the river is currently the focus of a major controversy. The city council of Springdale wants to use the Chatham for its water supply, a supply that Springdale needs very badly given the town's incredibly rapid growth. But our local farmers are objecting. They're worried that if Springdale takes water out of the Chatham, there will not be sufficient moisture for them to irrigate their crops.

"Friends of the Chatham," Marie continues, "sides neither with the residents of Springdale nor with the farmers. We have a different view. We want the Chatham to remain relatively untouched and unspoiled, with sufficient water in its banks to be of value to many different people: people who fish in it, hike along it, and boat on it. Now, I understand that I am speaking to a group of ethics students, so I want to explain our reasons.

"We have two arguments for wanting to protect the river. You might call our first argument 'humanitarian.' We believe that the Chatham river ecosystem is *instrumentally* valuable because it serves a wide variety of uses, including fishing, boating, and camping. These people enjoy the Chatham for its aesthetic and even spiritual characteristics. Our second reason is aligned with these spiritual considerations. You might call it the 'intrinsic' argument. The Chatham River ecosystem is *intrinsically* valuable, even if people do not use it or benefit from it, simply because it is relatively wild, untouched by human hands. According to the second line of argument, it would be morally wrong to pollute or use up or waste water from the Chatham simply because doing so would tend to undermine the wildness, stability, beauty, and integrity of the river's wonderful ecosystem."

Rich shakes his head in disagreement.

"Therefore," Marie continues, "I have been involved in actively opposing the Springdale town council's attempts to steal the Chatham's water. They are trying to drain the river of its lifeblood, an action that would not only harm farmers and recreational users of the river but also kill the Chatham river *itself*, and it would hurt those of us who love its wild beauty. Please join us. Nature has rights, and destroyers of nature must be stopped!"

Marie is clearly basing her case on an important controversy within environmental ethics. She has claimed that the Chatham River has a special sort of value, intrinsic value,[1] in addition to its instrumental value, or the fact that it contributes to some other good, such as an activity that humans enjoy.

She is also identified as a friend of the Chatham River, suggesting that she is not just looking at how the river can be used for human benefit but also is exploring how to take seriously the idea that the river can be looked at in terms of friendship and trying to determine what is in its own best interests. When we talk about instrumental value, we are trying to determine what something is *good for*. Money has instrumental value because it is good for buying things that we desire. A river may have instrumental value because it can be used to irrigate crops, serve as a source of water that people need, and for many other reasons. However, terms like "intrinsic value" or "for its own good" introduce another complication: the idea that something's value and how it should be factored into moral judgments sometimes go beyond instrumental value. It is a way of saying that there are things that are morally right or wrong independently of how useful they are for a specific purpose.[2] It is important to keep in mind (1) that something can have *both* instrumental and intrinsic value and (2) that intrinsic value is not always more important than instrumental value. If I am alone in a burning building and trying to escape, it would make sense for me to leave my child's fingerpainting that is hanging on the refrigerator (which has intrinsic value), and pick up a wallet containing cash and credit cards, which have "only" instrumental value.

Rich shakes his head in disagreement because, although he understands the distinction that Marie is trying to make, he has not heard good arguments to show that something, especially something like a river, can be evaluated in terms such as "intrinsic value," let alone why we need to distinguish it from instrumental value. He may even think back to the Four Cs and worry about clarity, and how whatever is supposed to have intrinsic value can be identified. What, exactly, is referred to by "the Chatham River"? The actual river bed and the water that flows over it? That, plus the immediate surroundings? The species of animals and plants that grow in or around it?

Marie does not help her case when she describes "our [Friends of the Chatham] different view" because she starts with the view that the river should remain relatively unspoiled and have sufficient water to sustain fishing, boating, and other activities, and these sound like instrumental criteria. It is similar to claims that we ought to preserve the tropical rainforests because some undiscovered plant might help us find a cure for cancer. Even when she appeals to aesthetic and spiritual values, it is important to ask whether these are truly intrinsic—whether they apply to the Chatham regardless of what people in the area think of it—or whether these too are instrumental values, even though they are noneconomic and something we think we cannot or should not put a price tag on.

In order to establish that something has intrinsic value, you need to explain and justify the claim that this thing has a good of its own that must be taken into account. For that reason, her explanation that the Chatham has intrinsic value because it is "relatively wild, untouched by human hands" is a real argument, even though it might be true. *Why* does a "relatively wild" river have intrinsic value? Since "wild" is meant to contrast it with other, tamer, or more artificial rivers—let's say the Avon River that flows through Stratford, Ontario—*why* would it be true that the Chatham has a sort of value that the Avon lacks? You might also note that it seems inconsistent to claim that she wants the river to be valued because people fish in it and so on, while at the same time claiming that is valuable because it is "untouched by human hands."

Since she has given a reasonably clear distinction between instrumental and intrinsic value, Marie has done a good job of explaining the distinction in general. Rich is, we hope, more worried about the apparent lack of *argument* to show that the Chatham River has intrinsic value. Keeping an open mind, he listens carefully to what Marie has to say next to see whether she offers an argument for her position that stands up to the Four Cs discussed in Chapter 3.

Turn to Appendix A and perform Exercise 5.A before continuing with this chapter.

DISCUSSION OF ISSUES

The position defended by Marie the environmentalist is introduced as the *noninstrumentalist* position that "nature has rights." Marie herself describes her position as an appeal to a "natural" reason, namely that the Chatham River is intrinsically valuable. In doing so, she echoes the arguments of several important authors and schools in environmental ethics who base their moral claims on the thesis that species, the ecosystem, or nature itself has intrinsic value. Some, like J. Baird Callicott, appeal to a deeper ecological understanding, of the sort advocated by Aldo Leopold as a means of demonstrating the intrinsic value of a properly functioning ecosystem.[3] Others embrace the concept of *deep ecology*, a term introduced by Arne Naess and popularized by Devall and Sessions. Finally, many branches of ecofeminism demand a personal involvement with nature as a way of justifying an "ethics of care" for nature. To understand and evaluate Marie's position, we must analyze which of these theories, if any, is implicit in her appeal, and which, if any, offers support for her defense of the Chatham River. The common denominator in all these approaches is that they all make reference to the intrinsic value of nature or some part of it that is not an individual sentient creature: a mountain, river, species, ecosystem, 'wilderness,' or nature as a whole.

Although appeals to intrinsic value figure prominently in environmental ethics, note that not all philosophers take this route. In the previous case, "Gordon the lawyer," environmentalists advanced the claim that nature can be harmed and that it is a moral affront to do so, but it is not clear that this requires an appeal to the rights or intrinsic value of nature. At least one prominent figure in environmental ethics, Bryan Norton, would agree with Gordon but would base his argument on pragmatic considerations, theories about the scientific bases of assessing ecosytemic health, and concern for future generations. Since these are also influential positions in environmental ethics, we will want to consider theories that do not appeal to the intrinsic value of nature. In the last section of this chapter, I defend a conclusion similar to Norton's—that appeals to intrinsic value are not an effective argument in favor of environmental ethics—but the reasons I give are importantly different from Norton's.

How are these claims best understood? What are their implications for real-life applications, specifically environmental policy? Do good supporting arguments exist that could be offered in their defense, and what are some problems that might afflict those arguments? The main goal of this chapter is to address these questions, but first, I put the debate in a broader context with the help of a familiar real controversy.

Before proceeding any further, turn to Appendix A and try Exercise 5.B.

BASIC CONCEPT: INTRINSIC VALUE

Statements such as "the Chatham River has rights" and "the Chatham River has intrinsic value" are often used interchangeably. They both tend to be invoked as "trump cards" designed to put a halt to utilitarian-based cost benefit analyses. The connection between rights and intrinsic value is far from necessary in either direction, but a full discussion of rights would take us too far afield. Although a discussion of different concepts of rights falls outside the scope of this chapter, a closer look at the idea of intrinsic value is essential to any critical analysis of arguments in environmental ethics. We can make significant progress in this area while setting aside, for the purpose of this discussion, a careful examination of the possibility of attributing rights to nature. In embarking on this task, I begin with a caveat. Many philosophers have distinguished between the concepts of *intrinsic value* and *inherent value* (or inherent worth). However, two problems arise with the distinction: (1) many writers use the terms interchangeably, and (2) those who draw the distinction often do so in ways that differ from other writers who want to use both terms. In what follows, I use the term *intrinsic value* to cover both ideas, but as you pursue further reading in environmental ethics, be alert for nuanced distinctions. Also, discussions of value tend to focus on things that are good, but values can be negative as well as positive.

The first feature to notice about attempts to define the concept of intrinsic value is that almost all these attempts are phrased negatively: intrinsic value might be defined as nonanthropocentric, nonrelational, or noninstrumental.[4] The most basic of these characterizations is the last, the contrast between intrinsic and instrumental value, so I start there.

As the name implies, saying that something has instrumental value is to say that it is good because it serves some *further* purpose. A $20 bill, for example is instrumentally good because it can be used to purchase food, buy a ticket to a concert, or contribute to the Nature Conservancy. Claims of intrinsic value, on the other hand, are intended to highlight ends in themselves: To say that something has intrinsic value is to say, in effect, "the buck stops here" or "this is simply a good thing." As the "trump card" analogy was intended to suggest, it puts a halt to the demand for justification. It seems reasonable to ask "What is a $20 bill good for?" or "Why is it good to contribute to the Nature Conservancy?" but inevitably the point comes at which such questions have only the answer "because it's good." Marie is claiming that having a clean, free-flowing river is such a stopping point: A healthy river is good in itself; it has intrinsic value.

The contrast between intrinsic and instrumental value also clearly shows why any ethical theory has to recognize *some* concept of intrinsic value. The theory needs a foundation, something that instrumental goods ultimately aim toward. Without such a foundation, our system of values (sometimes referred to as an *axiology*) would at best be circular and at worst be so chaotic that our choice of values would fall into an arbitrary set with little room for giving arguments, justification, or reasons for attributing values.

Utilitarianism, as we saw in Chapter 3, must specify which basic values, such as happiness, must be maximized. These would be intrinsic or inherent values. Anything that is not a basic value will be judged on its instrumental value, the way it contributes to a basic good. However, deontological or rights-based theories will often argue that some things have intrinsic value whether or not they contribute to human happiness. A clear example of an appeal to intrinsic value within utilitarianism that deliberately intends to avoid that sort of arbitrariness can be found in the classical utilitarianism of Jeremy Bentham and John Stuart Mill. Both argued that happiness was an intrinsic good, valued for its own sake, and that anything else was good only instrumentally, only insofar as it maximized happiness or reduced pain and suffering (which are intrinsically bad). But clearly these values require a subject capable of experiencing pain or happiness. This brings us to the second important question about intrinsic value: Does a "valuer" have to be in the picture?

If we continue for a moment to restrict ourselves to the simple case, if happiness is the only thing that is intrinsically good, if there were no people (or other individuals capable of feeling pleasure or happiness) *nothing* in such a world would have intrinsic value.[5] This approach to intrinsic value is commonly called *anthropocentric*, or, by at least one writer (Callicott) *anthropogenic*. The prefix *anthropo*, implying the need for a human valuer, is traditional but can be misleading. Many philosophers who are directly involved in the debate—Bentham, Singer, Callicott—argue that any sentient creature is by definition a source, determinant, or definer of value. No recognized philosophical term exists for this sort of theory; perhaps a term such as *protecentrism*, from the Greek verb *to choose*[6] would be a useful addition to our standard vocabulary. Protecentrism falls between anthropocentrism and biocentrism. It shares with anthropocentrism the emphasis on the need for valuers but does not assume that only humans are capable of valuing things. On the other hand, living things that do not have preferences—for example, the simplest forms of animals, and plants—would be included in biocentrism but not in protecentrism. Biocentrism ascribes intrinsic value to all living things, regardless of whether they are sentient or have preferences. Biocentrism is popularly associated with Albert Schweitzer and has been given a sophisticated analysis and defense by Paul Taylor. Biocentrism is also the first point of NASA's statement of bioethical principles on animal research. All three differ from ecocentrism in that they all focus on individuals rather than groups, systems, or species.

A subtle difference exists among philosophers who advocate anthropocentric, and by extension, protecentric views; this difference is, unfortunately, often overlooked. Some, like Bentham, use the term to mean that only certain sorts of experiences (happiness) can be intrinsically valuable. A tree can have instrumental value if it contributes to happiness, but it cannot itself be intrinsically valuable. I call this view *strong anthropocentrism* or *strong protecentrism*, depending on whether only human interests and experiences are considered or the experiences and satisfaction of all sentient creatures are included. Others cast their net more widely: anything can have intrinsic value if it is (correctly) valued for its own sake. Thus, on this second variation, a tree can have intrinsic value if some valuer correctly recognizes it as being good "in itself" rather than for some other purpose. I call this second view *weak anthropocentrism* or *weak protecentrism*. Dr. Wright's definition of anthropocentrists seems to cover only strong anthropocentrists, and my point is that anthropocentrism is more complex than his brief remarks might suggest. However, it is important to remember that in both variations, a world without valuers is a world without value; for that reason, both weak and strong anthropocentrism are variations within the general category of anthropocentrism and protecentrism.

In contrast to anthropocentric and protecentric views are nonprotecentic views— biocentrism and ecocentrism—and these have played a particularly important role in environmental ethics. As the names imply, these theories hold that something can be intrinsically valuable even if no sentient creature is available to recognize that value.[7] In environmental ethics, the most influential nonanthropocentric view is *ecocentrism*, the view that certain nonliving things—rocks, mountains, rivers—as well as some sorts of groups or systems—such as species, ecosystems, nature, wilderness, Gaia— have intrinsic value. It is this ecocentric view that Marie seems to have in mind when she contrasts her natural reason with humanitarian arguments.

A useful and popular thought experiment to determine whether a particular person or theory is appealing to a nonanthropocentric understanding of intrinsic value has come to be known as the "last person argument." If you were the last person (or last sentient individual or valuer) on earth and were about to die, would there be anything morally wrong with cutting down the last redwood tree or destroying the Grand Canyon? If a thesis, argument, or theory entails that such destruction would still be morally wrong, then it presupposes a nonprotecentric understanding of value.

The final issue surrounding intrinsic value, especially nonanthropocentric versions, is that a fully complete account will have to explain how we should justify and evaluate claims that something is intrinsically valuable, and thus how to settle, or at least make progress on, disagreements about whether, for example, the Chatham River ecosystem is intrinsically valuable. We have already seen that such an explanation cannot *merely* fall back on an appeal to instrumental value, but issues and questions still remain to be addressed. Generally, they are epistemological: How do we know, or justify our belief that, X has or lacks intrinsic value?[8] However, the questions might be metaphysical in nature: What qualities must a thing have to be intrinsically valuable? To address these questions fully would take us deep into disputed areas of abstract metaethical theory, so I must set them aside for now, but toward the end of the chapter, when I address the question of whether an ecosystem can have intrinsic value, I suggest some principles that can be applied. This practice is quite common in environmental ethics: Starting with a specific case or type of case and using it as a paradigm from which one can try to extract more general principles. This is certainly the case for Aldo Leopold, the father of ecology in America, and since Marie directly invokes his view, that would be a logical place to start the move from general concepts to specific positions in environmental ethics.

LEOPOLD'S LEGACY

The works of Aldo Leopold have often been cited as offering a good criterion for intrinsically valuable ecosystems. Marie's claim is based on a direct quotation from "The Land Ethic," an essay included in *A Sand County Almanac*: "a thing is morally right when it tends to support the stability, integrity, and beauty of the land, and it is wrong when it tends otherwise" (262).[9] When he used the term *land*, Leopold meant the entire ecosystem: animals; plants; water systems; even the soil.[10] This formula entails that the fundamental good, that which is intrinsically good, is a stable, integrated, beautiful ecosystem; something is instrumentally good if and only if it promotes such ecosystems.

Thus, two key themes in contemporary environmental ethics have their roots in Leopold's writing: a holistic approach and a rejection of the idea that the value of land is to be judged solely in terms of what it can produce that is useful for humans. Trained as a forester, Leopold was particularly opposed to forestry practices that turned mixed, thriving ecosystems into monocultures of a particular sort of "valuable" tree, planted in rows and managed "like cabbages" (p. 259). The following text looks at each of those in more depth and then asks whether the criterion is an effective way of determining which things are intrinsically good.

As noted previously, Leopold claims that the moral worth of any individual thing, event, or action—that is, whether it is morally good or bad—is determined by its effect on the land or the ecosystem. A forest fire might be good or bad, depending on whether it is necessary for the continued stability of the ecosystem (as it is, for example, on prairies) or whether it destroys the balance in an irreparable way or disrupts the stability of the ecosystem. The contemporary term for determining the value of a thing or event in terms of its effect on the ecosytem is *ecoholism*.

Although Leopold did not use the term, the emphasis on effects of actions would suggest that he based his judgments on utilitarian grounds. He applied the same standards to selective logging, the introduction of new species or removal of others, and hunting and fishing. Thus, when an overpopulation of deer exists, Leopold would support controlled hunting: Even though the individual deer will suffer, the ecosystem as a whole will be better off. Leopold himself was an avid hunter for a good portion of his life.

As these examples indicate, Leopold did not think that the land must remain untouched by humans, or that human interference was necessarily morally bad. In this way, he differs from most contemporary ecoholists who view any human impact on the environment as suspicious. A related question about Leopold's views is whether they are truly nonanthropocentric as they are sometimes portrayed. The obvious observation here is that "beauty," one of the three criteria, is obviously based on human evaluation.[11] To say this is not to condemn it as a reasonable criterion for evaluating ecosystems; it is merely to observe that Leopold may not be as committed to contemporary visions of ecoholism as some have portrayed him. However, this observation raises the question of what Leopold's criteria are intended to do. Do they give us necessary and sufficient conditions for determining the health of an ecosystem? Do they tell us whether an ecosystem has value, or how we ought to evaluate an action that will have an impact on ecosystem? The three criteria, remember, are "integrity, stability, and beauty." I believe taking them up in reverse order is useful.

An appeal to beauty is the most obvious barrier to the claim that Leopold is nonanthropocentric; of the three criteria, this one most clearly cries out for a sentient being to recognize beauty. Unless one begs the question, however, requiring a valuer does not necessarily disqualify it as a good or justifiable criterion. In many ways, beauty lies at the core of Leopold's approach, which has led some philosophers to dub his theory a "land aesthetic" rather than a "land ethic." Using the distinction introduced earlier, you can see that, at least with his appeal to beauty, Leopold falls into the category of weak anthropocentrism. This is important because it reveals that one can be both anthropocentric and an ecoholist. It follows from this that ecoholism and ecocentrism are not the same position, and the first does not entail the second.

The concern about an aesthetic basis for intrinsic value is that it initially seems too capricious to provide a meaningful basis for a rational environmental ethic. After all, some of us like mountains, others prefer the ocean, and still others the lights of Broadway; some would prefer to preserve an area of wetlands whereas others would rather see the "swamp" developed into a convenient Wal-Mart. However, just as people can hold educated and uneducated judgments about music and painting, so can people have educated and uneducated judgments about the beauty of an ecosystem. Leopold was quick to point out that the beauty of the land can best be judged and appreciated by those who understand how the ecology of the area works. An insect or plant that might seem boring and unattractive to the untutored eye may be recognized as a valuable gem by the sensitive ecologist. This factor does not guarantee the elimination of all disagreement, any more than a demand for informed rationality could eliminate all disputes in ethics, but it can provide a norm by which different views can be compared and evaluated.

The criterion of stability has necessarily undergone reinterpretation in order to remain a viable candidate for evaluating ecosystems. Ordinarily, *stability* suggests a sort of permanence, a lack of change; on this understanding, the surface of the moon would be perhaps the most stable ecosystem we know. However, contemporary ecology emphasizes the fact that ecosystems must constantly change and adapt to internal as well as "outside" influences (more about the reason for putting *outside* in quotation marks momentarily). Thus, a contemporary reading of Leopold's criteria would take a stable ecosystem to be one that is capable of responding to a wide variety of changing conditions while remaining a balanced system in equilibrium.

One must be careful here: Claiming that an ecosystem is unstable just because it is unable to maintain any sort of equilibrium in the face of bulldozers and concrete would be silly. Rather, we expect a stable ecosystem to respond appropriately to *normal* or *natural* forces: internal forces such as increased squirrel population or the growth of a taller and thicker forest canopy, as well as external forces such as lightning strikes. This expectation raises a further difficulty: what counts as normal or natural? As noted, Leopold allows that humans can have a positive effect on an ecosystem, that human actions that change an ecosystem can be morally good. Therefore we cannot assume that *normal* or *natural* can be defined as anything like *without human interference*.

Even if we could answer these difficult questions, we must deal with the context-relative nature of stability. That is, how stable a system is depends on the scope of our survey, both in size and time. A system that seems chaotic right now may, within a period of a month, year, or decade, return to a state of balance. Similarly, a tidal pool or flood plain might seem quite unstable as changing conditions result in dramatic changes in flora and fauna, but these systems can also be seen as part of a larger system that is stable in part because of the activity within that one small area. Conversely, an apparently stable system might turn out to be nothing more than a dormant stage in a long history of instability. For example, one concern about various attempts to "reclaim" or "rehabilitate" a system is that these efforts may produce only a temporary fix, one that will disintegrate as soon as the engineers pull out. This issue may seem empirical—what is the proper perspective from which to do scientific investigations of ecosystems most effectively?—but it brings us directly to Leopold's third criterion, integrity.

Integrity, according to the *New Shorter Oxford English Dictionary*, has as its first two definitions: (1) The condition of having no part or element taken away or lacking; undivided state; completeness; (2) The condition of not being marred or violated; unimpaired or uncorrupted condition; original state; soundness.

Given what we have already said about Leopold, it follows that his notion of integrity must concentrate on "having no part taken away" and "completeness." Although "soundness" is important to Leopold, it would already be covered under the criterion of stability. But those concepts require that an ecosystem have a definite identity if we are to have good reasons for judging it to be complete or incomplete, of having or lacking its essential parts. This brings us back to the motive for putting "outside" in quotation marks in the preceding discussion of the criterion of stability.

One of the most basic messages that ecology conveys is that "no ecosystem is an island." The Chatham River is affected by its surrounding river basin and the runoff from it. The runoff is obviously affected by land use in the immediately surrounding area (such as fertilizers and pesticides used, types of vegetation, how much land is developed and paved, and so on). Perhaps less obviously, much broader climatic effects at the global level or perhaps at the level of the solar system can also have an impact. In short, the hope of determining the boundaries of an ecosystem, what it encompasses and what lies outside it, seems more and more misguided as we gain a better understanding of both the theoretical and empirical issues involved in such an enterprise.

What, then, shall we say about the theoretical cogency of Leopold's criteria? On the one hand, at least the concepts of stability and integrity raise difficult empirical issues, important questions on which to consult the best available ecological sciences. However, no such thing as a purely value neutral evaluation exists; notions of stability, integrity, and ecosystem health[12] all involve choices about what is valuable and what the appropriate perspective is. Therefore, these issues are not merely a matter of "getting the science right." One needs to be clear about the values one is importing, even if only implicitly. What *kind* of stability, and on what scale, do we value and why? When we talk about the integrity of a system, how do *we* draw the unavoidably arbitrary boundaries? In short, we need to dig deeper into the criteria proposed by Leopold.

The second question just posed serves as a reminder of the basic metaphysical problem that became explicit in the discussion of integrity: The identity of the ecosystem. What constitutes an ecosystem and where does it begin and end, both in space and time? Some philosophers have tried to finesse the issue by insisting that nature as a whole is "the ecosystem,"[13] but doing so brings problems of its own, as you can see when I discuss ecoholism further.

If the basic concepts on which Leopold relies turn out to require more basic value judgments or to be irreparably unclear or arbitrary, we would do well to continue further in our search for a basis for attributing intrinsic value to an ecosystem and/or some of its components. Therefore, I next discuss how his ideas have fared in contemporary environmental ethics.

CONTEMPORARY HOLISTIC APPROACHES

Environmental ethics has become increasingly recognized as a legitimate part of moral theory and hence has spawned a variety of approaches. Many discussions of environmental ethics proceed within the traditional frameworks of philosophical, legal, or social ethics or become an extension thereof; I examine those more closely later, in the section "Environmental Pragmatism." Some philosophers and activists have found this allegiance to tradition an unsatisfactory way of defending the environment, however. These other philosophers, who are attracted at least to Leopold's basic approach and find the traditional approach lacking in some way (such as being too anthropocentric), have proposed what are presented as totally new approaches, or approaches grounded in other traditions not generally acknowledged in Western ethical theory. I examine two of the most influential.

ECOHOLISM

Ecoholism is sometimes thought of as another name for "deep ecology," a term coined by Arne Naess, made popular in philosophical circles by Bill Devall and George Sessions, and often associated with the "Earth First" movement. I would suggest that deep ecology is better understood as one type of ecoholism, for reasons that the following definitions should make clear.

Ecoholism, briefly, is the doctrine that a fundamental source of value, perhaps the most fundamental, is the ecosystem as a whole. Parts of the ecosystem, whether individuals, subspecies, or species, derive their value from the contribution they make to the welfare of the ecosystem. As noted earlier, ecoholism tends to view ecology as teaching us that the environment or nature as a whole is some sort of almost organic unity, thus avoiding the "identity problem" mentioned in conjunction with Leopold. In this way, ecoholism tries to combine ecocentrism with empirical claims about the environment.[14] Two influential proponents of ecoholism are J. Baird Callicott and Holmes Rolston, both of whom (particularly Callicott) see their arguments as having their foundations in Leopold.

Opponents of ecoholism often point out that this view entails an abrogation of individual rights. Tom Regan had at one time referred to this view as "ecological fascism" as a way of drawing attention to the fact that individuals or groups might be sacrificed for the good of the "state", that is, the system. It is also important to remember that this is a view about moral values, not merely the empirical claim that the various parts of the ecosystem are so interconnected that affecting any one part of it may well have an effect on all the others. Even the most radical anthropocentrist generally is becoming increasingly more aware of the truth of the latter claim, but that does not make him or her an ecoholist. As long as someone insists that the foundation of moral good rests with the good of individual humans, or individual animals, or even individual species, one is not an ecoholist. Only when one argues that all of these goods must inherit their moral value from their impact on the ecosystem as a whole or on nature does one get to the defining characteristic of ecoholism.

The previous discussion of intrinsic value reveals an intuitive conceptual appeal to a holistic approach. If intrinsic value is, as stated earlier, a trump card or a way of saying "the buck stops here," then one cannot go much further back than nature as a whole. However, we must weigh that idea against the serious concern underlying Regan's rather contentious label of "environmental fascism": According to Regan's interpretation of ecolhoisim, nothing that is merely good for an individual without benefiting the environment, will count as having moral value. Thus, according to ecoholism, donating to a charity such as a college scholarship fund, or the American Heart Association, or even saving the life of a drowning child would not be a morally good thing (unless we can figure out some way in which it benefits nature as a whole). At best, it would be morally neutral or permissible; it might even be morally bad.

Ecoholism also faces a deep and as yet unresolved conceptual issue, which can be summarized by the following question: "Are humans part of nature?" The question raises a dilemma:

(1) If humans are a part of nature, then what we do to the environment is natural and therefore acceptable.
(2) If humans are not a part of nature, then *nature* must be defined as the part of the world that does not include human interference.
(3) Either anything we do is acceptable, or else ecoholism applies only to those parts of nature that are pristine free from human interference.

The force of this dilemma is driven home when we realize that probably *no* place on earth has not been touched by humans, either directly or indirectly. To escape the dilemma, ecoholism must provide a satisfactory definition of *nature* that allows it to slip between the horns. So far, such a definition has not been forthcoming.

Deep ecology represents a radical version of ecoholism in that it adds what it refers to as a spiritual dimension, a personal experience and connection. It is also a call to action, including very far-reaching policy changes that will require population reduction and "basic economic, technological, and ideological structures . . . The ideological change will be mainly that of appreciating life quality (dwelling in situations of inherent value) rather than adhering to an increasingly higher standard of living."[15]

Environmental ethics, then, is no longer a purely rational, academic discipline for the deep ecologist. Deep ecologists often put more emphasis on wilderness than on other sorts of ecosystems, perhaps in the hope of weakening the force of the dilemma noted in ecoholism. Unfortunately, the term *wilderness* has exactly the same problems as *nature*, perhaps even compounded.[16] Moreover, deep ecology has often been accused of promoting misanthropic and elitist attitudes: misanthropic because human interests are almost always discounted; elitist because, although it is fine for comfortable academicians in the developed countries to renounce increasingly higher standards of living, it seems grossly unjust to make that a general recommendation, including to those struggling in third-world situations of extreme poverty. To sum up: Deep ecology inherits all the difficulties of ecoholism in general and creates more problems of its own.[17]

One of the interesting things about Aldo Leopold is that although he was always committed to some version of ecoholism, he seemed to move back and forth between a sort of deep ecology and a much more pragmatic sort of ecoholism, with ample room for humans. His way of escaping the dilemma would probably be to deny the first premise: Humans are a part of nature (I noted previously that he did not think human use of natural resources, or even changing an ecosystem, was necessarily bad) but humans have a proper place. Our actions are morally wrong when we ignore that sense of place. Unsurprisingly, an understanding of proper place sounds, in Leopold's writing, very much like an aesthetic appreciation coupled with an understanding of ecological principles.

ECOFEMINISM

Ecofeminism is often associated with ecoholism, and indeed many ecofeminists are ecoholists. A few even endorse deep ecology, although many ecofeminists criticize deep ecology for its emphasis on self realization rather than relationships with others, and for its failure to pay sufficient attention to social factors (see, for example, Cheney and Plumwood; for a response, see Fox). Even the association between ecofeminism and ecoholism, however, ignores several important theoretical and practical distinctions, not the least of which is the range of conclusions defended by ecofeminists. As you can see presently, an ongoing debate exists about the incompatibility of ecofeminism and deep ecology. First, however, I try to identify some common characteristics of various forms of ecofeminism.

If ecofeminism has any defining characteristic, it would seem to be an emphasis on the issue of domination. More specifically, current attitudes toward nature are seen as part of a pattern of domination: men dominating women, whites dominating people of color, colonialists dominating native people, humans dominating nonhumans, and so on. As Karen J. Warren, a leading ecofeminist, notes:

What all ecofeminist philosophers *do* hold in common, however, is the view that there are important connections between the domination of women (and other human subordinates) and the domination of nature *and* that a failure to recognize these connections results in inadequate feminisms, environmentalism, and environmental philosophy. (Warren, 1996, p. x)

Domination is often linked to *dichotomized* thinking: Separating the world into *us* and *other* is often a first step toward the view that *we* are more valuable. Thus, many ecofeminists emphasize connections and community rather than a search for differences.

A second theme often found in ecofeminism and in feminist ethics more generally is the claim that traditional moral theories overemphasize rationality and impartiality and undervalue feelings, connectedness, and personal relations. This theme has its roots in Carol Gilligan's critique of Kohlberg's formulation of stages of moral development in which one progresses to "higher" stages of moral thinking by engaging more and more in abstract, impersonal reasoning and setting aside specific personal connections and context. Gilligan argued that the latter was an equally important component of morality, and, adopting Gilligan's terminology and expanding on it, is sometimes referred to as "an ethics of care." Disagreement exists, of course, about how much such impartial rationality is too much; Wendy Donner, for example warns against a too thorough rejection of these traditional values as well as the strongly individualistic sense of self they imply. However, particularly since an emphasis on rationality and "coolness" is often linked to practices of domination and dichotimizing, this theme is also quite important to ecofeminism.

Both the emphasis on community and the location of attitudes toward nature within a larger pattern of domination that includes intrahuman domination make it unsurprising that many ecofeminists have paid close attention to "real world" social and political situations, especially in a global context, with special attention to third-world issues. Vandana Shiva is only one example of a philosopher who addresses this important part of ecofeminism.

As noted earlier, some natural affinities do seem to exist between deep ecology and ecofeminism: a rejection of the domination of nature; the tendency to think of nature as a connected whole; and the insistence that disinterested rational analysis by itself is not a sufficient grounding for environmental ethics. Serious disagreements have arisen, however, among some proponents of each theory (usually accompanied by the charge that the opposition is "shallow"). Much of the dispute revolves around the issue of how much our attitudes toward nature must be understood as part of a pattern that includes domination of other people, as well as the social and political forces that shape our interactions. Ecofeminists may charge deep ecologists with ignoring the particularized, cultural, and political forces that affect human relations as well as interaction with the environment, oversimplifying when they talk about "human" attitudes as if they were univocal and ahistoric. Deep ecologists sometimes object to feminists' attention to what some see as traditional human-centered concerns. Again, this is an area in which substantial disagreement arises in both camps, but it serves as a useful reminder not to oversimplify either position and reduce it to a caricature.[18]

Whatever the differences, all ecoholists and most ecofeminists advocate a nonextensionist approach: that is, the appropriate attitude toward nature (and other groups) is not merely a matter of retaining our individualistic approach to moral ethics while perhaps extending to more individuals (sentient creatures, or living things). Thus, if any of these alternative approaches is to be plausible, we must be able to make sense of nature as a suitable object of moral concern and duty, and do so in a way that is irreducible to concerns for individuals. To see whether we should move in this direction, we must first identify the alternative, individualistic theories that holism finds wanting before turning directly to the question of whether nature can be harmed.

ENVIRONMENTAL PRAGMATISM

The term *environmental pragmatism* has two related roots. The first, as its name implies in ordinary usage, eschews very abstract theoretical analysis in favor of issues that translate more directly into policy. The second, more closely tied to philosophical terminology, suggests a tie with the school known as *American pragmatism*, associated with philosophers such as William James, John Dewey, and Charles Sanders Pierce. Most contemporary writers representing environmental pragmatism, notably Bryan Norton, Andrew Light, Eric Katz, and Paul Thompson, combine elements of both these orientations.

Norton has written quite extensively on the first of these themes. One of his repeated targets is the concern with inherent or intrinsic value, which he finds to be a distraction rather than a help to increased clarity about environmental ethics. He argues that we can make more progress via alternative routes, particularly by addressing questions of scope and scale. With respect to temporal scale, for example, he argues that if we are sufficiently cognizant of the fact that environmental policy must consider short-term, medium-range, and long-term, multigenerational effects, we can arrive at sound policies without getting bogged down in debates about kinds of values that may have no clear implications for policy at all.

As noted earlier, the phrase *environmental pragmatism* is also used to invoke a connection with the philosophical tradition known as *American pragmatism*. The dual use of the phrase is not coincidental, since American pragmatists approached classic issues in philosophy by searching for more common-sense and useful answers to questions about knowledge or reality as well as ethical questions. Moreover, they generally championed a closer connection between philosophy and the natural sciences. This connection has two aspects. First, humans are viewed as part of nature, not distinct from it. Second, our understanding of the natural world, like our understanding of reality in general, is viewed by pragmatists as being to at least some extent relative to context and interest rather than a predetermined given.

Although I have described environmental pragmatism in general terms, more subtle differences exist among proponents of this movement to which I cannot do justice here. An anthology edited by Andrew Light and Eric Katz, *Environmental Pragmatism*, offers an excellent collection of essays on this topic. It is interesting to note the speculation that Aldo Leopold, with whom I began this chapter, may have been influenced by a pragmatist (in both senses) orientation.

One potential problem with both versions of environmental pragmatism is a concern about the lack of a theoretical foundation and the subsequent risk of begging questions about which policies should be adopted. This approach seems to assume that people can reach agreement at a policy level as long as we pay proper attention to the scientific data—that is, we are well grounded in ecology—and common sense. All the participants in the Chatham River debate, however, seem to meet both criteria and no consensus is in sight. Other real-world cases also suggest that pragmatism may be too optimistic about the way in which consensus will be generated.

EXTENSIONALIST THEORIES

Philosophers who are skeptical about a theory's ability to overcome the serious difficulties discussed in the previous section, who find Leopold's approach overly romantic and lacking in rigorous analysis, or who simply remain unconvinced that we need a "new environmental ethic" may argue in favor of a more individualist approach: anthropocentrism, protecentrism, or biocentrism. Such philosophers often agree that environmental ethics should extend the boundary of traditional ethical concerns but believe that such an extension can proceed from already familiar ethical and legal foundations.[19] Christopher Stone, for example, argued that the legal notion of rights currently in use can and should be extended to trees. Paul Taylor presents his thesis as a natural and logical extension of the Kantian perspective on respect and rights. Gary Varner has defended an extensionalist approach by arguing that proper attention to our duties to individual animals, whether grounded in a Singer-type utilitarianism or the rights-based theory defended by Regan, will also entail protection of the environment, including hunting some species (see the section "A Triangular Affair," later in this chapter, for more detail about this issue). I have argued that our duties to protect endangered species and ecosystems is best understood as a recognition of the interests of sentient creatures and the intrinsic aesthetic value of individuals (including plants and "places" as the latter term is often used in the phrase "sense of place.").

The previous list by no means covers all the possibilities: Nothing is said about appeals to natural law or the Continental tradition in environmental ethics.[20] Indeed, since individualistic approaches in traditional ethical theories are so varied, and since most such approaches can be extended to some degree to apply to environmental ethics—if only negatively—attempting to survey all the possibilities would be pointless.

The main issue to keep in mind when evaluating individualistic approaches is whether anything is lost if we no longer consider "ecosystems," "species," or "nature itself" as a direct object of moral concern. So, I turn to that question next.

CAN NATURE BE HARMED?

One way of asking whether nature can be an object of moral concern is to consider whether it can be harmed in a morally significant way. It puts the issue in more concrete, familiar terms, while still serving as a reflection of the more basic issue of whether nature itself has intrinsic value, as discussed shortly.

Although reading the question as "Can nature be *harmed*?" is natural, I propose to put the emphasis elsewhere: "Can *nature* be harmed?" That is, is nature the sort of thing to which it makes sense to apply concepts such as benefit or harm? The two versions of the question should lead to the same answer, but the emphasis will help guide our attempt to answer it. To do so, we must first get clear on what is meant by *harm* in these contexts. I therefore stipulate that in the following discussion, *harm* and *benefit* are to be understood as moral concepts; to harm someone or something in this sense is to do something that is *prima facie* morally bad.

There are two ways in which something or someone can be harmed in a morally significant way: one harms X if one fails to respect X's interests or if one's actions result in a net decrease of X's intrinsic value (or the intrinsic value of some Y of which X is a part).[21] These two forms of harm are not equivalent but they are not exclusive, either. So, in order for X to be the sort of thing that can be harmed, it must either have interests or intrinsic value; we can safely set aside the parenthetical condition in the case of nature. So, is nature the sort of thing that can be harmed in one of these ways?

First, it is clear that nature cannot have interests in the narrow sense, cannot have something it "cares about." It is not sentient, even though some parts of it are: sentient individual animals, which *may* include humans, depending on how nature is defined. To have interests in the narrow sense, one must care about something, and in order to care about something, one must be sentient. (These are necessary, but perhaps not sufficient, conditions.) There is a broader sense of interest, sometimes called *welfare interest*, in which X has an interest in anything that contributes to its faring ill or well. Appeals to this broad sense, however, beg the question unless we have independently established what it means for nature to fare ill or well and why that is morally significant, since having interests satisfied or thwarted does not always create moral obligation.

The second alternative is the only possibility, then. Is it possible for some action we might take (or fail to take) to make nature less valuable? In considering this question, it is important to keep in mind that we are talking about intrinsic value, or what might make nature valuable "in and of itself" rather than instrumentally valuable for some other purpose that we or other interest-holders might have; that was the reason for the change in emphasis in the first sentences of this section in order to focus attention on the question understood as "Can *nature* be harmed?" as opposed to "Can nature be *harmed*?"). The first alternative (whether our actions can make nature less valuable) might seem to involve an obvious answer; after all, it seems clear that water pollution, clear cutting, erosion caused by all-terrain vehicles (ATVs) in fragile environments, and allowing an endangered species to become extinct are all examples of how we might harm nature itself. But initial appearances may be superficial; we need to look deeper.

I am not disputing that all the examples just cited are bad things. It is not clear, however, that they are bad because they diminish the intrinsic value of nature, if such value exists, once we distinguish damage to nature itself from weak anthropocentrism or protecentrism. After all, we don't just think that water pollution is bad because it kills fish and other life that depends on the water supply, we actually *define* what counts as pollution by its effects on plants, animals, and people. Similarly, we decry clear-cutting of trees (but not always of corn[22]), but the reasons we cite tend to be both aesthetic (clear-cutting destroys the beauty of the forest) and very specific about its effects on sentient beings (depriving animals of necessary habitat, allowing runoff that will negatively impact stream life, and other effects). In short, our objection is not fundamentally that nature *itself* is harmed but that the properties of nature that creatures who have an interest in nature being one way rather than another are changed in ways that affect those creatures.

Of course, life is not that simple: What is helpful to one individual may harm another. In those cases, the natural impulse is to sum and compare those benefits and harms, or to appeal to the rights of those affected, both of which are classic individualistic approaches. That is to say, even assuming that we have a clear understanding of what *nature* means, we see it as important for the interests of individuals but not necessarily intrinsically valuable.

All this presents the following challenge to those who would attribute intrinsic value to nature itself:

(1) If nature has intrinsic value, harming nature directly by decreasing or destroying its intrinsic value must be possible.
(2) The examples surveyed rest on harms to individual sentient creatures rather than direct harms to nature.

(3) Absent a new counter example, so-called harms to nature are more fundamentally understood as harms to individual sentient creatures or diminution of instrumental value as judged by those creatures.
(4) No convincing argument exists at present that nature can be harmed.
(5) No convincing argument exists at present that nature has intrinsic value.

In addition to the problem of attributing intrinsic value to nature, we must also consider another issue that has already been mentioned: What do we mean by *nature*? This is not really a new question, since we ask about what we mean by nature in order to figure out what would be a harm to *nature* as opposed to something else, but considering the question in this form reinforces the previous argument. The question was raised earlier when I considered the dilemma of how humans fit into nature. There are three possible answers: (a) no human interference is natural—that is, any environment affected by human interference is non-natural; (b) humans are simply part of nature, and any effect they have is therefore natural; (c) some effects of humanity are natural but others are not. Unfortunately, no convincing arguments have been given for (a); (b) seems naive and certainly unhelpful to environmental ethics; and (c) requires a criterion, *with justification*, for determining which effects are natural and which are not; and such a criterion is not forthcoming.

So far, I have considered the issue of whether nature can be harmed, but the other side of the coin should also be recognized. Another question to be considered is whether we can identify anything that is *good* for nature, in and of itself. Here again, the same issues arise: We have not yet arrived at a good definition of nature, and what we tend initially to gravitate toward as examples of things that are good for nature seem, upon closer examination, to turn out to be good because they benefit individuals. To give one example: Is prairie restoration good for nature? Certainly, prairies are good things for many reasons. They provide excellent habitat for many species, they are good for enriching the soil and curbing erosion, they remind humans of their past, they are beautiful (although the recognition of their beauty may require a better understanding of ecology than is generally present), and so on. But it is not clear that any of these reasons show that such activities are good for nature itself. Moreover, by definition, prairie *restoration* is a human activity; it is, on one interpretation of *nature*, unnatural.

In short, attempts to answer the question of whether nature itself can be harmed reveals two fundamental problems with any attempt to argue that it can be harmed: (1) whether we have a clear enough sense of what *nature* means to know what the question refers to; and (2) whether nature is the sort of thing that really has intrinsic value—that is, whether beneficial or objectionable actions are judged to be beneficial or objectionable because they affect nature itself, or rather on the basis of their effects on individuals within the natural world. These two issues carry over directly to questions about specific ecosystems, such as the Chatham River, and species.

CAN ECOSYSTEMS OR SPECIES HAVE INTRINSIC VALUE?

As I have discussed, attributing intrinsic value demands two linked, necessary (but not sufficient) conditions: a clear definition of identity and a criterion for *harm for X* that is independent of instrumental concerns or concerns about how those harms are fundamentally harmful for someone or something else. These conditions must be met by anyone who argues that an ecosystem, such as the Chatham River, or a species, such as the spotted owl, has intrinsic value. Can these conditions be met? I argue that they cannot; each fails to satisfy at least one of the conditions. I start with the issue of whether species can have intrinsic value.

Although biologists have frequently revised the definition of species, the question of whether two individuals belong to the same or different species is rarely a serious issue. Exceptions to this general rule do occur with subspecies (different subspecies of trout), and species that once had been geographically isolated but have expanded into overlapping ranges (for example, different types of juncoes), but these exceptions rarely have an impact on environmental debates. One notable exception is the dusky seaside sparrow, which, when reclassified as a color variation of the seaside sparrow, was no longer protected as an endangered species. This enabled Disney World to be built on the last known natural habitat of the sparrow. The last dusky seaside sparrow died in captivity in Disney World. However, concerns about protection of endangered species often do not parallel the biological definition of species. Protection of subspecies such as the Mt. Graham ground squirrel, or even domestic breeds such as the Norwegian Fjord horse, is not uncommon. Thus, although the "identity condition" can be scientifically met on the species level, those distinctions do not seem to capture what environmentalists are often concerned about. At the subspecies level, the identity criterion becomes more problematic but perhaps not an insurmountable barrier.

The real problem for species is the second condition: Can a species be harmed? On the face of things, the question seems silly; the whole idea of a species being endangered or going extinct seems an obvious sort of harm. But if we probe a bit deeper and ask who or what is harmed when a species is endangered or becomes extinct, the issue becomes more difficult. Obviously, individual animals are harmed if their habitat is destroyed or they are killed, whether through hunting by humans or other animals or by the destruction of their habitat—again, whether through human impact or natural causes.

Humans may be harmed if they are deprived of the opportunity to see a beautiful snow leopard or an awe-inspiring stand of old-growth Douglas firs. A common line of reasoning about the rain forest is that it contains hundreds of species that we haven't identified yet and one of them could prove to be a cure for cancer. This argument also is strongly anthropocentric. Other arguments tend to focus on the fact that a species may occupy an essential niche in the environment. If we eradicate black flies, we will unknowingly interfere with the pollination of wild blueberry bushes. This is persuasive if we like blueberries.

Thus, when we examine the question of whether species can be harmed, species do not seem to be the right level at which to identify any harm done by the endangerment or extinction of a species unless we try, as Rolston does, to argue that the species itself is some sort of "superorganism." Certainly, harms is done to individuals: humans, members of the species in question, and other individual plants or animals that depend on the threatened species. In some cases, the ecosystem may be altered, perhaps irreparably damaged, although in many cases, such as that of the California condor or Prezwalskis' horse, little or no impact on the ecosystem occurs. If we want to argue that preserving the condor or Prezwalsi's horse is something we ought to do, an appeal to the effect on the ecosystem of the extinction of a species will not always work. Until a new suggestion is forthcoming, it would seem that individuals and ecosystems can be harmed but that *species* themselves are the wrong place to locate the harm done by extinction or endangerment. If so, species may have instrumental value but do not seem to meet the conditions for intrinsic value.

Neither Marie's nor Leopold's argument, however, focuses on individual species but rather on ecosystems as a whole. Thus we must consider whether ecosystems can have intrinsic value, judged according to the same two criteria of identity and harm. I consider them here in reverse order.

Many philosophers have suggested that the concept of "ecosystem health" is a value-neutral measure of harm and good to an ecosystem. Dale Jamieson and I have both argued that for a variety of reasons this is not so.. Although my argument was intended primarily as a rebuttal, however, of the claim that the concept is value-neutral (or, as sometimes claimed, *purely* scientific or *purely* objective, as if those three terms were synonymous), it also highlights the ways in which someone might argue that an ecosystem can be harmed. That is to say, an ecosystem is harmed if its health is diminished.

Ecosystem health has been defined in a number of ways, but two key and related features are natural diversity and a dynamic sort of resilience. *Resilience* here is intended to replace an outdated, static sort of stability. It refers to a system's ability to maintain itself by adapting to natural internal and external changes while still supporting a high level of biodiversity. Conversely, biodiversity is often what allows an ecosystem to be resilient: A larger gene pool is more likely to contain life forms with the ability to adapt to new conditions. Of course, some of the problems with Leopold's concept of stability noted earlier are inherited by its successor, but ecologists such as Costanza and Rapport, who have tried to define the concept of *ecosystem health*, have made significant progress in these areas.

The other criterion for intrinsic value is the identity condition: Can we specify with reasonable precision *what* is supposed to have intrinsic value? It is necessary to identify the *bearer* of value, or else we are left with free-floating value, a nonsensical idea. Here is where the attempt to ascribe intrinsic value to ecosystems encounters serious obstacles.

An ecosystem is shaped by native and invasive animals and plants, migration, water, soil, climate, and air, just to name the main components. None of these has clear boundaries, so it follows that the ecosystem that comprises them will not have identifiable boundaries, either. But without boundaries, without any way of determining what is or is not part of the ecosystem, no way seems to exist of meeting the identity criterion. So, the main challenge to anyone who wishes to argue that ecosystems have intrinsic value would be to develop a coherent account of how to identify and delineate an ecosystem.

CONFLICTING INTERESTS

So far, we have identified certain values that should guide environmental ethics. It would be easy to decide what to do if we could just run down a checklist of these considerations and calculate the values involved. Unfortunately, in the real world, demands and values often come into conflict. Whether or not we have a satisfactory way of resolving the conflict, we should at least be cognizant of the most common areas in which such problems arise. This section surveys some of the most important areas of conflict. Together, these issues remind us that environmental ethics does not occur in a vacuum; it must be developed and evaluated in a context of ethical concerns about humans, nonhuman animals, and social justice.

HUMAN FLOURISHING VERSUS WILDERNESS

As noted earlier, deep ecology places special emphasis on the importance of wilderness but it is certainly not the only theory in environmental ethics to do so. In fact, arguments for protection of wilderness areas are often the central focus of environmental ethics. However, protection of the wilderness poses a theoretical problem about how to define *wilderness*, as well as at least three potential areas of conflict with what we might generally call *human flourishing*.

What, exactly, is a wilderness and why is it singled out for special recognition and (sometimes) protection? In the strictest sense, a wilderness is an ecosystem that is unaffected by human interference, and if that is the standard we choose, there are almost certainly no wilderness areas on this planet. More typically, *wilderness* designates an area that is set aside and protected from development or uses that would change the area's "natural" condition. This leads back to the question considered earlier of what *nature* or *natural* means. There is no need to go over the theoretical issues again, but it might be useful to identify some practical issues that arise specifically with wilderness areas.

It would seem obvious that a basic prerequisite for wilderness preservation is the absence of major human interference, either in the form of development or interference with natural processes. However, wilderness areas have, at least until recently, been actively managed. Such management can sometimes yield apparently devastating results, at least in the short run, such as the fires in Yellowstone (for a unique perspective on human interference in Yellowstone, see Chase). Second, wilderness areas have usually been so designated long after the natural balance of life has been disrupted, with special impact on large predators. Without such predators, other species exhibit an impact: Prey species such as deer or elk may multiply far past the carrying capacity of the land, and other species such as coyotes may partially, but not fully, expand to cover an empty niche. What should be done about these situations? Third, wilderness areas are isolated islands surrounded by developed areas, and this, too, poses problems. One current example is the fate of bison in Yellowstone: Their natural behavior is to migrate to lower ground in winter to obtain better grazing, but current policy—dictated by ranchers who are concerned about the possibility of buffalo infecting domestic cattle with brucellosis—allows, and sometimes demands, that they be shot as soon as they stray outside park boundaries.

These are major, dramatic issues affecting wilderness, but sometimes the seemingly minor issues are more illuminating from a philosophical standpoint. The basic theoretical problem is how much interaction between humans and wilderness should be allowed or encouraged. One of the reasons often given for maintaining wilderness areas is that they provide a valuable opportunity for people to get back to nature, but that means that people must have access to those areas. Deciding what sort of access is appropriate creates a host of practical problems.

Assume for the moment that all of the previous major problems have been resolved. A less dramatic question would be whether to have marked trails and campsites in a wilderness area, and if so, what sort and how many. Obviously, such things are unnatural but they are part of almost every managed wilderness area. Paths, trails, and established campsites are intended to control human impact by concentrating it in limited areas: Hikers should use the trails rather than go bushwhacking through fragile underbrush. But they also change the ecosystem; they are more susceptible to erosion; nonnative plants are more easily established at such sites; and the local wildlife is quick to adapt to the "easy pickings" at a campsite or portage head. So by making the wilderness experience accessible to more people, and by trying to protect the land from damage imposed by human use, we seem to be making the area less and less of a wilderness. But if we do nothing, very few people will be able to enjoy the wilderness, and those who do will have an uncontrolled impact on the ecosystem.

The previous discussion, although relevant to U.S. wilderness areas, represents only a small part of the issue. When most Americans think of wilderness, we do so from the perspective of an affluent society with the leisure time and resources to enjoy, or at least contemplate, the rain forest, Siberian tigers, and unpolluted streams. From the perspective of a developing nation, setting aside nature preserves to protect tigers while ignoring the needs of people who depend on that land for the basic necessities of survival might well be viewed as imperialist or elitist. Ramachandra Guha and Anil Agarwal have been particularly eloquent in analyzing this topic and identifying the presuppositions built into a glorification of wilderness.

As the previous issues imply, an emphasis on wilderness runs the risk of inviting us to overlook other pressing environmental concerns. After all, very few people have any direct contact with wilderness areas, and most of the world's population lives in an urban environment. Dale Jamieson has discussed the relevance of environmental ethics to cities in an illuminating way, serving as a reminder that preservation of wilderness is hardly the only, or even most pressing, issue in environmental ethics. Thus, any fully developed and effective ethical theory ought to address urban concerns, and insofar as a "wilderness focus" occupies a glamorous, romantic role, it runs the risk of distracting attention from environmental issues that have a far greater real-world impact.

A TRIANGULAR AFFAIR

The connection between environmental ethics and animal rights, or animal libera- tion movements, has always been murky. Tom Regan has accused ecologists of "envi- ronmental fascism" because they sacrifice the rights of individual mammals (clearly, sentient creatures) in order to preserve species and ecosystems, even to the point of killing sentient animals to preserve plant species. In 1980, J. Baird Callicott published an article titled "Animal Liberation: A Triangular Affair" in which he argued that "animal liberation," or more generally an ethical stance that emphasized the moral considerability of individual sentient animals, is often incompatible with environmen- tal ethics. Although he later modified his own interpretation, his initial challenge still poses an important problem for environmental ethics.

As just suggested, there are at least three potential conflicts between environmental ethics and concerns about animal welfare: Environmental ethics tends to value the welfare of a wild species over that of equally sentient domestic animals; might justify sacrificing the welfare of individual animals for the good of the species; and sometimes justifies harm to a sentient creature in order to preserve a nonsentient but endangered species. In all three cases, it is difficult to see how to reconcile conflicting demands. Callicott is not the only philosopher to highlight these conflicts; Aldo Leopold was an avid hunter, Mark Sagoff has responded to Callicott's attempt to retreat from his early position, and, as noted earlier, Regan has characterized some versions of environmental ethics as "fascism."

With regard to the relative status of domestic versus wild species in moral deliber- ation, it is clear that any ethical theory that judges the worth of something in terms of its contribution to an ecosystem will automatically dismiss almost all concerns about domestic animals. Callicott refers to them as "artifacts," and claims that

> ... a herd of cattle, sheep, or pigs is as much or more of a ruinous blight on the landscape as a fleet of four-wheel-drive off-road vehicles. There is thus something profoundly incoherent and insensitive as well in the complaint of some animal liberationists that the "natural behavior" of chickens and bobby calves is cruelly frustrated on factory farms. It would make almost as much sense to speak of the natural behavior of tables and chairs (p. 30)

Despite Callicott's later renouncement of this view, there is a clear tension here between the emphasis on sentience and the ability to suffer emphasized by animal projectionists, and moral considerability based on contributions to an ecosystem.

Second, environmentalism tends to focus on species rather than individuals. This focus has several consequences. An endangered species should be given more weight, morally speaking, than a common one. Individuals can and should be sacrificed for the good of their own or other species (this is, presumably, what provoked Regan's label of environmental fascism, since fascism demands sacrifices of individual rights for the good of the state). This may involve active hunting or trapping to control populations, encouraging predators, and allowing animals to die of starvation or accident even when rescue could be effected.

Finally, environmental ethics typically downplays the importance of sentience. If forced to choose between killing a plant or killing a sentient creature, the decision should be made on the basis of what each contributes to the ecosystem, not on which will suffer more pain.

In pointing out these conflicts, I do not intend to condemn environmental ethics or to suggest that it has overlooked something vital. Indeed, environmentalists may be on the right side of all these issues. But they are real issues and must be recognized and analyzed as carefully as possible.

ENVIRONMENTAL RACISM

Environmental ethics has sometimes been portrayed as less concerned than it should be with issues of justice, particularly when those issues have their greatest impact on minorities or Third World, non-European nations. We have already touched on this issue in the discussion of urban environments, since, at least in the United States, cities tend to have a higher minority population than rural areas. However, other ways exist to raise the question of environmental racism.

In this country, local sources of pollution tend to concentrate in poorer, often minority, areas, for two reasons. First, people with money are more apt to have their voices heard in the political systems that decide on zoning—what industries to allow in an area or even promote. Rich and poor both have the right to say "not in my back yard," but politicians generally pay more attention to the former. Second, undesirable types of land use, ranging from landfills and large hog farms to nuclear waste disposal sites, make surrounding property less desirable and less valuable. Those who can afford to do so may choose to relocate, but those who don't have the same means will be forced to live with the environmental hazard or might find such areas the only location for affordable housing. In both cases, the motive may not be racist, but the effect is.

Environmental racism also exists on a global scale. The same issues that exist intranationally—using other places as dumping grounds—contain other areas of potential abuse as well, usually grounded in a practice of discounting developing, usually nonwhite countries, or viewing them from our own perspective of already having "made it." I have also already mentioned the issue of wilderness protection and endangered species. Cries to protect the rain forest often fall into the same category. Deforestation and loss of biodiversity are supposed to be global problems, but all too often the local population is expected to bear the apparently invisible economic and social burden that such efforts entail. People frequently expect areas of Africa, Central and South America, and India to be set aside for nature preserves, and some laudable efforts are made to ensure that such projects actively involve and respect native human populations. Other campaigns fail to realize how much havoc an elephant can cause, or how few alternatives are readily available to those who depend on a slash-and-burn style of agriculture.

Similarly, calls for controls on ozone-depleting chemicals or other practices that pollute the air or water can sometimes ring hollow. Developed countries have depended for years on such practices and are now in a position to understand the environmental impact they have had. However, issues of justice arise when attempts are made to hold developing nations to the same standards of restricting resources and their use that they, in their current affluence, are able to consider.

Environmental ethics is a rich and rewarding field, posing everything from deep theoretical problems about the nature of value and moral standing to very practical decisions. This essay could only hope to scratch the surface of some of those issues, but I hope that it has provided some guidance about how we can make a bit more progress in our thinking about them.

REFERENCES AND ADDITIONAL READING

Agerwal, Anil. "Human-Nature Interactions in a Third World Country", *The Environmentalist*, v. 6 (1986).

Armstrong, Susan J. and Richard G. Bolzer, eds., *Environmental Ethics: Divergence and Convergence* (New York: McGraw Hill, Inc., 1993).

Baier, Annette. "For the Sake of Future Generations", in Regan, *Earthbound*.

Bookchin, Murray, and David Foreman; Steve Chase, ed., *Defending the Earth*, (Boston: South End Press,1991).

Callicott, J. Baird. *In Defense of the Land Ethic: Essays in Environmental Philosophy* (Albany: SUNY Press, 1989).

Callicott, J. Baird. "The Value of Ecosystem Health," *Environmental Values*, v. 4, 345–362.

Chase, Alston. *Playing God in Yellowstone* (New York: Harcourt Brace Jovanovich, 1987).

Cheney, Jim. "Ecofeminism and Deep Ecology," *Environmental Ethics*, v. 9 (1987), 115–145.

Cheney, Jim. "Intrinsic Value in Environmental Ethics: Beyond Subjectivity and Objectivity"

Comstock, Gary. "An Extensionist Environmental Ethic," in N. Cooper and R. C. J. Carling, eds., *Ecologists and Ethical Judgments* (London: Chapman & Hall, 1996), 43–55.

Comstock, Gary. "Do Agriculturalists Need a New, an Ecocentric, Ethic?" *Agriculture and Human Values* 12 (Winter 1995), 2–16.

Comstock, Gary. *Vexing Nature? On the Ethical Case Against Agricultural Biotechnology* (Boston: Kluwer, 2001), 199–220.

Costanza, Robert, Bryan Norton, and B. Haskell, eds., *Ecosystem Health: New Goals for Environmental Management* (Washington: Island Press, 1992).

Davion, Victoria. "How Feminist is Ecofeminism?"

Devall, Bill and George Sessions. *Deep Ecology: Living as If Nature Mattered* (Layton, Utah: Gibbs M. Smith, Inc., 1985).

Donner, Wendy. "Self and Community in Environmental Ethics", in Warren, ed., 1997, 375–389.

Elliot, Robert, ed. *Environmental Ethics*. (New York: Oxford University Press, 1995).

Elliot, Robert. "Intrinsic Value, Environmental Obligation, and Naturalism," *The Monist*, v. 75 (1992), 138–160.

Ferry, Luc. *The New Ecological Order* (Chicago: University of Chicago Press, 1995).

Fox, Warwick. *Towards a Transpersonal Ecology* (Boston: Shambala, 1990).

Guha, Ramachandra. "Radical American Environmentalism and Wilderness Preservation: A Third World Critique," *Environmental Ethics*, v. 11 (1989), 71–83.

Hargrove, Eugene. *Foundations of Environmental Ethics* (Englewood Cliffs, NJ: Prentice Hall, 1989).

Hargrove, Eugene. "Weak Anthropocentric Intrinsic Value," *The Monist*, v. 75 (1992), 183–207.

Jamieson, Dale. "Some Preventative Medicine," *Environmental Values*, v. 4, 333–344.

Jamieson, Dale. "The City Around Us," in Regan, *Earthbound*.

Katz, Eric. "Searching for Intrinsic Value: Pragmatism and Despair in Environmental Ethics," in Light and Katz., 307–318.

Leopold, Aldo. *A Sand County Almanac* (New York: Ballantine, 1966; originally copyrighted by Oxford University Press in 1949).

Light, Andrew, and Eric Katz, eds., *Environmental Pragmatism* (New York: Routledge, 1996).

Naess, Arne. "Self Realization in Mixed Communities of Humans, Bears, Sheep, and Wolves", *Inquiry*, v. 22 (1979), 231–241.

Naess, Arne. "The Shallow and the Deep, Long Range Ecology Movements: A Summary," *Inquiry*, v. 16 (1973), 95–100.

Norton, Bryan. "Integration or Reduction: Two Approaches to Environmental Values," in Light and Katz, 105–138.

Norton, Bryan. "Epistemology and Environmental Values," *The Monist*, v. 75 (1992), 208–226.

Norton, Bryan. *Why Preserve Natural Variety?* (Princeton, NJ: Princeton University Press:, 1987).

O'Neill, John. "The Varieties of Intrinsic Value," *The Monist*, v. 75 (1992), 119–137.

Plumwood, Val. "Nature, Self, and Gender: Feminism, Environmental Philosophy, and the Critique of Naturalism," in Warren, 1996.

Rapport, David. "Ecosystem Health: Exploring the Territory," *Ecosystem Health*, v. 1, (1995), 5–13.

Regan, Tom. *All That Dwell Within* (Berkeley: University Of California Press, 1982).

Regan, Tom, ed.. *Earthbound: New Introductory Essays in Environmental Ethics* (New York: Random House, 1984).

Regan, Tom. "Does Environmental Ethics Rest on a Mistake?" *The Monist*, v. 75 (1992), 161–182.

Regan, Tom. *The Case for Animal Rights* (Berkeley: University Of California Press, 1983).

Rolston, Holmes, *Philosophy Gone Wild: Essays in Environmental Ethics* (Buffalo, NY: Prometheus Books, 1986).

Routley, Richard. "Is There a Need for a New, an Environmental Ethic?" *Proceedings of the Fifteenth World Congress of Philosophy* (Sophia, Bulgaria: Sophia Press, 1973, 205–210).

Russow, Lilly-Marlene. "An Objective Evaluation?" *Environmental Values*, v. 4, 363–369.

Russow, Lilly-Marlene. "Prairies Do Not Have Intrinsic Value" *The Ag Bioethics Forum*, v. 6, no. 4 (1994), 1, 6–7.

Sagoff, Mark. "Animal Liberation and Environmental Ethics: Bad Marriage, Quick Divorce," *Osgoode Hall Law Journal*, v. 22 (1984), 306.

Schweitzer, Albert.

Sessions, Robert. "Deep Ecology versus Ecofeminism: Healthy Differences or Incompatible Philosophies? in Warren, 1996.

Shiva, Vandana. "Women in Nature" in *Staying Alive: Women, Ecology and Development in India*, (London: Zed Books, 1989), 38–54.

Singer, Peter. *Practical Ethics* (New York: Cambridge University Press, 1979).

Stone, Christopher. *Should Trees Have Standing? Toward Legal Rights for Natural Objects* (Los Alos, CA: William Kaufmann, 1974).

Taylor, Paul. *Respect for Nature* (Princeton, NJ: Princeton University Press, 1986).

Thompson, Paul. "Pragmatism and Policy: The Case of Water," in Light and Katz, *Environmental Pragmatism*.

Thompson, Paul. *The Spirit of the Soil: Agriculture and Environmental Ethics*. (New York: Routledge, 1995).

Varner, Gary. *In Nature's Interest?* (New York: Oxford University Press, 1998).

Varner, Gary. "Do Species Have Standing?" *Environmental Ethics*, v. 9 (1987), 57–72.

Warren, Karen J., ed., *Ecofeminism* (Bloomington, IN: Indiana University Press, 1997).

Warren, Karen J., ed., *Ecological Feminist Philosophies* (Bloomington, IN: Indiana University Press, 1996).

Westra, Laura, and Peter Wenz, eds., *Faces of Environmental Racism* (Lanham, MD: Rowman and Littlefield, 1995).

NOTES

1. Some philosophers draw a distinction between *intrinsic* and *inherent* values, but that is not necessary in order to understand Marie's basic point. For the most part, we can treat *intrinsic* and *inherent* as having the same meaning.
2. There are strong connections between utilitarianism and instrumental value and between intrinsic value and deontological theories (see Chapter 3), but it would be oversimplistic just to assume that these connections hold.
3. Specific references for selected examples of Callicott's work, and other authors or schools mentioned in the text, can be found in the bibliography/suggestions for further reading at the end of the chapter.
4. Apparent exceptions to this are the arguments claiming that intrinsic value has some sort of objectivity that instrumental values lack. This claim is highly controversial; see Cheney (1992) and O'Neill for further discussion.
5. This does not contradict my earlier claim that any system of values requires some foundation in intrinsic value. There would be nothing valuable in such a world, and indeed the whole talk of ethical theory or moral principles would make no sense.
6. My thanks to James Stephens for this suggestion.
7. This phrasing is a bit convoluted but necessary for accuracy. The second variation on anthropocentrism, if coupled with moral realism, might agree that something

could have intrinsic value even if nobody *in fact* valued it for its own sake: The valuers might simply be wrong about their moral judgments.

8. Generally anthropocentric views will have more options here: They might, for example, appeal to a moral sense, intuition, or natural emotion. Some of these options may be open to the nonanthropocentric, but they fit less well together.

9. All references in this section are to *A Sand County Almanac*. See bibliography for full citation.

10. However, in other essays, *land* is used as a pejorative term: "There is much confusion between land and country. Land is the place where corn, gullies, and mortgages grow. Country is the personality of land, the collective harmony of its soil, life, and weather" (Leopold, 177).

11. Indeed, I have argued that it is often the most important consideration, if understood correctly.

12. See Jamieson (1995) and Nelson and Russow (1995) for discussions of how the concept of ecosystem health is value laden, and see Norton (1995) and Rapport for dissenting views.

13. Callicott has taken this even further, pointing out that the land ethic "could fairly be called a case of earth chauvinism" (262)—although he thinks it's a perfectly justified form of chauvinism.

14. However, Gary Comstock (1995, 2001) has raised serious doubts about the scientific currency of this interpretation of ecology.

15. This is taken from Naess's "Eight Points," originally formulated in Naess (1987) but widely reprinted in anthologies, including Armstrong and Boltzer, from which this is taken.

16. It is interesting to note in this regard that the largest designated "wilderness area" in the United States, the Boundary Waters Canoe Area (BWCA), was in fact fairly extensively logged through the early twentieth century.

17. One response to some of these issues has been a split between deep ecology and "social ecology," the latter most closely associated with Murray Bookchin. See Chase (1991).

18. For a more detailed discussion, see the essays by Sessions and Plumwood in Warren (1996).

19. Hence Gary Comstock's (1996) term for such theories: "extensionist environmental ethics."

20. For a detailed discussion and a scathing critique of the latter, see Ferry.

21. The second condition is not quite precise enough: X must somehow contribute to the intrinsic value of Y, such that our effect on X diminishes the value of Y. That is, there must be a direct causal link between what we do to X and Y's diminution of intrinsic value; it can't just be a coincidence.

22. Although arguments in favor of conservation tillage often follow similar lines of argument as those adduced against clear-cutting of trees.

6

FOOD

Hugh LaFollette and Larry May

CASE STUDY: DHRUVA THE DESTITUTE

During the discussion with Marie the environmentalist, Emily notices Rich becoming increasingly upset. She is surprised, however, by Doug's reaction. She asks Doug several days later what is bothering him. He says that the environmentalists fail to realize how important agriculture is. "If we can't use water to feed ourselves," he complains, "we won't have the luxury to be worrying ourselves about philosophical niceties regarding the moral status of nonhuman entities. We'll be dead. Agriculture helps us to feed ourselves, and efficient farming provides us with goods so that we can do valuable things such as study philosophy."

At the next class, Doug raises this objection to Dr. Wright, who points out that the reading for the next class period addresses that very issue. That night, Doug calls Emily.

"Have you read the Peter Singer essay yet?" he asks.

"Yeah. What'd you think?" Emily replies.

"I really liked his argument that we ought to do everything we can to help people."

"Why?"

"Well, first, I just think that people have a right to be fed if it's possible to feed them. But I also think that Singer makes the case for the superiority of farmers over environmentalists. Our first job is to take care of people's basic needs, which includes the production of food. That's a big job and it requires using our natural resources in a way that maximizes production. There are millions of malnourished people out there! Who's going to feed them if not farmers?"

Emily has been thinking about this question ever since seeing a story on TV about the state of children in certain developing countries. The story told about a ten-year-old boy named Dhruva in one such country, a boy who was abandoned at birth because he was very sick and who now made a living by begging in one of the country's major urban areas. He does not eat well and his prospects are not good.

"I see your point," answers Emily. "I'm all for feeding people when it is possible. But I'm skeptical about giving my money to relief agencies because they spend it all on these big huge salaries of their executives. And even if they didn't, I wonder whether there is anything I can do that would actually help to feed strangers halfway across the globe."

"You put your finger on it," answers Doug. "It seems we have a duty to help others. But what if there is no mechanism by which we can actually meet this duty? And how far does the duty extend? Should we give to relief agencies until we are impoverished? If we do so, then we will be contributing to the problem, because we ourselves will need someone else's assistance."

"Right," responds Emily. "But, it's possible to overstate the difficulty. Certainly each of us is capable of doing something that will help to alleviate malnourishment in developing countries. The mere fact that we could impoverish ourselves if we gave too much is no excuse for not giving anything. Consider this argument. If we give something, we will save some people. If we give nothing, some people will die. Now what's the difference between murdering someone by attacking and killing them and murdering someone by failing to send five dollars to a relief agency that will give them bread when bread is all they need?"

Turn to Appendix A of this book to find discussion questions for this case study (Exercise 6.A).

DISCUSSION OF ISSUES

Emily, moved by the plight of Dhruva, is not alone in caring for the hungry, and it is a rare person who does not share her urge to help the starving. The faces of the malnourished are compelling, and it would not seem difficult to motivate people to assist them. For children are the real victims of world hunger: At least 70% of the malnourished people of the world are young people. By best estimates, seventy-five of every one thousand children in developing countries will die before they are five years old (United Nations Development Program, 2000:189). Children do not have the ability to forage for themselves and their nutritional needs are exceptionally high. Hence, they are unable to survive for long on their own, especially in lean times. Moreover, they are especially susceptible to diseases and conditions that are the staple of undernourished people: simple infections and simple diarrhea (UNICEF, 1993: 22). Unless others provide adequate food, water, and care, children will suffer and die (WHO, 1974: 677, 679). This fact must frame any moral discussions of the problem.

And so it does—at least pre-philosophically. When most of us first see pictures of seriously undernourished children, we want to help them, we have a sense of responsibility to them, we feel sympathy toward them (Hume 1978: 368–71). Even those who think we needn't or shouldn't help the starving take this initial response seriously: They go to great pains to show that this sympathetic response should be constrained. They typically claim that assisting the hungry will demand too much of us, or that assistance would be useless and probably detrimental. The efforts of objectors to undermine this natural sympathetic reaction would be pointless unless they saw its psychological force.

We want to explain and bolster this sympathetic reaction—this conviction that those of us in a position to help are responsible to the malnourished and starving children of the world. We contend that we have this responsibility to starving children unless there are compelling reasons that show that this sympathetic reaction is morally inappropriate (Ibid., 582). Anyone wishing to demonstrate this responsibility must, among other things, seek some "steady and general point of view" from which to rebut standard attempts to explain away this instinctive sympathetic response. By showing that assistance is neither too demanding nor futile, we think more people will be more inclined to act upon that pre-philosophical sense of responsibility. And philosophically championing that sense of responsibility will make most people feel more justified in so acting.

VULNERABILITY AND INNOCENCE

Our initial sense of responsibility to the starving and malnourished children of the world is intricately tied to their being paradigmatically vulnerable and innocent. They are paradigmatically vulnerable because they do not have the wherewithal to care for themselves; they must rely on others to care for them. All children are directly dependent on their parents or guardians, and children whose parents cannot provide them with food—either because of famine or economic arrangements—are also indirectly dependent on others: relief agencies or (their own or foreign) governments. Children are paradigmatically innocent because they are neither causally nor morally responsible for their plight. They did not cause drought, parched land, soil erosion, and overpopulation; nor are they responsible for social, political, and economic arrangements that make it more difficult for their parents to obtain food. If anyone were ever an innocent victim, a child who suffers and dies from hunger is.

Infants are especially vulnerable. They temporarily lack the capacities that would empower them to acquire the necessities of life. Thus, they are completely dependent on others for sustenance. This partly explains our urge to help infants in need. James Q. Wilson claims that our instinctive reaction to the cry of a newborn child is demonstrated quite early in life.

> As early as ten months of age, toddlers react visibly to signs of distress in others, often becoming agitated; when they are one-and-a-half years old they seek to do something to alleviate the other's distress; by the time they are two years old they verbally sympathize . . . and look for help. (Wilson 1993: 139–40)

Although this response may be partly explained by early training, available evidence suggests that humans have an "innate sensitivity to the feelings of others" (Wilson, 1993: 140). Indeed, Hans Jonas claims the parent-child relationship is the "archetype of responsibility," in which the cry of the newborn baby is an ontic imperative "in which the plain factual 'is' evidently coincides with an 'ought'" (1983: 30).

This urge to respond to the infant in need is, we think, the appropriate starting point for discussion. But we should also explain how this natural response generates or is somehow connected to moral responsibility.

THE PURPOSE OF MORALITY

The focus of everyday moral discussion about world hunger is on the children who are its victims. Yet the centrality of children is often lost in more abstract debates about rights, obligations, duties, development, and governmental sovereignty. We do not want to belittle either the cogency or the conclusions of those arguments. Rather, we propose a different way of conceptualizing this problem. Although it may be intellectually satisfying to determine whether children have a right to be fed or whether we have an obligation to assist them, if these arguments do not move us to action, then they are of little use—at least to the children in need. So we are especially interested in philosophical arguments that are more likely to motivate people to act. We think arguments that keep the spotlight on starving children are more likely to have that effect.

Moreover, by thinking about hunger in these ways, we can better understand and respond to those who claim that we have no obligation to assist the starving. For we suspect that when all the rhetoric of rights, obligations, and population control are swept away, what most objectors fear is that asking people to assist the starving and undernourished is to ask too much. Morality or no, people are unlikely to act in ways they think require them to substantially sacrifice their personal interests. Thus, as long as most people think that helping others demands too much, they are unlikely to provide help.

John Arthur's critique of Peter Singer highlights just this concern. Arthur objects to moral rules that require people to abandon important things to which they have a right.

Rights or entitlements to things that are our own reflect important facts about people. Each of us has only one life and it is uniquely valuable to each of us. Your choices do not constitute my life, nor do mine yours . . . . It seems, then, that in determining whether to give aid to starving persons . . . [agents must assign] special weight to their own interests. (1977: 43)

Thus, people need not assist others if this requires abandoning something of substantial moral significance. Since what we mean by "substantial moral significance" has an ineliminable subjective element (Arthur 1977: 47), some individuals may conclude that sending *any* money to feed the starving children would be to ask too much of them. Arthur thereby captures a significant element of most people's worries about assisting the needy. The concern for our own projects and interests is thought to justify completely repressing, or at least constraining, our natural sympathies for children in need.

We suspect that the issue at bottom is the proper conception and scope of morality. Some philosophers have argued that morality should not be exceedingly demanding; indeed, one of the stock criticisms of utilitarianism is that it is far too demanding. On the other hand, some theorists, including more than a few utilitarians, have bitten the proverbial bullet and claimed that morality is indeed demanding and that its demandingness in no way counts against its cogency (Parfit, 1984; Kegan, 1988; Cullity, 1996). On the former view, morality should set expectations that all but the most weak-willed and self-centered person can satisfy; on the latter view, morality makes demands that are beyond the reach of most, if not all, of us.

We wish to take the middle ground and suggest that morality is a delicate balancing act between Milquetoast expectations that merely sanctify what people already do and expectations that are *excessively* demanding and, thus, psychologically impossible—or at least highly improbable. Our view is that the purpose of morality is not to establish an edifice that people fear but rather to set expectations that are likely to improve us, and—more relevant to the current issue—to improve the lot of those we might assist. Morality would thus be like any goal that enables us to grow and mature: the goal must be within reach, yet not easily reachable (LaFollette, 1989: 503–6). Of course, what is within reach changes over time; and what is psychologically probable depends, in no small measure, on our beliefs about what is morally expected of us. So, by expecting ourselves to do more and to be more than we currently do and are, we effectively stimulate ourselves to grow and improve. But all that is part of the balancing act of which we speak.

Thus, we frame the moral question in the following way: What should responsible people do? Our initial sympathetic response is to help the starving children. Are there any compelling reasons for thinking that our compassion should, from some "steady and general point of view," be squelched? We think the answer is "No." Are there additional reasons that bolster this initial reaction? We think the answer is "Yes." In short, we think that our initial conviction that we are responsible to malnourished children not only is undefeated but also rationally justified.

Moral Responsibility

We "instinctively" respond to the needs of starving and malnourished children. But are we, in fact, morally responsible for their plight? Two different questions are, of course, intermingled here: 1) Are we *causally* responsible *for* their condition—did we, individually or collectively, cause their hunger or create the environment that made their hunger and malnourishment more likely? 2) Are we *morally* responsible *to* these children, whether or not we are causally responsible for the conditions that make them hungry?

It is a commonplace moral argument that people are morally responsible to those whom they harm. If I run a stoplight and hit your auto, then I must pay any medical bills and either repair or replace your auto. If I trip you, causing you to break your arm, then I am expected to carry any resulting financial burden. The principle here is that we should respond to those whose cry for help results from our actions. If others are contributing causes to the harm, we may be jointly responsible to you (Hart and Honore, 1959: 188–229). Or, if my action was itself caused by the actions of some other agent—for example, if I shoved another person into you—then I am both causally and morally responsible for the harm. But, barring such conditions, a person is morally responsible for harms he or she causes.

Some commentators have argued that the affluent nations, especially colonial powers, are morally responsible to the starving because they created the conditions that make world-wide starvation possible and perhaps inevitable (O'Neill, 1993: 263–4). We find such claims plausible. But, such claims, although plausible, are contentious. Hence, for purposes of argument, we assume that we in affluent nations are in no way causally responsible for the plight of the starving. If we can show that we are (morally) responsible to the children even if we are not (causally) responsible for their plight, then our responsibility to them will be all the stronger if, as we suspect, these causal claims are true.

Shared Responsibility

If we are the cause of harm, then we are responsible *to* the "victim" because we are responsible *for* their condition. For instance, we assume that biological parents have *some* responsibility *to* children because they were responsible *for* bringing them into the world. However, being the cause of harm is not the only condition that creates a responsibility *to* someone. We are also responsible to those whom we have explicitly agreed or promised to help. For instance, by assuming a job as a lifeguard, I have agreed to care for those who swim at the beach or pool, even if they, through lack of care or foresight, put themselves into jeopardy.

More important for the current argument, responsibilities also arise from actions that, although not explicit agreements, nonetheless create reasonable expectations of care. For example, although *some* of the parents' responsibilities to their children are explained by their being the cause of the children's existence, this clearly does not explain the full *range* of parental responsibilities. For even when an agent is indisputably responsible *for* the harm to another, we would *never* think the agent is obliged to change the "victim's" soiled pants, to hold her at night when she is sick, or to listen patiently as she recounts her afternoon's activities. Yet we *do* expect this—and much more—of parents.

Our ordinary understanding of parental responsibilities makes no attempt to ground specific responsibilities *to* the child on any causal claims about the parents' responsibility *for* the child's condition. Rather, this understanding focuses on the needs of the child and the fact that the parents are in the best position to respond to those needs. This is exactly where the focus should be.

Although for any number of reasons these responsibilities typically fall to the child's biological parents, the responsibilities are not limited to the parents. Others of us (individually or collectively) have a responsibility to care for children whose parents die or abandon them. It matters not that we neither brought these children into the world nor voluntarily agreed to care for them. Rather, as responsible people, we should care for children in need, especially because they are paradigmatically vulnerable and innocent. This is our natural sympathetic reaction. "No quality of human nature is more remarkable, both in itself and in its consequences, than the propensity we have to sympathize with others" (Hume, 1978: 316).

This helps explain our shared moral responsibility to care for children who are not being cared for by their parents. Since the range of parental responsibilities cannot be explained either by the parents' being the cause of the child's existence or by their explicitly agreeing to care for the child, it should not be surprising that our shared responsibility likewise does not depend on an explicit agreement or an implicit assumption of responsibility. We assume that responsible people will, in fact, care for abandoned children. This shared responsibility springs from our common vulnerability and from our ability to respond to others who are similarily situated.

ACUTE NEED

Until now, we have written as though all starvation and malnutrition were created equal. They are not. The hunger with which people are most familiar—the hunger whose images often appear on our television sets—is hunger caused by famine. And famines tend to be episodic; often they are unpredictable. An extended drought or a devastating flood may destroy crops in a region so that the people of that region can no longer feed themselves. (Or, as is more often the case, these environmental catastrophes may not destroy all crops but primarily that portion of the crop that is used to feed the local population; crops used for export may be protected in some way.) In these cases, the problem may emerge quickly and, with some assistance, may disappear quickly. Such need is acute.

The nature of our responsibility to the starving arguably depends on the nature of their need. Peter Singer offers a vivid example of acute need and claims his example shows that we have a serious moral obligation to relieve world starvation.

> If I am walking past a shallow pond and see a child drowning in it, I ought to wade in and pull the child out. That will mean getting my clothes muddy, but this is insignificant when the death of the child would presumably be a very bad thing. (1971: 231)

This case, Singer claims, illustrates the intuitive appeal of the following moral principle: "If it is in our power to prevent something bad from happening, without thereby sacrificing something of comparable moral importance, we ought, morally, to do it." In the case in question, this is sage moral advice. If muddying my clothes saves the life of an innocent child, then it is time for me to send the cleaners some additional business.

Singer's example vividly illustrates our fundamental moral responsibility to meet acute need, especially the acute need of children—those who are paradigmatically vulnerable and innocent. In Singer's example, the child is in immediate danger; with relatively little effort we can remove her from danger. As we argued earlier, people have a shared moral responsibility that arises from our common vulnerability. None of us has complete control over our lives. All of us are vulnerable to circumstances beyond our control: floods, hurricanes, droughts, and so on. Through no fault of our own, our lives and welfare may be jeopardized. Admittedly, some acute need results from our ignorance or stupidity. Even so, others should assist us when feasible, at least if the cost to them is slight. After all, even the most careful person occasionally makes mistakes. When need is caused by natural disaster or personal error, we each want others to come to our aid. Indeed, we think they *should* come to our aid. If, upon reflection, our desire for assistance is reasonable when *we* are in need, then, by extension, we should acknowledge that we should help others in similar need. Shared responsibility and sympathy conspire to create the sense that we should go to the aid of those who cannot alleviate their own acute needs.

Although we are here emphasizing responsibility rather than justice (narrowly defined), it is noteworthy that the conditions that generate responsibility to help others in acute need resemble the conditions that Hume cites as generating our sense of justice: " . . . 'tis only from selfishness and confin'd generosity of man, along with the scanty provision nature has made for his wants, that justice derives its origin" (1978: 495). Our common vulnerability to circumstances and to the "scanty provision nature has made" leads us to seek ways to protect ourselves against misfortune and error. Natural disasters occur. They may occur where we live; they may not. Prudent people will recognize that we are all more secure, and thus better off, if we recognize a shared responsibility to assist others in acute need.

As we have suggested throughout this essay, this responsibility is all the more apparent when those in need cannot care for themselves and are in no way responsible for their plight. In short, the responsibility is greatest (and less contentious) when children are the victims. In fact, when children are in acute need, especially when many people are in a position to help, there is little moral difference between the responsibility of biological parents and others. If a child is drowning, then even if the parents (or some third party) tossed the child into the pond (and are thus singularly responsible for the child's plight), we should still rescue her if we can. Likewise, if a child is starving and her need is acute, then even if the child's parents and its government have acted irresponsibly, we should still feed the child if we can.

Arguably, the problem is different if the acute need is so substantial and so widespread as to require us to make considerable sacrifices to help those in need. In this case, our responsibilities *to* the children in acute need may resemble our responsibilities to children in chronic need.

CHRONIC NEED

Acute need arises once (or at least relatively infrequently). It requires immediate action that, if successful, often alleviates the need. But most hunger is not acute; it is chronic. Chronic hunger is the hunger of persistently malnourished children, where the causes of hunger are neither episodic nor easily removed. If the need can be met at all, it can be met only through more substantial, sustained effort and often only by making numerous (and perhaps fundamental) institutional changes, both within our countries and the other countries in need of aid.

That is why Singer's case is disanalogous with most world hunger. The drowning child is in acute need. Suppose, however, that Singer's fictional child lives on the edge of a pond where she is relatively unsupervised. We cannot protect this child by simply dirtying our clothes once. Rather, we must camp on the pond's edge, poised to rescue her whenever she falls or slips into the water. However, can we reasonably expect anyone to devote his or her entire life (or even the next six years) as this child's lifeguard? It is difficult to see how. The expectation seems even less appropriate if many children are living beside the pond.

Likely the only sensible way to protect the child from harm is to relocate her away from the pond. Or perhaps we could teach her to swim. But are we responsible for making these efforts? Do we have the authority to forcibly relocate the child or to erect an impregnable fence around the pond? Can we *require* her to take swimming lessons? Can we *force* her government to make substantial internal economic and political changes? In short, even though we are morally responsible to assist those in acute need (and especially children), we cannot straightforwardly infer that we must assist those (even children) in chronic need.

For instance, if we try to save a child from famine, we may have reason to think that quick action will yield substantial results. Not so with chronic hunger. Because we are less likely to see the fruits of our efforts, we may be less motivated to assist. Moreover, some have argued that we can alleviate chronic need only if we exert enormous effort, over a long period of time. If so, expecting someone to respond to chronic need arguably burdens that person unduly. Responsible people need not spend all their time and resources helping those in chronic need, especially if only a small chance of success exists. This is surely the insight in Arthur's view.

Consider the following analogy, which illuminates that insight. Suppose that an adult builds a house by the side of a river that floods every few years. After the first flood, we may help the homeowner, thinking that we should respond to someone who appears to be in acute need. However, after the second or third flood, we will feel that our continuing to help is asking too much. We would probably conclude that this adult has intentionally chosen a risky lifestyle. He has made his own bed; now he must sleep in it. Although this case may well be disanalogous to the plight of starving adults— because most have little control over the weather, soil erosion, or governmental policy—nonetheless, many people in affluent nations think it is analogous.

The case, however, is indisputably and totally disanalogous to the plight of children. Children did not choose to live in an economically deprived country or in a country with a corrupt government. Nor can they abandon their parents and relocate in a land of plenty or in a democratic regime. Hence, they are completely innocent—in no sense did they cause their own predicament. Moreover, they are paradigms of vulnerability.

Because they are the principal victims of chronic malnutrition, it is inappropriate to refuse to help them unless someone can show that assisting them would require an unacceptable sacrifice. That, of course, demands that we draw a line between reasonable and unreasonable sacrifice. We do not know how to draw that line. Perhaps, though, before drawing the line, we should ask the following: If the child who was starving were our own, where would we want the line to be drawn?

A DOSE OF REALITY

Evidence suggests, however, that this whole line of inquiry is beside the point. Although determining how to draw the line between reasonable and unreasonable sacrifices would be theoretically interesting, this is not a determination we need make when discussing world hunger. Doomsayers such as Garrett Hardin claim that we have long since crossed that line: Feeding starving children requires more than we can reasonably expect even highly responsible people to do; indeed, Hardin claims that such assistance effectively is suicide (1974). The doomsayers are mistaken, however. Current efforts to alleviate hunger have fallen far short of efforts that would require a substantial sacrifice from any of us. Nonetheless, even these relatively measly efforts have made a noticeable dent in the problem of world hunger. And these successes have been achieved with smaller than anticipated growth in population. According to the FAO:

- The number of chronically undernourished people in developing countries with populations exceeding 1 million is estimated at 786 million for 1988–90, reflecting a decline from 941 million in 1969–71 and a lowering of their proportion of the population from 36 to 20 percent. . . . (FAO, 1992b: 1)
- During the same period, the average number of calories consumed per person per day went from 2,430 to 2,700—more than a 10% increase. (FAO 1992b: 3)

Because the relatively meager efforts to assist the starving have made a noticeable dent in the incidence of world hunger, then, although enormous problems clearly remain, we have good reason to think that heightened efforts—efforts still *far* short of those requiring substantial sacrifices from the affluent—could seriously curtail, if not completely eliminate, world starvation. If so, we do not need to decide where the line should be drawn. We are still some distance from that line. Put differently, many of the world's poor are not like the unsupervised child who lives on the side of the lake. Even though their need may be chronic, that need can be met short of the enormous efforts that would require us to camp next to the pond for the remainder of our days. To that extent, our responsibility to chronically starving children is, despite first appearances, similar to our responsibility to children in acute need.

HOW TO ACT RESPONSIBLY

Many people are already motivated to help others (and especially children) in need. Indeed, this fact helps to explain the influence and appeal of Singer's essay more than two decades after its publication. Thus the claim that we have a shared responsibility to meet the needs of others in acute need is psychologically plausible. Even so, it is often difficult to motivate people to respond to others in chronic need. Many in affluent nations feel or fear that aid just won't do anything more than line the pockets of charitable organizations or corrupt governments. Doubtless, some money sent for aid does not reach its intended source. But that reality may simply reflect our inability to determine which relief agencies are most effective. Moreover, even if some aid does not reach those in need, it is even more obvious that most relief aid *does* reach its desired target. That is what the statistics cited in the last section demonstrate.

We suspect that the strongest barrier to helping those in chronic need is more psychological than philosophical: Most people just don't feel any connection with someone starving halfway around the world (or, for that matter, in the ghetto across town). As Hume noted, most of us do tend to feel more sympathy for what we see than for what we do not see. This at least partly explains why many of us are less willing to help starving children in foreign lands—we don't see them and thus don't feel a tie or connection to them. As we have argued throughout this essay, this is the core insight in Arthur's view: Moral obligations that require us to abandon what is important to us, especially in the absence of some connection with those in need, will rarely be met by many people—and thus, will make no moral difference. Someone might argue, on more abstract philosophical grounds, that we should not need that link. Perhaps that is true. But whether we should need to feel this connection, the fact is that most people do need it. Our concern in this essay is how to help meet the needs of the children. Thus, we need to know what will *actually* motivate people to act.

Of course, just as we should not take our initial sense of responsibility *to* children as *determining* our moral obligations, neither should we put too much weight on the unanalyzed notion of "normal ties." Doing so ignores ways in which our moral feelings can be shaped for good and for ill. So perhaps the better question is not whether we have such feelings but whether we could cultivate them in ourselves and perhaps all humanity and, if so, whether that would be appropriate. We suspect, though, that many of us cannot develop a sense of shared responsibility for *every* person in need. More likely, we must rely on a more limited sense of shared responsibility; certainly that is not beyond the psychological reach of most of us. Indeed, it is already present in many of us. Thus, working to cultivate this sense of responsibility in ourselves and others would increase the likelihood that we could curtail starvation.

Because people have a natural sympathetic response to the cry of children, the best way to cultivate this connection is to keep people focused on children as the real victims of starvation and malnutrition. If we keep this fact firmly at the fore of our minds, we are more likely, individually and collectively, to feel and act upon this sense of shared responsibility.

But even if we acknowledge this responsibility, how should we meet it? Should we provide food directly? Perhaps sometimes. But this direct approach will not solve chronic starvation. More likely we should empower the children's primary caretakers so that they can feed and care for their children. To this extent our shared responsibility to hungry children is mediated by the choices and actions of others. Thus, it might be best conceptualized as akin to (although obviously not exactly like) our responsibility to provide education. Our responsibility is not to ensure that each child receives an education (although we will be bothered if a child "slips through the cracks"). Rather, our responsibility is to establish institutions that make it more likely that all children will be educated. By analogy, because feeding children directly is virtually impossible, our responsibility is not to particular children but rather to changing the circumstances that make starvation likely.

Changing those circumstances might occasionally require that we be a bit heavy-handed. Perhaps such heavy-handedness is unavoidable if we wish to achieve the desired results. OXFAM, for example, provides aid to empower people in lands prone to famine and malnutrition to feed themselves and their children. If the recipients do not use the aid wisely, then OXFAM will be less likely to provide aid again. This is a bit draconian, but perhaps not so much as to be morally objectionable.

CONCLUSION

Both in cases of chronic and of acute need, we must remember the children who are the real victims of world hunger. The suffering child is paradigmatically vulnerable and innocent. Because we can, without serious damage to our relatively affluent lifestyles, aid these children, we should help. We share a responsibility *to* them because we are well placed to help them and because we can do so without substantially sacrificing our own interests. This is so even if we in *no way* caused or sustained the conditions that make their hunger likely.

However, if the stronger claim that we *caused* their starvation (or created the conditions that made their starvation more likely) can be defended—as we think it probably can—this responsibility becomes a stronger imperative. Thus, if the views of Sen, Crocker, and Balakrishnan/Narayan (all in Aiken, 1996) are correct—and we suspect that they are—then most of our responsibility is to cease supporting national and international institutions that cause and sustain conditions that make hunger likely. And *this* responsibility could be explained much more simply as a responsibility not to harm others.

We should also mention that the issue of hunger is deeply connected to the issue of animals and the environment, discussed by Varner and Russow. Here's how. According to agricultural scientist Paul Waggoner, "a vegetarian diet for 10 billion could be furnished by present agricultural production . . . (1994: 15). That is, by changing our diets, we could have enough food to feed not only everyone currently alive but also everyone predicted to be alive at mid-century. How can that be? Simple. Animals raised for food consume far more human-edible protein than they yield. If that food went to feed humans rather than farm animals, we could quickly meet any foreseeable human demand for food. And we could meet that demand without further damaging our environment.

Thus, although the arguments for vegetarianism and the environment are rather different from the arguments for feeding the hungry, their solutions are mutually supportive. By changing our eating habits, we have a way to diminish animal and human suffering without gobbling up more land, further polluting our rivers, cutting more trees, or destroying more plant species. Morally, that is a happy coincidence.

REFERENCES

Aiken, W and H. LaFollette. *World Hunger and Moral Obligation*, Englewood Cliffs, NJ: Prentice-Hall, Inc.

Arthur, J. "Rights and the Duty to Bring Aid." In W. Aiken and H. LaFollette, *World Hunger and Moral Obligation* (Englewood Cliffs, NJ: Prentice-Hall, Inc, 1977).

Brown, L. *The State of the World 1994: A World Watch Institute Report on Progress Toward a Sustainable Society* (New York: W.W. Norton, 1994).

———*In the Human Interest* (New York: W.W. Norton. 1974).

Food and Agricultural Organization (FAO). *World Food Supplies and Prevalence of Chronic Undernutrition in Developing Regions as Assessed in 1992* (Rome: FAO Press, 1992a).

———FAO News Release (Rome: FAO Press, 1992b).

———*World Hunger* (Rome: FAO Press, 1989).

Hardin, G. *Exploring a New Ethics for Survival* (New York: Penguin, 1975).

"Lifeboat Ethics: The Case Against Helping the Poor," *Psychology Today* (1974) 8: 38–43, 123–6.

Hart, H. and Honore, A. *Causation in the Law* (Oxford: Oxford University Press, 1959).

Hume, D. *A Treatise of Human Nature*, L.A. Selby-Bigge, ed. (Oxford: Oxford University Press, 1978).

Jonas, H. *The Imperative of Responsibility* (Chicago: University of Chicago Press, 1984).

Kegan, S. *The Limits of Morality* (Oxford: Oxford University Press, 1988).

LaFollette, H. "World Hunger," in R. Frey and C. H. Wellman, eds., *Blackwell Companion to Applied Ethics* (Oxford: Blackwell Publishers, 2002).

LaFollette, H. "The Truth in Psychological Egoism," in J. Feinberg, ed., *Reason and Responsibility* (Belmont, CA: Wadsworth Publishing Co., 1989).

May, L. *Socially Responsive Self* (Chicago: University of Chicago Press, 1996).

———*Sharing Responsibility* (Chicago: University of Chicago Press, 1992).

Mesarovic, M. and Pestel, E. *Mankind at the Turning Point* (New York: Signet Books, 1974).

O'Neill, O. "Ending World Hunger," in T. Regan, ed., *Matters of Life and Death* (New York: McGraw Hill, 1993).

Parfit, D. *Reasons and Persons* (Oxford: Oxford University Press, 1984).

Singer, P. "Famine, Affluence, and Morality," *Philosophy and Public Affairs* (1972) 1: 229–43.

United Nations Children's Fund (UNICEF). *The State of the World's Children 1993* (Oxford: Oxford University Press, 1993).

United Nations Development Program. *Human Development Report 2000* (Oxford: Oxford, 2000).

Waggoner, Paul. *How Much Land Can Ten Billion People Save for Nature?* (Ames, IA: Council for Agricultural Science and Technology, 1994).

Wilson, J. *The Moral Sense* (New York: The Free Press, 1993).

World Health Organization (WHO). *Health Statistics Report* (Geneva: World Health, 1974).

NOTES

We wish to thank William Aiken, John Hardwig, and Carl Wellman for helpful comments on earlier drafts of this paper.

This paper was originally published as "Suffer the Little Children," in *World Hunger and Morality*, William Aiken and Hugh LaFollette, eds. (Upper Saddle River, NJ: Prentice-Hall, Inc. 1996). For further development of a view similar to ours see Robert Goodin (1985), Protecting the Vulnerable, University of Chicago Press. Unfortunately, Goodin's work came to our attention after our article appeared.

Chapter 7

ANIMALS

Gary Varner

CASE STUDY: MISHA THE COW

Dr. Wright asks the class to read Tom Regan's book, *The Case for Animal Rights*. While they are working through it, Doug invites Dawn out to his farm and, on a Saturday morning, she is helping him with the milking. She notices that he calls each cow by name, so she asks him what he thinks about the early chapters of the book, in which Regan discusses the mental abilities of animals.

"Well, you know what?" Doug says. "I think Regan is right about one thing: Each of these cows is 'a subject of a life.' Holsteins have been a part of my life for as long as I can remember, and I think I know them pretty well. A lot of people think cattle look stupid, I guess, but when you watch them carefully you see that they have their own ways of solving problems. When they're confronted with something new, you can almost see them thinking about it: what they want and how to get it. And each one has its own personality, like Misha, who's shy and gentle but likes to hang around the rambunctious and ornery Daisy. I'm certainly not an animal rightist, but I think Regan is right that animals like cows are at least conscious and have preferences about their futures. What do you think?"

"I think he's right that many animals are conscious 'subjects of a life,'" Dawn replied, "but I wonder about the conclusions that he claims follow from that. I can see why animal rights activists would object to things like cramming five hens into a small cage, and maybe even to slaughter, even if it's painless. Because like Regan says, if you

believe that animals have rights, then you can't justify slaughtering them, even painlessly, just because humans like the taste of meat. But Misha seems to live a very nice
life, and she sure doesn't seem to mind being milked."

Doug thinks a moment. "The objectionable part is probably not the milking, or how
they spend their days. You're right about that. Our cows spend most of their time on
that ten acres we have fenced in for them. And if we go to a rotational grazing system,
then they'll even be getting almost all of their food by grazing rather than us bringing
them forage. We never mistreat them and would not allow any of the people we hire to
help with milking to mistreat them either."

"So then, what's wrong with dairy farms according to someone like Regan?"

"Well," Doug continues, "the whole point of breeding dairy cows is to maximize the
amount of milk they will produce over their lifetimes. We breed our cows so that they
will first give birth when they are about two years old, and we continue to breed them
so that they calve about once a year after that. Then, when they're no longer producing
milk at an efficient rate, we cull them."

"Cull them?"

"Send them to slaughter. Years ago, we'd cull cows when they were seven or eight
years old; now we cull them at about three or four years. So each year, we ship about a
quarter of our two hundred milkers to market. We have to keep fifty heifers to replace
them, but only one out of every two cows born is female. The rest are male; they become veal calves, which are slaughtered after a few months."

"I see," Dawn replies. "If every cow gives birth each year and only fifty of those
calves can be used to replace retiring dairy cows, then you're producing 150 unneeded
calves each year."

Doug looks at the ceiling, thinking. "Well, that's not quite right," he says. "Since the
fifty milkers who are being retired aren't bred, we have only 150 calves at most, and
since fifty of them end up replacing the retiring cows, that leaves only one hundred."

By this time, everyone working in the milking barn is cleaning up and putting
equipment away.

"Okay," Dawn continues, "so one problem an animal rights activist might have with
dairy farms is that they end up depending on a lot of slaughter, like maybe a number
equal to half of the milking herd every year. But just how bad is it, being slaughtered?"

"Well, actually, when my dad and I have gone to that slaughterhouse out in the
county, it hasn't seemed so bad. The cows go through a walled pathway—called a
'race'—just like we have here on the farm for getting them into the device that restrains
them during veterinary care. The races at the slaughterhouse are curved so that they
can't see far ahead, and that helps keep them calm. And my dad and I can get them to
move forward through the races without using cattle prods or anything. You just have
to get into the right place in their field of vision and they'll move on down the races trying to stay away from you."

"But don't people say that animals know when they're about to be killed?"

"The ones we watched sure didn't seem to. The slaughterhouse is ventilated so that air is drawn in through the kill chute, so you can't smell anything, and like I said, they can't see ahead. So if you don't get them excited, they don't get scared."

"So how are they actually killed? In my junior high English course we read a book called *The Jungle*, by Upton Sinclair, and it described guys beating cows to death with big hammers."

Doug laughs. "It's nothing like that anymore. They use a pneumatic gun that fires out a bolt about the size of your pinky finger, and when it's placed correctly, the cow just slumps unconscious immediately. You can even see a grey mist, which is the brain being obliterated, so the cows must lose consciousness instantly and permanently."

"Well, that's not a pretty sight, but you're right, it doesn't sound like they suffer."

"Yeah, but my dad's seen chickens slaughtered, and he said it was different. The chickens get hung from their feet, fully conscious, on a conveyer belt. They are killed by having their heads sliced off, but if things work right, they are stunned electrically first."

"But remember that expression, 'running around like a chicken with its head cut off'?" Dawn asks. "Maybe chickens are not subjects of a life, and maybe they don't have rights because they can't think in even the simple ways Misha can. I mean, how far do 'animal rights' go? If chickens have rights, then do fish? What about cockroaches?"

Before Doug can respond, his mother rings the dinner bell on the back porch. "We better hurry and get cleaned up for lunch," he says, and they head for the house.

Turn to Appendix A and perform Exercise 7.A; then, continue reading this chapter.

DISCUSSION OF ISSUES

Do nonhuman animals have rights? The answer to that question depends not only on whom you ask but also on how you ask them.

Philosopher Bernard Rollin of Colorado State University in Fort Collins has extensive experience talking to cattle ranchers about the well being of their animals and about the animal rights movement. The ranchers are understandably suspicious of the animal rights movement. After all, they are in the business of raising beef, and a well-known slogan of the animal rights movement is that "Animals are not ours to eat, wear, or experiment on." So if an animal rights activist were to ask them whether their animals "have rights," the ranchers might well answer "no." But philosopher Rollin finds that if you ask ranchers whether animals "have rights" and you make it clear that a "yes" answer to that question does not entail that ranching is wrong, they will often say "of course," citing what they take to be an obvious duty they have to treat their animals well.

Clearly, then, to understand what is at issue in debates over various uses of animals, it is critical that we get more clear about what it means to "have rights." This chapter begins with an overview of various ways that rights claims are used. Subsequent sections survey the ways that the main contributors to the contemporary debate over animal rights have interpreted rights claims and the implications that those differing interpretations have for a range of human uses of animals.

THE LOGIC OF RIGHTS CLAIMS

If it sounds odd for cattle ranchers to acknowledge that their animals "have rights" while denying that it is wrong to kill them for food, it is only because "having rights" means a number of different things in different contexts.

When a rancher says that cattle "have rights," she or he might mean only that there are right and wrong ways of treating cattle, that cattle are not mere objects that you can use any way you please without concern for their well being. A cow, the rancher might say, is a sensitive being who can suffer if mistreated and who therefore deserves to be treated well; but the rancher might maintain that humane slaughter, in which cows do not suffer, is okay. For an animal rights activist, however, to say that a cow "has rights" probably means something much stronger; it might mean that we would be wrong to use the cow as a means to our ends, even if we treat it very well in the process. For this reason, an animal rights activist might oppose dairy farming, even if all the animals were allowed to live out their natural lives rather than be sent to slaughter eventually.

These dramatically different views of "animal rights" reflect one of the most basic divisions in modern ethical theory: between consequentialist and utilitarian thinking in ethics on the one hand, and nonconsequentialist and rights-based thinking on the other.

CONSEQUENTIALISM AND UTILITARIANISM

Consequentialism is the view that the morality of actions or institutions is a function of their consequences. By *actions* is meant particular actions of individuals, whereas by *institutions* is meant social practices, laws, customs, and so on. With a consequentialist approach, you evaluate an action or institution in terms of the effects that flow from it. To the extent that the consequences are good on the whole, the action or institution is good.

The best-known version of consequentialism is utilitarianism. As a general school of thought, utilitarianism is often described as advocating "the greatest good for the greatest number," but a more precise definition would be as follows: To think in utilitarian terms is to think that the right thing to do is maximize aggregate happiness. It is important to note that utilitarianism is defined in terms of "*aggregate* happiness." To speak of "aggregate" happiness is to speak of the total or average happiness of the affected group rather than the agent's own happiness individually. So utilitarianism is not to be confused with egoism: Whereas an egoist would hold that each individual is obligated to maximize his or her own happiness, the utilitarian holds that each individual is obligated to maximize the total or average happiness of the group. That might well call for individual sacrifice, as when a parent forgoes something that would make him or her happy because the family as a whole would benefit from another use of available money.

Because utilitarianism is defined in terms of maximizing aggregate happiness, any utilitarian obviously owes us an analysis of this key term. Most utilitarians have fallen into one of two camps on this issue: *hedonistic* utilitarians define happiness in terms of feeling pleasure and avoiding pain; *preference* utilitarians define happiness in terms of some kind of integrated satisfaction of one's preferences—that is, of one's projects, plans, desires, hopes, dreams, and so on. Although most contemporary utilitarians endorse the latter, the classical utilitarians—Jeremy Bentham, John Stuart Mill, and Henry Sidgwick—all endorsed the former.

It is Mill who wrote that "It is better to be a human being dissatisfied than a pig satisfied; better to be Socrates dissatisfied than a fool satisfied." But Mill explicitly endorsed a purely hedonistic conception of happiness, writing that "By happiness is intended pleasure, and the absence of pain; by unhappiness, pain, and the privation of pleasure." He noted that such a definition of happiness "excites in many minds . . . inveterate dislike. To suppose that life has (as they express it) no higher end than pleasure—no better and nobler object of desire and pursuit—they designate as utterly mean and groveling; as a doctrine worthy only of swine." Mill replied to this "doctrine of swine objection" by arguing that the pleasures associated with various intellectual endeavors are qualitatively superior to pleasures associated with bodily functions, and that because only human beings are capable of the more intellectual endeavors (such as science, philosophy, art, and so on), a happy life for a human being will involve the exercise of those capacities.[1]

Commentators sometimes argue that in responding this way, Mill was implicitly abandoning his professed hedonism by defending the qualitative superiority of intellectual pleasures, arguing that humans strongly *prefer* a life that includes some of them. Whether or not that is the best way to interpret Mill, we can understand why many philosophers are dubious that a purely hedonistic view of happiness is adequate, at least in regard to humans, and why, therefore, most contemporary utilitarians endorse a preference-based conception of human happiness.

Many people think that happiness for an animal consists in merely avoiding pains and enjoying pleasures, and that only when it comes to humans is premature death really "tragic." Suppose, for instance, that tonight you are killed painlessly in your sleep. How are we to describe the harm you have suffered? We are assuming that you did not suffer consciously while being killed, so how do we describe the harm? To many people, it seems inadequate to describe it in terms of missed opportunities for pleasure in the future. When talking about animals, many people think that is an adequate way of describing what has been lost when they die young, but when a human dies young, people commonly think that the death is tragic in a way that no animal's death is. The tragic nature of a person's premature death might be explained in terms of its precluding the fulfillment of various long-term projects—hopes, dreams, or plans—that she or he had (for example, to raise a family, to succeed professionally, and so on) or, in the case of the very young, that she or he would eventually have developed.

This is why most contemporary utilitarians define happiness in terms of some kind of integrated fulfillment of one's preferences. For when one's happiness is conceived in these terms, it is easy to see how death could be tragic even when one passes away painlessly in one's sleep. However painless, death forecloses all possibilities for accomplishing all the things one hopes to do, for achieving one's goals, prosecuting one's projects, chasing one's dreams. Describing premature death this way seems to capture more adequately the "tragedy" that such a death is.

Obviously, a preference utilitarian has much more to say about happiness. In particular, what does achieving an "integrated fulfillment of one's preferences" mean? Are all preferences on a par, or do individuals have hierarchies of preferences, some of them more important and fundamental to one's happiness than others? If so, how do we decide which are the more and less fundamental ones? Can the satisfaction of many less fundamental ones outweigh, in the aggregate, the frustration of a more fundamental one?

Enough has been said here, however, to make the basic outlines of utilitarian thinking clear. Utilitarians are consequentialists; they evaluate actions and institutions in terms of their effects or consequences. Specifically, utilitarians look at effects on the aggregate happiness of the affected individuals. Although some utilitarians define happiness in terms of individuals simply feeling pleasure while avoiding pain, others say that individuals are happy to the extent that they achieve an integrated satisfaction of their preferences.

Nonconsequentialism and Rights-Based Thinking

The nonconsequentialist alternatives to utilitarianism are best introduced by identifying what is widely thought to be the central problem for utilitarian thinking in ethics, returning to the overarching topic in this section of the chapter. A succinct way of expressing the fundamental problem that many philosophers have seen in utilitarianism is as follows: Because it evaluates actions and institutions in terms of aggregate happiness, utilitarianism fails to respect individual rights.

To see why someone would think this, consider the institution of slavery. Presumably, slavery makes each slave dramatically less happy than she or he could be, but if each slave serves several free people, then perhaps the total added happiness for the slaveholders more than outweighs the aggregate unhappiness of the slaves. If so, then wouldn't slavery be justified on utilitarian grounds? Some utilitarians have responded that it is unrealistic to think that the dramatic sufferings of the slave class would really be outweighed by the increased leisure time and other benefits enjoyed by the slave-holders. But critics of utilitarianism argue that even if no real-world cases of slavery can be justified on utilitarian grounds, something is still wrong with a theory that eval-uates institutions such as slavery in terms of aggregate happiness. That utilitarianism would endorse human slavery, even just in principle, is a critical objection in many thinkers' eyes. This objection is commonly expressed in terms of rights claims: If slav-ery violates fundamental human rights, then even if the slaves' unhappiness were out-weighed by the slaveholders' happiness, slavery would still be wrong.

To criticize utilitarianism for neglecting individual rights in this way is to invoke the kind of stronger notion of "having rights" that I earlier attributed to animal rights ac-tivists and contrasted with "rights" of the kind that ranchers are willing to acknowl-edge that animals have. A utilitarian certainly thinks that individuals "have rights" in a sense. After all, because utilitarians believe that we ought always to maximize aggre-gate happiness, they ought to take into account effects on all individuals whose happi-ness is affected by an institution such as slavery, and they do just that in taking the slaves' unhappiness into consideration. But the rights that critics of utilitarianism have faulted it for neglecting are rights in some stronger sense. Just as we imagined an ani-mal rights activist objecting to various uses of animals—even perfectly humane uses—on the grounds that animals ought not to be treated as means to our ends, some philosophers have criticized utilitarians for treating individual humans as means to others' ends. One of the most famous opponents of utilitarian thinking in ethics was the eighteenth century Prussian philosopher Immanuel Kant. Kant formulated a very different way of thinking about right and wrong, which he called "the categorical im-perative." Although Kant claimed repeatedly that only one categorical imperative ex-ists, he gave several different formulations that are not obviously equivalent. One of them invokes the distinction between treating people "as ends rather than means": "Act in such a way that you treat humanity, whether in your own person or the person of any other, never simply as a means but always at the same time as an end." Kant claimed that precisely the same conclusions follow from this version of the categorical impera-tive as would follow from this very different formulation of it: "Act only on that maxim through which you can at the same time will that it should become a universal law." It is not obvious how to apply either of these criteria, nor is it obvious that they would imply exactly the same things as Kant claimed.[2]

Without going into any further detail about Kant's specific way of opposing utili-tarianism, I can sum up what I have covered about the logic of rights claims in this section. Utilitarianism is the consequentialist view that right actions and institutions maximize aggregate happiness. Utilitarians recognize that individuals "have rights" in the limited sense that every individual's happiness ought to be taken into account in deciding what to do. Critics of utilitarianism charge that it ignores a stronger sense of "having rights." If an individual "has rights" in this stronger sense, then that individual deserves to be treated as more than a utility receptacle. That is, if an individual "has rights" in the stronger sense, then it would be wrong to harm that individual unless some non-utilitarian justification could be given for doing so.

This of course leaves open many questions. First and foremost, how do we decide which individuals "have rights" in this stronger sense? Kant believed that only human beings are due this kind of respect, but obviously animal rights advocates who oppose even the most humane uses of animals think that the animals do as well. And even after the question of who has rights is settled, it remains to be said how we should decide whose rights to violate when rights come into conflict. Both of these important questions are discussed in more detail in the next section.

THE CONTEMPORARY DEBATE OVER "ANIMAL RIGHTS"

The battle lines in the current debate over "animal rights" can be drawn pretty clearly in terms of utilitarianism versus rights views, with hedonistic utilitarian thinking about "animal rights" corresponding to animal welfare. (Note: From here on, I put "animal rights" in quotation marks when I intend for the term to include both animal welfare philosophies and true animal *rights* philosophies.) The preceding discussion of rights claims clarifies how such an animal welfare position might lead to dramatically different conclusions than a true animal *rights* view. Before looking at such views in greater detail, however, it is important to contrast both with what is often called the neo-Cartesian position.

NEO-CARTESIANISM

René Descartes, commonly designated "the father of modern philosophy," held that animals lack all consciousness because they lack language and reason. Descartes and his followers held that, because they are not conscious, animal experimentation raises no ethical issues at all. Modern study of animal cognition makes it implausible to hold that animals entirely lack reason, and some studies suggest that they can master at least the rudiments of language. Moreover, Descartes never clearly explained why an organism that lacks language and reason must necessarily lack all consciousness whatsoever, including consciousness of pain, which is generally regarded as a less cognitive phenomenon. After all, newborn babies can neither reason nor use language, yet we believe that they can feel pain—otherwise, why would circumcising infants without using anesthetics be controversial?

Still, some contemporary philosophers hold that animals may entirely lack consciousness. The best-known example is Peter Carruthers of the University of Sheffield, England. In a widely discussed article called "Brute Experience," Carruthers argued that although animals clearly experience their environment, their experiences might all be nonconscious, the same way a driver distracted by conversation experiences and unconsciously avoids traffic on the road but can't recall anything about doing so later. Carruthers proposed that an experience is conscious only if it is available for reflection. That would explain why the preoccupied driver would claim not to remember anything about miles and miles of road she has just traversed, even though she was clearly conscious of other things during the same period. With that criterion for consciousness in mind, Carruthers questioned whether animals are ever conscious. In Carruthers' view, unless we can show that they think about at least some of their own experiences, we have no evidence that animals are conscious at all.

"Blind sight" is the term for a well-studied phenomenon in humans that can help us understand Carruthers' claim about animals. "Cortical blindness" results when part of the primary visual cortex in the brain is damaged. If the damage is limited, the blindness will be limited to part of the subject's visual field. In the damaged area, the patient has no conscious vision; these patients claim not to see anything at all, they steadfastly maintain that they are not conscious of, and do not think about, anything in the affected area of their visual field. Nevertheless, when forced to guess what is being presented there, they answer correctly with far higher-than-chance accuracy (in some situations, approaching 100 percent), proving that visual information is still being processed, that they really do see, but without that vision being available to consciousness.

Carruthers' suggestion about animals is that they may be, in effect, blind sighted in all respects; that all their experiences—including their experiences of pain—may be nonconscious experiences. Just as it would not seem to matter to a blind-sighted patient whether he or she is presented with a beautiful scene or a gory one in the area of her visual field of which she has no conscious awareness, it may not matter to an animal whether it feels pleasure or pain. Nonconscious pains don't have any conscious "feel" at all, so they can't feel *bad*, and seemingly the fact that our pains feel bad to us is what makes us think pain is a bad thing.[3]

Although a few contemporary philosophers are neo-Cartesians like Carruthers, most hold that at least some nonhuman animals are conscious, at least in certain ways, and this is certainly a feature of our common-sense world view. But if at least some animals are conscious, then the question of their moral status arises, for if animals can feel pain and therefore suffer, they apparently can be harmed in morally significant ways. Contemporary proponents of animal welfare and of animal rights argue for two different ways of taking the conscious suffering of animals into consideration, ways that map roughly onto the hedonistic utilitarian perspective and some kind of stronger claim about animal rights.

THE ANIMAL WELFARE POSITION

In response to the emergence of the contemporary animal rights movement (roughly following the release of Peter Singer's book *Animal Liberation* in 1975), agriculturalists, medical researchers, and others targeted by the movement began using the term "animal welfare" to describe their position on the moral status of animals. They used the term to emphasize that although they took seriously the moral standing of animals, they did not reach the abolitionist conclusions of self-styled animal rights activists. To be an animal welfarist came to mean roughly that one takes animal suffering into account, but in roughly the way that a hedonistic utilitarian would. Although the implications of applying utilitarian thinking to our treatment of other humans had long been a source of objections to the view, the analogous implications for our treatment of nonhuman animals allowed animal welfarists to endorse things such as medical research on animals, which promised to save countless humans (and animals) from preventable suffering and death, while distancing themselves from the neo-Cartesians who would deny that animals are conscious at all.

In the philosophical literatures on animal rights and environmental ethics, the term "animal welfare" has become synonymous with the hedonistic utilitarian approach to thinking about "animal rights." However, in popular discussions of the issue, the "animal welfare/rights" distinction is commonly employed in a less philosophical and more political sense, in terms of the goals that various activists have in mind and the tactics they employ in trying to reach those goals. For instance, in media coverage of the "animal rights" issue, animal welfarists are usually portrayed as people who work within the system to revise certain problematic practices without advocating a total end to those practices. Animal rightists, by contrast, are portrayed as extremists bent on abolishing various practices and willing to use illegal means to accomplish their goals. It is also common for animal welfarists to be portrayed as calm, well-informed, and rational critics, with animal rightists portrayed as emotional, unreasoning, and poorly informed.

This more popular, political version of the animal rights versus animal welfare distinction, however, obscures the agreement that exists at the level of philosophical principle between self-described animal welfarists and the best-known "animal rights" philosopher, Australian ethicist Peter Singer. Singer is thoroughly a utilitarian and, although he actually employs a complex mix of hedonistic and preference utilitarianism, Singer thinks, as do most animal welfarists, that a hedonistic conception adequately captures the concept of happiness as it applies to many nonhuman animals. Singer acknowledges that in principle his utilitarianism implies that some medical research and some forms of slaughter-based animal agriculture would be justifiable, but in practice, he argues, even a hedonistic utilitarian stance implies that these practices should be abolished or all but abolished. So, Singer is popularly characterized as an animal rightist because he argues for abolitionist conclusions, even though much of his philosophical stance corresponds to that of self-professed animal welfarists.

In this section, I look at a few important details of Singer's philosophical position, noting more carefully where and how his views correspond to those of self-professed animal welfarists and noting certain general implications of Singer's views. I wait until the final section to look in detail at specific uses of animals so that I can first look at the very different, and explicitly rights-based view, of the other best-known animal rights philosopher, Tom Regan. I can then carefully compare and contrast the implications of thinking about animals in hedonistic utilitarian terms and in terms of rights more strongly construed.

Any thorough summary of Singer's utilitarian approach to "animal rights" must discuss at least four concepts:

- His principle of equal consideration of interests
- His claim that sentience is a sufficient condition for having interests
- The related notion of speciesism
- His distinction between animals that are morally "replaceable" and those that are not

Singer discusses the first three of these points in the first chapter of his book *Animal Liberation*, which is one of the most widely reprinted pieces on the subject of "animal rights."

Singer titles the chapter "All Animals are Equal: or why the ethical principle on which human equality rests requires us to extend equal consideration to animals too."[4] In it, he argues first that "Equality is a moral idea, not an assertion of fact" (p. 4). That is, when people say that "all humans are equal," we do not assert that they are in fact equal in intelligence, capabilities, size, or other features. Rather, we assert that every human, regardless of these variations, deserves equal consideration of his or her interests. For instance, the interests of white males should not be counted while those of black females are not; the interests of less intelligent people should not be discounted, and so on. The moral equality we insist on for all humans, Singer concludes, comes to this: "The interests of every [individual] affected by an action are to be taken into account and given the same weight as the like interests of any other" (Singer, 5).

But, Singer then argues, if some nonhuman animals have interests that are similar to some human interests, then it would be arbitrary to limit equal consideration of interests to humans. In particular, he argues, all animals that are capable of feeling pain have a basic interest in common with humans, namely an interest in avoiding pain. Although the term *sentient* can be used to refer to any kind of consciousness whatsoever, the term has come to be associated with consciousness of pain specifically, because Singer chose to use it that way. We can summarize Singer's argument so far, then, as follows:

1. The principle of equal consideration of interests is the basis of our recognition of human moral equality.
2. According to that principle, the similar interests of all individuals affected by actions and institutions ought to be given equal weight in evaluating those actions and institutions.
3. Sentience (conceived of as the capacity to feel pain) is a sufficient condition for having interests.
4. So, we ought to treat the similar interests of all sentient animals equally in evaluating actions and institutions.

In this sense, all (sentient) animals are equal.

One of the most basic misunderstandings people get from a superficial acquaintance with Singer's argument for animal equality comes from not noticing that, to the extent that different individuals have different interests, equal consideration of similar interests does not require equal treatment of different individuals. Singer himself emphasizes this with the following example:

> Precisely what our concern or consideration requires us to do may vary according to the characteristics of those affected by what we do: concern for the well-being of children growing up in America would require that we teach them to read; concern for the well-being of pigs may require no more than that we leave them with other pigs in a place where there is adequate food and room to run freely. (p. 5)

So, recognizing (sentient) animals' moral equality with humans in the way that Singer advocates would not entail providing horses with libraries or giving monkeys the vote. Exactly what it might entail is considered further in the final section of this chapter, but Singer thinks that the implications are sweeping, as his discussion of speciesism in the first chapter of *Animal Liberation* makes clear.

The term *speciesism* was coined by Richard Ryder, a British author of several popular books on our treatment of animals. Singer describes speciesism as "a prejudice or attitude of bias toward the interests of members of one's own species and against those of members of other species" (p. 6). Speciesism thus uses the same logic as racism or sexism: Each involves either ignoring or differentially weighting the similar interests of members of different groups. Racists ignore or differentially weight the similar interests of different racial groups, and sexists ignore or differentially weight the similar interests of different sexes. A speciesist would be anyone who ignores or differentially weights the similar interests of different species.

In *Animal Liberation*, Singer rests his case against animal research, agriculture, hunting, and so forth on two kinds of arguments, both of which rely on the concept of speciesism. The first strategy can be called the argument from marginal cases. By *mar-*

ginal cases is meant human beings who lack many, most, or all of the intellectual traits distinctive of normal adult humans, which would include, for example, profoundly retarded persons, very young children, fetuses, and newborns, the enfeebled elderly, and the irreversibly comatose. (The label for this form of argument may be unfortunate but the label is widely used in the contemporary literature on animal rights.) An argument from marginal cases proceeds thus:

1. Identify a reason that most people think would suffice to make treating such "marginal" humans a certain way wrong.
2. Show how that same reason applies to animals who are commonly treated the same way.
3. Conclude that we are being speciesist when we treat the animals differently than the "marginal" humans.

Here is an example in Singer's own words:

> [I]f we consider it wrong to inflict pain on a baby for no good reason then we must, unless we are speciesists, consider it equally wrong to inflict the same amount of pain on a horse for no good reason. (p. 15)

Singer's other argument strategy, which we can call the generalization argument, works the other way around:

1. The advocate of a practice cites some reason for using certain nonhuman animals.
2. The reason given is that the animals lack some characteristic.
3. For any characteristic that the advocate cites, some human beings can be found who lack this same characteristic.
4. Therefore the advocate should feel justified in doing the same thing to those humans for the same reason.
5. But the advocate does not, and this shows that he is a speciesist.

Here is an example of this type of argument from Singer:

> [I]f we use [the argument that certain experiments would cause less pain if performed on nonhuman animals than if performed on normal human beings] to justify experiments on nonhuman animals[,] we have to ask ourselves whether we are also prepared to allow experiments on human infants and retarded adults … (p. 16)

Singer argues that "Most humans are speciesists" because most humans "take an active part in, acquiesce in, and allow their taxes to pay for practices that require the

sacrifice of the most important interests of members of other species in order to promote the most trivial interests of our own species" (p. 9). Because we would never sacrifice similarly important interests of human beings for the sake of trivial benefits, our
acceptance of various agricultural, scientific, educational, and recreational practices is
speciesist.

In *Animal Liberation*, Singer relied on such arguments rather than explicitly endorsing utilitarianism. He did this to avoid embroiling himself in philosophical controversies that were unnecessary to settle in his opinion. He thought that our acceptance of
various uses of animals is so clearly speciesist that one needn't even take sides on the
utilitarianism versus rights issue to oppose those practices, and he thought that a popular, less philosophical book would be more effective in generating opposition to them.
However, throughout his professional philosophical work, Singer endorses utilitarianism,
and in concluding my overview of Singer, I should note that Singer defends a complex
mix of hedonistic and preference utilitarianism.

In my general discussion of utilitarianism, I noted that different utilitarians employ
different conceptions of happiness. With the hedonistic conception, one is happy to
the extent that one feels pleasure and avoids pain, whereas the preference conception
identifies happiness with the "integrated fulfillment of one's preferences." Many people
find the latter to better capture the way premature death is tragic for a human being.
In *Practical Ethics*,[5] Singer agrees that the death of an individual who (like any normal,
adult human being in decent circumstances) has a robust sense of his or her future,
who has projects, plans, desires, hopes, dreams, and so on—which he identifies as being "self-conscious"—is a greater harm than is the death of an individual who lacks
such a robust sense of self and future.

However, Singer denies that only human beings are self-conscious. He cites research
that, he claims, clearly shows that some other primates are self-conscious (specifically chimpanzees, gorillas, and orangutans, Singer, 111–16, 118, and 132). He admits than many animals, including fish and chickens, may not be (pp. 95 and 133), but he claims that a case
can be made, though with varying degrees of confidence, on behalf of whales, dolphins, monkeys, dogs, cats, pigs, seals, bears, cattle, sheep and so on, perhaps even to
the point at which it may include all mammals (p. 132).

This is an important qualification of Singer's view because he concludes that the
morality of killing is very different between animals who are self-conscious and those
who are not.

Specifically, Singer claims that what he calls "the replaceability argument" applies
to the latter but not to the former:

> When we come to animals who, as far as we can tell, are not rational and
> self-conscious beings, the case against killing is weaker. . . . Even when the
> animal killed would have lived pleasantly, it is at least arguable that no

wrong is done if the animal killed will, as a result of the killing, be replaced by another animal living an equally pleasant life. (pp. 132-33)

Singer argues that this reasoning applies to individuals who are self-conscious in his sense, but not to those who are not. The following chart illustrates Singer's reasoning (see table 7.1). Suppose that a farmer has three happy chickens, kills one without causing it to suffer in the process, and replaces it with an equally happy chicken. Singer's claim is that in this case, the same total happiness exists in the world before and after, as illustrated in situation A. But this argument assumes that the opportunities for future pleasure forgone by the terminated individual (#1) do not count as negatives in the moral ledger after that individual is replaced by an equally happy individual (#4). By comparison, if the unfulfilled preferences of a self-conscious individual *do* count as negatives in the moral ledger after its death, then things are as represented in situation B, where aggregate happiness is not maintained by replacing the terminated individual (#1) with an equally happy individual (#4). According to Singer, this explains why killing an individual who is self-conscious is a more serious affair, morally speaking.

Singer's replaceability thesis is controversial, but for the purposes of this chapter it is interesting primarily because it shows where his philosophical views coincide most closely with those of self-professed animal welfarists. The animal welfarists think similarly to hedonistic utilitarians in regard to all animals. Singer does so only in regard to animals who are not self-conscious. This is not to say that for Singer we are never justified in harming a being that is self-conscious; as a utilitarian, he must admit that at some point the aggregate utility of doing so justifies the harm to that individual. However, in Singer's view, the painless death of an individual who is not self-conscious is easier to justify than is the painless death of a self-conscious one, and Singer believes that many of the animals involved in agriculture and scientific research, including birds and fish, are not self-conscious. As noted previously, he explicitly gives chickens and fish as examples, but we may surmise that he intends to include other birds, as well as all reptiles and amphibians.

TABLE 7.1. Illustrating the Replaceability Thesis

Situation A	Before	After	Situation B	Before	After
Individual #1	+1		Individual #1	+1	−1
Individual #2	+1	+1	Individual #2	+1	+1
Individual #3	+1	+1	Individual #3	+1	+1
Individual #4		+1	Individual #4		+1
Aggregate happiness	+3	+3	Aggregate Happiness	+3	+2

THE ANIMAL RIGHTS VIEW

I have shown how the views of the best-known "animal rights" philosopher, Peter Singer, correspond in significant ways to those of self-professed animal welfarists. In this section, I examine the views of the other best-known "animal rights" philosopher, Tom Regan of North Carolina State University. Regan defends what he calls "the rights view" and claims that it, rather than Singer's utilitarianism, is the philosophical foundation of the more abolitionist conclusions that self-professed animal rights activists reach. To adequately summarize Regan's view, I need to discuss three things:

- His rights-based opposition to utilitarianism
- His use of an argument from marginal cases
- His "miniride" and "worse-off" principles

All these features of his view are developed in his 1983 book *The Case for Animal Rights*.[6]

As the label he chooses for his view—"the rights view"—suggests, Regan rejects utilitarian justifications for harming individuals who have moral rights; that is, Regan agrees with the standard criticisms of utilitarian views, such as the slavery objection. Whereas Singer responds to this objection the way I described earlier—he argues that realistically, slavery would never actually maximize aggregate happiness (see *Practical Ethics*, p. 23)—Regan objects to the fact that a utilitarian could admit that something like slavery would ever be justified, even in principle. To do so, he argues, is to treat human beings as if they were mere "utility receptacles" (pp. 205–6, 236). Regan reasons that the respect we normally think is due our fellow humans, and that makes slavery wrong even in principle, is the kind of respect described earlier as "having rights" in the stronger sense. Regan concludes that if an individual "has rights" in this sense, it is wrong to harm him or her on the grounds that doing so maximizes aggregate utility— to do so would be to treat him or her as a mere "utility receptacle."

Thus far, Regan's view is a very traditional response to utilitarianism. It is the next step in Regan's argument that leads him toward a nontraditional attitude toward non-human animals: Regan uses an argument from marginal cases to defend extending such basic respect, or rights in the strong sense, from humans to include many animals. Specifically, Regan argues that the most plausible basis for attributing rights to humans, what he calls "the subject of a life criterion," implies that many animals, and at the very least all "normal adult mammals of a year or more" have rights in the same sense.

Regan describes a "subject of a life" as any individual who has

beliefs and desires; perception, memory, and a sense of the future, including [one's] own future; an emotional life together with feelings of

pleasure and pain; . . . the ability to initiate action in pursuit of their desires and goals; a psychological identity over time; and an individual welfare in the sense that their experiential life fares well or ill for them, logically independently of their utility for others and logically independently
of their being the object of anyone else's interests. (p. 243)

Notice how similar Regan's notion of being a "subject of a life" is to Singer's notion of
"self-consciousness." In both cases, what is meant, basically, is that the individual in
question has a fairly robust sense of his or her (or its) own future and preferences (desires, goals, plans) about that future.

Regan argues that the subject of a life criterion best explains the judgments we
make about respect being due our fellow human beings, because not only are normal
adults but also many "marginal" humans (including the quite profoundly retarded,
very young children, and the very enfeebled elderly) subjects of a life. But if their having the capacities listed previously is what qualifies all these humans for the special
kind of respect that moral rights in the stronger sense carry, then we ought to recognize that many animals have rights in this sense, too. In the early chapters of *The Case
for Animal Rights*, Regan presents empirical evidence that he thinks proves that at least
"all mentally normal mammals of a year or more" (p. 78) are subjects of a life.

Some commentators have misinterpreted this claim to mean that "mammals and no
other forms of life" have rights on Regan's view.[7] But Regan makes the point clear that he
is restricting the reference of "animals" in this way only to avoid the controversy over
"line drawing." He does so by making his arguments in the book "refer [to] individuals
well beyond the point where anyone could reasonably 'draw the line' separating those who
have the mental abilities in question from those who lack them" (Regan, 78). At various
places in the book, Regan reminds us that his meaning is not that birds, for instance, do
not have rights but only that the case for saying that they are subjects of a life is not in-
controvertible the way he believes it is for mammals (see, for example, Regan, 349).

So, to summarize Regan's use of the argument from marginal cases: He argues that
if most humans' being subjects of a life qualifies them for moral rights in the strong
sense, then so too does it qualify many nonhuman animals, including at least all mammals, probably also birds, and maybe even some other animals. In the final section of
this chapter, I say a bit more about which animals are subjects of a life as well as about
which ones feel pain (the two categories might not be co-extensive). To finish up this
summary treatment of Regan's view, however, I must first say something about his
"miniride" and "worse-off" principles.

These principles are Regan's way of answering the question "How can we decide
whose rights to violate when rights come into conflict?" Regan recognizes that such
conflicts will arise, but realizes that to decide on utilitarian grounds in such cases
would be to take back the basic respect for individuals that recognizing their moral
rights requires. He therefore proposes two non-utilitarian principles. The principles

apply in two different kinds of cases and require very different things. Here are convenient summaries of the two principles (based on what Regan says about them in section 8.10 of *The Case for Animal Rights*):

> *The miniride principle* applies where comparable harms are involved and requires that we override the rights of the few rather than the many.

> *The worse-off principle* applies where the harms involved are not comparable, and requires us to avoid harming the "worse-off individual."

Before applying these principles to various situations, I need to clarify two key concepts.

One is the concept of a "noncomparable" harm. To understand how Regan conceives of harm, remember that what qualifies individuals for rights is their being what he calls "subjects of a life," and what he means by this is roughly that they have conscious preferences and desires about their own futures. So, harm to a subject of a life consists in thwarting actual preferences and desires or in diminishing the individual's capacity to form or satisfy such preferences and desires in the future. What would be an example of "noncomparable harms" so construed? An obvious example would be losing a finger or a leg versus losing one's life. The former would diminish one's capacity to form and satisfy desires, at least temporarily as one adjusted to the loss of the digit or limb. It might even make ever pursuing certain previously valued goals impossible. But in contrast, losing one's life completely destroys the capacity to form and satisfy desires. As I discuss later, Regan does not think that his critique of animal agriculture and other uses of animals requires us to make any more fine-grained distinctions than this. He thinks that the harms animals suffer in these practices are just as obviously not comparable to the harms humans would suffer by eliminating the practices.[8]

The other concept crucial to applying Regan's two principles is that of "the worse-off individual." The best way to explain this concept is by considering an abstract example. In table 7.2, assume that each row represents a choice between two options in a given situation, that each negative number under an option represents the harm that would befall one individual if that option is chosen in that situation, and that a –10 harm is "noncomparably worse" than a –1 harm.

By "avoid harming the worse-off individual" Regan means avoiding harming whoever would suffer a harm that is noncomparably worse than the harm anyone would suffer under the alternative option. In situation A, this is the individual who, under option #1, would suffer a harm that is noncomparably worse than the harm any individual would suffer if option #2 were chosen. In situation B, each of the individuals under option #1 is "the worse-off individual" in relation to the individuals under option #2.

Regarding the implications of Regan's principles, note that in some cases, his principles will imply the same conclusions as the principle of utility, for instance in situation B (in which the aggregate harm under option #1 is –50, versus –5 under option

#2). Here the worse-off principle applies because noncomparable harms are again involved and the worse-off principle requires us to avoid harming "the worse-off individual" who is, as previously noted, any of the individuals under option #1 in relation to those individuals under option #2. But sometimes, as in situation A, the implications will be radically different. In situation A, the utilitarian would choose option #1 because less aggregate harm would occur (–10) in comparison to option #2 (–25). However, Regan's worse-off principle implies that we should forego maximizing aggregate utility in this case in order to respect the rights of the worse-off individual. This example illustrates how, according to Regan, the rights view rules out slavery even in principle. The relatively trivial harms each beneficiary of the institution would suffer via its abolition (the –1s under option #2 in situation A) do not justify, even in the aggregate, causing noncomparable harm to even one slave (the –10 under option #1).

In situation C, however, notice that everyone who would be harmed by either option #1 or option #2 would be harmed to the same degree (–10). So, "comparable" harms are involved and the miniride principle applies, requiring us to override the rights of the few rather than the many. Thus in situation C, we are required to choose option #2. Notice that this is the same option a preference utilitarian would choose, because the aggregate harm under option #2 (–50) is less than under option #1 (–250). Regan insists, however, that the miniride principle is not utilitarian, because its application does not turn on minimizing aggregate harm but rather on minimizing

TABLE 7.2. Applying Regan's two principles

	Option #1	Option #2
Situation A	–10	–1 –1 –1 –1 –1 –1 –1 –1 –1 –1 –1 –1 –1 –1 –1 –1 –1 –1 –1 –1 –1 –1 –1 –1 –1
Situation B	–10 –10 –10 –10 –10 –10 –10 –10 –10 –10 –10 –10 –10 –10 –10 –10 –10 –10 –10 –10 –10 –10 –10 –10 –10	–1 –1 –1 –1 –1
Situation C	–10 –10 –10 –10 –10 –10 –10 –10 –10 –10 –10 –10 –10 –10 –10 –10 –10 –10 –10 –10 –10 –10 –10 –10 –10	–10 –10 –10 –10 –10

the overriding of rights (Regan, 305–306). Where *comparable* harms are involved, Regan claims, respecting individuals equally means counting each rights violation equally and minimizing the total number of individual rights violations. His reasoning is, roughly, that when comparable harms are involved, the rights violations are equally serious, whereas when noncomparable harms are involved, they are not; so, in the former case but not in the latter, it is the number of rights violations that is crucial.

ANIMAL RIGHTS AND AGRICULTURE

This section compares and contrasts the animal welfare position, which employs hedonistic utilitarian thinking about society's treatment of animals, with a view that, like Regan's, attributes rights in a stronger sense to many nonhuman animals. A wide range of practices, both traditional and new, have come under the moral microscope since animal rights views achieved new prominence in the last quarter of the twentieth century. One of the most prominent targets of animal rights groups has been animal agriculture, but even within this area a range of practices have come under fire, raising a variety of issues. Here I can discuss only a few general practices: beef, dairy, and poultry and egg production, but the differences among these three practices provide the means for critically comparing and contrasting the implications of Singer's and Regan's views.

One difference is that, as we saw in the preceding section, Singer's and Regan's views might imply different things about the two types of animals involved. Singer, for instance, holds that the replaceability argument applies to animals that are not self-conscious. Singer refers to chickens as an example of animals that might be sentient without being self-conscious. For Regan, only animals that are subjects of a life "have rights" in his strong sense. So, the questions "Which animals are self-conscious?" and "Which animals are subjects of a life?" are crucially important for understanding the implications of these two views.

But how do we answer such questions? All attributions and denials of mental states or consciousness[9] concerning other animals, including our fellow human beings, involve arguments by analogy. For even when it comes to our fellow human beings, we can never directly observe one another's conscious states (pains, desires, beliefs, and so on). We are always reasoning by analogy.

Formally, arguments by analogy have the following structure:

1. a, b, and c are all known to have properties P and Q.
2. a and b are known to have property R as well.
3. So, probably c has property R as well.

Suppose, for instance, that you see a stranger accidentally place his hand on a hot stove. When he screams and jumps away, nursing his hand, you are certain that he is in pain, even though you do not see or feel his pain. How can you be so certain? In the case of a fellow human being, you know that he has similar neurophysiology and behaviors to yours (think of these as P and Q in the preceding argument form). Because you know that you would feel pain in his situation (R), you think it likely (indeed, *extremely* likely) that he does. For nonhuman animals, the situation is no different except that the analogies usually are not as strong. For humans, innumerable analogies exist under the general categories of behavior and neurophysiology. However, depending on the animal and the type of conscious state in question, the case for saying that the animal has the conscious state in question may be stronger or weaker, depending on the number and type of analogies scientific research or simple observation provide.

Bear in mind, however, that the strength of an argument by analogy is not simply a matter of *how many* analogies one can find. To see why, consider this obviously weak argument by analogy:

1. Chickens, turkeys, pheasants, and cattle are all animals and they are all eaten by humans.
2. Chickens, turkeys, and pheasants are all born from eggs.
3. So, probably cattle are born from eggs.

Obviously, one could come up with more and more analogies to list off (for example, chickens, turkeys, pheasants, and cattle all have hearts, all have lungs, all have bones . . .). The preceding argument is weak because it ignores a crucial *disanalogy*: that cattle are mammals, whereas the others are all birds, and we have very different theories about how the two classes are conceived, gestated, and born. So, in assessing an argument by analogy, we do not look just at the raw number of analogies cited but also at the significance of the analogies cited and whether there are any relevant disanalogies.

Regarding consciousness of pain, specifically, the authors of the four most detailed examinations of the available analogies all reach the conclusion that although all vertebrates probably can feel pain, most invertebrates probably cannot (the notable exception being the cephalopods, which are octopi, squid, and cuttlefish). They reached this conclusion by considering behaviors such as avoidance of previously harmful stimuli and situations and the favoring of injured limbs, along with neurophysiological considerations such as whether various animals respond to known analgesics and have natural pain blockers (endogenous opioids) in their systems. The four studies note that whereas vertebrates uniformly score high on such comparisons, invertebrates (with the exception of cephalopods) score low.[10]

On the matter of being "self-conscious" or a "subject of a life" or both, however, there is much more disagreement, both about what analogies are relevant and which animals fit those analogies. When scientists have studied the concept of "self-consciousness," their focus has usually been on animals' use of either mirrors or language. Regarding both, studies have usually shown that only a few of the great apes behave in the ways the scientists have assumed that a "self-conscious" being would. For instance, chimpanzees and gorillas have been taught crude versions of human language, and their uses of these languages have suggested to some scientists that these animals might be self-conscious. Other scientists have thought that a self-conscious organism would recognize itself in a mirror, as opposed to treating its mirror image as an unknown conspecific (member of its own species), and would investigate parts of its body that it would otherwise never be able to look at (such as its teeth and forehead). But by this criterion, only chimpanzees, orangutans, and possibly gorillas would appear to be self-conscious, because only these animals have been shown to recognize changes in their own bodies (for example, a paint spot applied to the forehead during anesthesia) after acclimation to mirrors, and many animals (such as baboons) can live with mirrors for years and never stop acting aggressively (as if to an unknown conspecific) toward their own mirror image.[11]

As discussed in "The Animal Welfare Position," however, Singer seems to think that all mammals might be self-conscious in his sense. But all he means by the term "self-conscious," and all Regan means by the term "subject of a life," is (roughly) having conscious preferences about one's own future, which might not require the kind of capacities discussed in the preceding paragraph. So what scientific work would be relevant to determining which animals have this more limited capacity? Elsewhere I have discussed certain research on basic learning strategies that suggests that although mammals, birds, and herpetofauna (reptiles and amphibians) probably have such forward-looking desires, fish may not, and invertebrates (with the exception of cephalopods) probably do not. The studies in question involve very simple learning strategies that would seem to be indicative of thinking consciously about the environment and how to respond to it in ways that will get you what you want, rather than "mindlessly" repeating behavior patterns that you have been habituated into. Insofar as most invertebrates and even fish lack these basic learning strategies, the case for thinking that they have conscious preferences is weak relative to mammals, birds, and herpetofauna, which employ these strategies.[12]

It is much more difficult to say what counts as clear evidence for conscious preferences. Therefore, in light of the much less controversial evidence for consciousness of pain in all vertebrates, answers to the questions "Which animals are self-conscious in Singer's sense?" and "Which animals are subjects of a life in Regan's sense?" are controversial. Also, the best answers may not be the same as the answers to the question "Which animals can feel pain?"

For present purposes, then, assume, as Singer at least seems willing to do, that although all mammals are in his sense "self conscious," birds are not. A crucial difference in his analysis of the beef and poultry-and-egg industries would follow, because the replaceability argument would apply to chickens but not cattle. A sufficiently humane form of poultry and egg production, in which animals lived happy lives, died humane deaths, and were replaced by similarly happy animals, would be justified, especially in light of the nutritional and culinary benefits of having eggs and chicken to eat. Singer acknowledges precisely this in *Practical Ethics* (Singer, 132–33). Previously, I noted that self-professed animal welfarists differ from Singer in denying the assumption at the start of this paragraph, namely that cattle are self-conscious animals whose well being cannot be adequately captured in purely hedonistic terms, so that, according to these animal welfarists, the replaceability argument does not apply to them. This difference clarifies why an animal welfarist might conclude that a sufficiently humane form of cattle production is also morally acceptable.

Singer admits that the replaceability argument might justify a sufficiently humane form of slaughter-based agriculture for those animals to whom it applies. He then immediately argues, however, that even if it applies to birds, the argument doesn't justify the contemporary poultry industry, which is based on "factory farming, where animals do not have pleasant lives" (Singer, 133). It is easy to see why someone thinking from an animal welfare perspective would be critical of the contemporary U.S. poultry industry. Today, most laying hens live in small, crowded cages, and most broilers (chickens raised for meat) are raised in very large numbers in large barns where crowding becomes extreme as the birds mature to slaughter weight. Economies of scale and other factors drove the egg industry toward confinement early: By 1990, 90 percent of all laying hens in the United States were caged. One of the main reasons is that intensive confinement systems require far less labor to maintain. For instance, although building a single facility in which 300,000 hens live while laying eggs might cost $2.5 million, only three laborers might be needed to run such a facility.[13] Little labor is needed because feeding, watering, and sanitation are mechanized and farmers can afford to forego monitoring for underproductive hens and simply replace the entire flock when average production falls below a certain level (which usually occurs every twelve to fifteen months).[14] With several confined to a small wire cage, today's laying hens cannot forage, fly, dust-bathe, or nest. Poultry are still exempt from U.S. federal humane slaughter regulations, and poultry slaughter is a relatively indelicate affair, with fully conscious birds hung from their legs on conveyor belts before being electrically stunned and then mechanically beheaded. Poultry slaughter is on the rise. As beef consumption fell during the late 1980s and early 1990s, chicken slaughter grew at a rate of almost 300 million animals a year, reaching a yearly rate of almost six *billion* chickens in the United States alone.

By contrast, relatively few cattle—approximately thirty million per year—are slaughtered.[15] Of these, many have lived a portion of their lives loose on Western ranges, and conditions of the slaughter procedure have improved dramatically since the scenes described in the early 1900's in Upton Sinclair's novel *The Jungle*. Large-scale, state-of-the-art facilities today are capable of slaughtering as many as four hundred to six hundred cattle per hour, but (probably contrary to popular belief) systems of this kind, when well designed and operated, can be the most humane. The "races" or walled paths approaching the stunning chute can be designed to look just like those through which cattle have traveled previously for routine veterinary care, and experienced handlers can move animals along without prodding. Although inexperienced or poorly trained handlers may prod every animal who passes by, doing so is unnecessary. By simply shaking a pom-pom on a stick in their "flight zone," an experienced handler can use cows' natural herding instinct to move them forward without ever touching them in most cases. Also contrary to popular belief, cattle do not "smell blood in the chutes." Any unfamiliar or unusual object or fluid will arouse the animals, but a well-designed slaughterhouse will be ventilated so that air flows into the building through the kill chute, thus making it impossible to smell anything from outside the building. The kill chute itself in state-of-the-art facilities today allows cattle to see nothing but the buttocks of the animal immediately in front of them because they are supported on either side of their briskets by a double-track conveyor system just prior to being hit with the "stun gun." Finally, "stunning" is a misnomer for what happens in the chute, since a properly placed shot with a stun gun obliterates the animal's brain, making it impossible for it to remain conscious or ever regain consciousness. For longer than an hour at a time, I have watched cattle being killed at a slaughterhouse on the Colorado plains. I observed very few animals struggling at all while entering the double-track conveyor system, and almost always the first shot of the stun gun sent a mist of grey matter into the air.[16]

Not all self-professed animal welfarists are critics of the U.S. poultry industry, but from a hedonistic utilitarian perspective, the beef and poultry industries can look very different. This perspective can also explain why the poultry industry in Scandinavian countries has been a special target of legislative reform and welfare-oriented animal science.

From a viewpoint like Regan's, which holds that animals "have rights" in a stronger sense, the beef and poultry industries would fare equally badly. For even if the animals involved live perfectly happy lives and die completely painless deaths, Regan argues that animal slaughter violates the worse-off principle, at least in developed nations such as the United States and the Scandinavian countries. Regan admits that the worse-off principle would justify humans in killing animals for food if this were the only way to survive. For, he argues, "the harm that death is, is a function of the opportunities for [preference formation and] satisfaction it forecloses" (Regan, 324). Therefore, because the range of preference formation and satisfaction open to a human being is dramatically greater than that open to any of the animals commonly consumed for food, the harm that death is to a human being is noncomparably worse than the harm that death is to any animal. However, he argues, in developed Western nations, people cannot plausibly claim that they have to kill animals to survive; vegetarian diets can suffice and even be superior in terms of nutrition and taste.

The foregoing argument also shows us why an animal rights advocate such as Regan might oppose the dairy industry, at least as it exists today in a country such as the United States. Today's dairy industry is heavily dependent on slaughter. Milking cows spend only about three to four years in production, after which they are slaughtered for relatively low-grade beef. Also, dairy farmers maintain high milk production by breeding their milkers to calve about once per year.[17] The resulting calves are removed from their mothers immediately or within days, and although as many as one-half to two-thirds of the female calves become replacement milk cows, the remaining heifers and all the males become beef or veal. Altogether, the dairy industry produces around one-seventh of the cattle slaughtered in the United States.[18]

So, someone who, like Regan, objects to animal slaughter might also object to the dairy industry. It would be economically infeasible in today's market for dairy farmers not to slaughter older cows who give relatively little milk, to cease breeding their milkers yearly, or not to send unused animals into the beef trade. So assuming that a strict vegetarian diet can be nutritionally adequate, humans can live without dairy products as well. That assumption is controversial because some nutritionists doubt that a vegan diet, which excludes all animal by-products such as milk and eggs, can be nutritionally adequate, at least for people with high metabolic needs such as growing children and pregnant and lactating women. Others believe that a vegan diet can be perfectly healthful for these individuals, and some have even claimed that heavy reliance on dairy products actually increases the incidence of osteoporosis or brittle bones in the elderly. These nutritional issues are complex,[19] but the tie between the dairy and beef industries at least clarifies why animal rights advocates might be seriously concerned with the dairy industry even if the animals involved are treated very well on a day-to-day basis.

Regan concludes that if we recognize that the animals involved have rights in the strong sense, then we ought to be committed to "the dissolution of commercial animal farming as we know it" (Regan, 353). You may know people who are dependent on the beef, dairy, or poultry industries and wonder how even an animal rights philosopher such as Regan could call for this. Millions of families are economically dependent on animal agriculture, either as farmers, processors, or retailers, so even if consumers could be convinced to give up meat, dairy, and eggs, wouldn't it be better to preserve these industries because so many families are dependent on them? After all, wouldn't the harms be very serious, especially if you take into consideration lost opportunities for education and other important things, so that Regan's own worse-off principle would imply that we should preserve these industries in order to avoid these harms, some of which might be noncomparably worse than what an animal suffers through humane slaughter?

Regan considers this objection and responds by arguing that the miniride and worse-off principles do not protect individuals who voluntarily participate in competitive, risky enterprises (Regan, 339). He acknowledges that we have a duty to prevent their dependents from being made worse-off (because they did not choose to be dependent on animal agriculture). But, he argues, "it is not the consumers, in their capacity as consumers, who have this responsibility" (Regan, 341). At most, we as a society have an obligation to protect these innocent victims of economic forces just as we have obligations to the dependents of people unemployed for other reasons; we have no duty to buy animal products.

In conclusion, you may know people who are dependent on animal agriculture and you may disagree with Regan's response to this objection. However, my discussion of Regan's strong, abolitionist stance on animal rights at least demonstrates that the popular, political characterization of animal rights people as uninformed and unreasoning is a caricature. Regan's work on animal rights shows that an abolitionist stance can follow from a carefully argued application of a traditional ethical theory, the theory that individuals are due a kind of respect that utilitarianism denies them.

NOTES

1. Quotations from John Stuart Mill, *Utilitarianism* (New York: Macmillan Publishing Company, 1985), 10, 11, 14.

2. Quotations from Immanuel Kant, *The Moral Law: Kant's Groundwork of the Metaphysics of Morals*, H.J. Paton transl. (London: Hutchinson and Company, 1948), 96 and 88. Readers desiring a more detailed treatment of Kantian thinking in ethics should see chapters eight and nine of Marcus Singer's *Generalization in Ethics* (New York: Atheneum, 1971). Singer makes it very clear how to apply the second formulation of the categorical imperative given above and why the "treat humanity as an end" formulation is not equivalent. (Marcus Singer is no relation to Peter Singer, the animal rights philosopher.)

3. For more on Carruthers' view, see his initial article "Brute Experience," *Journal of Philosophy* 1989, 258–69, and his later, book-length overview of related issues, *The Animals Issue: Moral Theory in Practice* (Cambridge University Press, 1992). In more recent work, Carruthers has entertained the hypothesis that pains that are nonconscious may nevertheless be morally significant. See "Sympathy and Subjectivity," *Australasian Journal of Philosophy* 77 (1999), 465–82.

4. Peter Singer, *Animal Liberation*, second edition (New York: Avon Books, 1990). In the text, page references are to this second, minimally revised edition.

5. Peter Singer, *Practical Ethics*, second edition (Cambridge University Press, 1993). Page references in the text are to this second edition.

6. Tom Regan, *The Case For Animal Rights* (Berkeley: University of California Press, 1983).

7. Eugene C. Hargrove, "Preface" to *The Animal Rights/Environmental Ethics Debate: The Environmental Perspective* (New York: State University of New York Press, 1992), x. Similarly, in his review of Regan's book, J. Baird Callicott ridicules Regan for titling it *The Case for Animal Rights*: "Regan insists on using the word *animal* throughout his discussion, even though what he really means, as he himself notes, is 'mammal,' on the grounds of 'economy of expression.' This is puzzling since both words contain six letters. Why wasn't the book called *The Case for Mammal Rights?*" J. Baird Callicott, review of *The Case for Animal Rights, Environmental Ethics* 7 (1985), 365–372, at 366.

8. Actually, there are two complications: first, in medical research the harms some human beings will suffer if disease cures are not found might actually be noncomparably worse, in Regan's view, than the harms that experimental animals suffer; and second, Regan invokes "special considerations" both regarding medical research and in response to certain objections to his abolitionist conclusions regarding animal agriculture.

9. Although here I use these two terms interchangeably, sometimes there is a distinction to be made. An obvious example would be Freud's use of unconscious mental states in the explanation of behavior. However, as was mentioned in my earlier discussion of Carruthers' views, it is only *conscious* mental states that seem to matter morally. For as Carruthers argues, if animals' mental states are all unconscious, then our treatment of them seems to raise no moral issues at all.

10. The four studies are (1) Jane A. Smith and Kenneth M. Boyd, eds., *Lives in the Balance: The Ethics of Using Animals in Biomedical Research* (New York: Oxford University Press, 1991); (2) David DeGrazia and Andrew Rowan, "Pain, Suffering and Anxiety in Animals and Humans," *Theoretical Medicine* 12 (1991): 193–211; (3) Patrick Bateson, "Assessment of Pain in Animals," *Animal Behavior* 42 (1991): 827–39; and (4) Margaret Rose and David Adams, "Evidence for Pain and Suffering in Other Animals," in Gill Langley, ed., *Animal Experimentation: The Consensus Changes* (New York: Chapman and Hall, 1989), 42–71. For a table summarizing the evidence, see Gary Varner, *In Nature's Interests? Interests, Animal Rights, and Environmental Ethics* (New York: Oxford University Press, 1998), 53. A version of that table is available via the World Wide Web at: http://www-phil.tamu.edu/~gary/awvar/lecture/pain.html.

11. For a recent overview of relevant research, see James R. Anderson and Gordon G. Gallup, "Self-Recognition in Nonhuman Primates: Past and Future Challenges," in Marc Haug and Richard E. Whalen, eds., *Animal Models of Human Emotion and Cognition* (Washington, DC: American Psychological Association, 1999), 175–94.

12. See "Localizing Desire," chapter two of Gary Varner, *In Nature's Interests? Interests, Animal Rights, and Environmental Ethics* (Oxford University Press, 1998).

13. Carmen R. Parkhurst and George J. Mountney, *Poultry Meat and Egg Production* (New York: Van Nostrand Reinhold, 1988), 194, 196.

14. Richard E. Austic and Malden C. Neshem, *Poultry Production*, 13th ed. (Philadelphia: Lea & Febiger, 1990), 56.

15. This statistic and the statistics in the foregoing paragraph are based on the U.S. Department of Agriculture's publications *Livestock Slaughter* and *Poultry Slaughter*, various dates.

16. I am indebted to Colorado State University animal scientist Temple Grandin for showing me the facility and for other details of cattle slaughter described in this paragraph.

17. Donald L. Bath, Frank N. Dickinson, H. Allen Tucker, and Robert D. Appleman, *Dairy Cattle: Principles, Problems, and Profits*, third edition (Philadelphia: Lea & Febiger, 1985), 325, 259.

18. Based on the USDA's *Livestock Slaughter*, various dates.

19. For an overview, see the special edition of the *Journal of Agricultural and Environmental Ethics*, volume 7 (1994), number 1, which contains relevant articles by several nutritionists as well as several philosophers.

Chapter

8

LAND

PAUL B. THOMPSON

CASE STUDY: GORDON THE LAWYER

"Remember Marie the environmentalist?"

Doug is looking at Emily, who is busy reading about property rights.

"Yes, of course."

"Remember she said that she was opposed to the Springdale town council's attempt to take water from the Chatham River?"

"Yes, why?"

"My brother is the city attorney in Springdale."

"Really?"

"Yup."

"So he's doing battle with Marie?"

"Yup."

"What's the story?"

"Well, Springdale borders the Chatham River and, like many rural communities, is desperately seeking economic growth. The Springdale town council is convinced that economic growth can be assured only if water can be made available for light industry and residential use. Herein lies the rub, for Springdale's wells are already producing at their limit. So, the council has proposed to divert large flows of water from the river into the city."

Doug's brother's name is Gordon. In a recent phone conversation, Gordon explained to Doug the challenges he faces in defending the council's plan. The town is opposed by two groups. The first group consists of farmers, who are riparian (water use) rights holders who have used Chatham water for more than a century. Farm water usage has expanded and contracted over the years, and at various times has included watering of stock and irrigation of crops as well as farm household uses. The State Water Board study of stream flow in the Chatham River indicates that in years of reduced flow, the town council's plan would prevent farmers from maintaining their current levels of use, much less expanding them. The farmers are certain, however, that their property rights as riparians entitle them to expanded use. In short, they think the town council is surreptitiously planning to steal water belonging to them and them alone. The farm group is rumored to be considering legal action to protect its rights.

"I have some sympathy with these guys," Gordon said. Like Doug, he was raised on the family farm and knows how difficult making a living there is. "But it is my professional duty, nonetheless, to defend the town council and defeat these farmers. Even in court, if it comes to that."

The second group, Gordon explained, is Marie's Friends of the Chatham, a loose-knit coalition of outdoors enthusiasts. The Chatham River supplies sustenance to fish and wildlife up and down its length, including an endangered species of salamander unique to Chatham River habitats. The town council's plan will threaten the Chatham ecosystem on which this wildlife depends during years of reduced flow. Harm to the Chatham ecosystem is offensive to the Friends of the Chatham for two reasons: Recreational users of the Chatham ecosystem come from around the state and have a strong attachment to the fishing, camping, and appreciation of nature that the ecosystem provides. Second, members of Friends of the Chatham who do not use the Chatham River for recreation nevertheless believe that harm to the wildlife and, indeed, the ecosystem itself, is a moral affront. The town council, they hold, has offended nature simply by proposing a plan that displays little apparent regard for the interests of nature. Like the farmers, the environmentalists are utterly opposed to the plan Gordon must defend.

See Appendix A and perform Exercise 8.A before continuing with this chapter.

DISCUSSION OF ISSUES

The dispute about the Chatham River in which Gordon is embroiled is like many disputes over water use. People use resources differently and those uses are not always compatible. For citizens of Springdale, the Chatham is a resource for industrial and residential development. For farmers in the area, the river is an input for their production process; they believe that they purchased the rights to the water when they bought their land. For hunters and outdoor recreationists, the water is a "leisure good"— critical for enjoyable activities, such as bird watching, hiking, water skiing, and hunting. For environmentalists, the river ecosystem is valuable in its own right, even if humans do not make use of it. In this view, humans should understand their use of water in terms of what nature demands. The members of each of these groups—the citizens, the farmers, the hunters, and the environmentalists—believe that they value the river correctly, and they are loath to be told otherwise. How should Doug understand the different values these people place on the water?[1]

Clearly, the groups have an economic or political interest in seeing the policy issue resolved in a way that permits the water to be used *their* way. If Gordon were a typical policy lawyer, he would not see this as a problem in ethics. Each interested party has that group's preferred solution to the Springdale water issue, and there is nothing to be said about whether some preferences are better than others are. Of course, Gordon does have his own ethical standards. He believes that as a lawyer he should negotiate a solution that leads to the greatest possible satisfaction of interests. But such a solution demands that he understand why the members of each interest group believe that they are in the right.

This realization leads Gordon to see that each group not only wants the dispute resolved in their favor but also has a philosophical framework in mind. This framework, or ethical worldview, provides criteria for saying which interests have moral priority—that is, which uses are compatible with what ethics demands. In real-world politics it is often difficult to tell what comes first, political interest or moral principle. Some people no doubt employ ethical arguments simply because those arguments happen to support their economic or political interests. Even if that fact makes us skeptical about their intentions, the merit of the arguments themselves does not depend on the motive for making them. And, of course, other people come to support political interests as a *result* of moral deliberation.

Gordon tells Doug that if he were to review the main arguments that are used to make a moral evaluation of soil and water use, he would discover four main types. One stresses the instrumental use of soil and water in producing food; a second argument stresses property rights. A third group of egalitarian arguments interpret land use as part of a larger problem, namely, social inequality. The last group of arguments understands a person's relation to land as an irreducible component of moral character. But before moving into these different viewpoints, we should remind ourselves of just how important soil and water actually are.

AGRICULTURE AND ENVIRONMENT

As Gordon knows, soil and water are crucial to agriculture. Indeed, agriculture has a more extensive spatial impact on the environment than any other single human practice. According to the Food and Agriculture Organization (FAO) of the United Nations, slightly more than one-third of the earth's land is used for agriculture. The remaining two-thirds are forests, deserts, tundra, swamps, wetlands, and savanna. Only a fraction of the earth's land mass consists of concentrated urban areas. In the United States, agricultural uses account for nearly half of the total land mass. American urban and recreational lands (including uninhabited deserts, swamps, and high mountain ranges) account for a mere 20 percent of the total.

Obviously, the way in which farmers use soil and water is critical to the health of our planet. Lester Brown, who founded the Worldwatch Institute and coined the phrase "sustainable development," has long believed that food production is the key to sustainable land use. Agriculture also uses the largest share of fresh water. Not surprisingly, a large percentage of the world's crops and pasture lands are located in areas of reliable rainfall. Rain-fed farms and ranches get first crack at this water, taking their share before rainwater enters ground or surface systems. Brown (1997, 31) writes, "There is a tendency in public discourse to talk about the water problem and the food problem as though they are independent. But with some seventy percent of all the water that is pumped from underground used for irrigation—the water problem and the food problem are in large measure the same."

The ethics of land and water use are first and foremost ethics for agricultural production. What environmental parameters should be considered in farming and ranching? Should food production operate within a framework of totally renewable resources, or should some consumption of nonrenewable energy, soil, and water resources be regarded as an acceptable trade-off for the production of food for hungry mouths? What does it mean for farmers to be good stewards of nature, and how do economic or policy incentives affect their stewardship? Only when these questions have been answered does it become meaningful for the 98 percent of U.S. population *not* directly involved in farming or ranching to ask how their consumption choices can be made on a more ethical basis.

SOIL AND WATER AS INSTRUMENTS OF PRODUCTION: THE UTILITARIAN VIEW

Farmers have long recognized duties of stewardship, duties to leave the land as they found it, at least as far as soil and water are concerned.[2] The moral foundation for these duties has been a mix of religious and secular obligations to Creation, to posterity, and to nature itself, often personified as having intent and purpose. The moral justification for converting soil and water assets in meat, milk, and grain commodities, however, has typically rested on the role that these goods have in satisfying human wants and in the contributions to wealth and prosperity that their production brings.

Gordon has heard agricultural economists explain how these facts provide a basis for understanding the value of land. *Asset theory* holds that price of land will reflect its relative capacity for bringing forth the goods that people want.[3] Soil and water figure prominently in the assets of any land put to farming purposes. The richer the soil and the more reliable the water supply, the more valuable land will be. The greater productive potential of fertile and well-watered land makes farmers willing to pay a higher price for it. They will make up for the higher cost of land by producing more per acre. Land prices can be distorted by other factors, but when distortions are absent, the market price of land reflects its asset value.

Of course, a non-farm buyer may place a higher value on soil and water. In most areas, industrial and residential users are willing to pay much more for water than farmers. Why shouldn't the water just go to the buyers who are willing to bid the most for it? Wouldn't that be consistent with the goal of allocating soil and water assets to their most valued use? Indeed, anyone who is inclined to decide soil and water use by comparing farming to other uses will soon conclude that farming should be done only in those places where no other use is profitable. That is a conclusion that supports the view of developers in places such as Springdale.

The idea of a "most valued use" is a moral norm that is often used in making social decisions about soil and water. It is a decision rule that derives from *utilitarian* philosophy. The basic pattern of utilitarian thinking is simple. Consider the options available and estimate the costs and benefits of each option. Then choose the option that has the best consequences, the best balance of cost and benefit. However, as utilitarians start to understand and compare cost and benefit, they inevitably find themselves making a series of additional assumptions. Some of these assumptions are summarized in the following premises, frequently adopted by those who tend to think of soil and water in terms of their asset value:

1. Soil and water are instruments for producing goods that are, in turn, instruments for the satisfaction of human wants and desires. They are *not* intrinsically valuable.
2. Value is attributed to a good by individual human beings. Something has value because at least one human being performs the subjective mental act of wanting or desiring it. To say that something has intrinsic value is simply to say that one cannot go any further in saying why someone wants or desires it.
3. The goodness of health or the psychological and emotional welfare of individual human beings requires no further justification. Benefit or harm to health and welfare thus represent intrinsic values.
4. Comparison of benefit and harm should consider *everyone* affected by the options under review. This principle has led Peter Singer to the view that benefit and harm to nonhuman animals should be included in the assessment.
5. Benefits and harms are quantitatively additive. One can simply "add up" the benefits and "subtract" the harms expected for each affected party. The total satisfaction (or utility) increases or decreases when the health and welfare of a given individual changes or when another individual is included in (or eliminated from) the affected group.

6. Monetary or material wealth is valuable because it is instrumental to intrinsic values. However, monetary wealth is so readily convertible into conveyances of health and welfare that abundance or lack of wealth is both an essential feature and a reasonable estimate of both individual and social welfare. Many utilitarians go further: Everything has a price.
7. Production of food and fiber commodities is justified up to the point that it contributes to individuals' ability to satisfy intrinsic wants and needs, and to the extent that exchange of these commodities contributes to individual and social wealth. Simply put, if it's profitable, it's morally right.

These propositions imply that soil and water are valuable precisely because they are inputs into the production of food and fiber commodities. In this view, the asset value of soil and water becomes equivalent to moral value and we should invest the asset value of soil and water so that all of human society, including posterity, receives the greatest total return of value, or as Bentham and Mill wrote "the greatest good for the greatest number."

Utilitarianism thinking reflects the way that many farmers, businessmen, government leaders and scholars of agriculture come to understand production. As already noted, utilitarianism often supports the idea of simply allowing the market to decide issues of resource use. In the Chatham River case, for example, it appears that the most valued use is development rather than farming. But we must be sure that one has included all the costs and benefits to all the affected parties in our comparison. Doing so requires considering the benefit and harm to future generations. Future generations are affected parties, too. Yet if we are not careful, the utilitarian analysis of future generations ends in a quixotic recommendation.

For example, suppose that the soil fertility and water availability on a given plot of land will return $100 in crop production every year forever, as long as it is farmed carefully. What would be an equivalent asset measured in dollars? The asset value of the land is the amount of money that would return the same amount in perpetuity. In order to keep the math simple, assume a constant interest rate of ten percent, making the asset value $1,000. Now (still keeping the math simple) suppose that the farmer can farm so that the annual return will be $110 but that the productivity of the soil and water will remain constant for one hundred years and then drop to zero.[4] At the end of 100 years, an extra $1,000 has been earned. The farmer's heirs can get $100 a year from the interest on that, and they will get the money without having to trouble themselves with farming. Under either scenario, ecologically sustainable or not, the farmer's heirs receive an "economically sustainable" $100 per year forever. Through the miracle of compound interest, if the farm family puts the extra ten dollars a year in an investment that yields ten percent, then the break-even point comes at only about twenty-five (rather than one hundred) years. If the productivity of soil and water hold out longer than that, it is economically foolish *not* to deplete them.

This reasoning provides a philosophical basis for the claim that farmers can be said to meet their obligations to posterity *even if* they farm in a way that is not ecologically sustainable. Utilitarian thinking appears to show that a soil-and-water-exhausting agriculture is morally acceptable.[5] Furthermore, if farmers consider only themselves and their heirs, it can seem to be a pretty compelling argument. Good reasons exist, however, to resist this advice. If all farmers farmed this way (and some utilitarians believe they should), posterity would be left with plenty of money in the bank and no capacity to produce food within a rather short period of time. We reach this conclusion by considering the farmer's practice in isolation. When we ask, "What if everyone did that?" we are asking for a more sophisticated way of understanding the total societal impact of individual production choices.

In fact, a utilitarian analysis of soil and water can provide a very illuminating analysis of ecological problems. Take, for example, the American Dust Bowl. Farmers in the 1930s tilled so many acres of fragile soils so extensively that drought caused not merely the loss of a crop but also wind erosion (and on a phenomenal scale). Dust was piled everywhere, devastating the productive capacity of all land, whether fertility or surface water had been conserved or not. The Dust Bowl is an example of the "tragedy of the commons" first described by Garrett Hardin. When many producers use a common resource, they follow a logic of "use it or lose it," resulting in a collapse of the resource's productive capacity. Why did Dust Bowl farmers have to "use it or lose it"?

The answer is complicated, but we can gain insight into individual farmer decision making by looking at the Prisoner's Dilemma model from game theory. In the Prisoner's Dilemma, two prisoners are being interrogated separately for a crime they committed together. If both confess, they will both be prosecuted for a felony. If neither confesses, the police can't make their case and both will be punished for a minor offense. The police offer each prisoner an attractive deal to confess and testify against the other in court. So each has an incentive to confess, but if both confess, the testimony will not be needed and the deals fall through.

Again with some assumptions to simplify the math, we can represent the choices and payoffs for both prisoners in a two-by-two matrix. Each prisoner has two choices: confess and don't confess. Figure 8.1 represents the payoff or expected value for each choice as years spent in jail. Negative numbers serve as a reminder that spending a year in jail is a bad outcome from the perspective of a prisoner. Payoffs for Prisoner Row are listed in the lower-left corner of each box, whereas payoffs for Prisoner Column are in the upper right. In the middle is the utilitarian or net social payoff, which is just the sum of payoffs for all affected parties. If the prisoners cooperate with one another (if they refuse to confess), they will both get off lightly, but since they are being held separately, such cooperation may be difficult to negotiate. In the meantime, if one prisoner thinks the other will not confess, he is tempted by the possibility of skipping all the prison time by testifying against the other. That is what a self-interested utility maximizer will do in that case. But if the other prisoner is thinking the same way, we expect her to confess for similar reasons. So what should one do if one expects one's counterpart to confess? One should still confess in order to gain a bit of leniency and save two years prison time (noncooperative prisoners are treated more harshly). It seems that the only circumstance in which one would not confess is when one could be confident that the other would not and when both are willing to endure the two-year sentence.

Substituting losses in fertility or water availability for years in prison, the economic logic of the Prisoner's Dilemma is the economic logic of the Dust Bowl or the tragedy of the commons. Farmers would be better off to cooperate and take either modest gains or minor losses, but the one who does not cooperate can reap rewards at the expense of others. This happens because the dust blows on everyone's land, conserver and exploiter alike. Because one farmer expects the other to be rationally self interested, the result is the worst-case scenario, in which dust rolls across the plains, fisheries collapse, and the fertility of rangelands plunges into a death spiral.

Some look at this logic and draw the obvious conclusion that farmers (or prisoners) should simply cooperate in order to get the payoffs represented in the lower right box. Indeed, that is what the utilitarian maxim recommends. But the Prisoner's Dilemma is important because it shows how individuals rationally pursuing their own ends can produce outcomes that are not only socially suboptimal but also obviously contrary to every individual's abiding interest. People will cooperate voluntarily in these situations only if they are confident in their fellows, a situation unlikely to occur when many users who are strangers to one another rely on a common resource. The alternative to

	Confess	Don't Confess
Confess	-4 (-8) -4	-6 (-6) 0
Don't Confess	0 (-6) -6	-2 (-4) -2

FIGURE 8.1
The Prisoner's Dilemma

voluntary cooperation is regulation: enforced cooperation that is truly "for one's own good." The Prisoner's Dilemma thus shows why "free market" solutions fail, and why society might find regulating practices that affect soil and water necessary. Sometimes coercion is the only way to achieve "the greatest good for the greatest number."

Soil and Water as Private Property: The Libertarian View

Regulation is not the only solution to the tragedy of the commons. Garrett Hardin proposed to solve the problem by placing soil and water resources into private hands. Divide up the commons and give each person a share to manage on his or her own, without threat from use by everyone else. Private ownership would, he thought, supply the incentives needed to conserve by ensuring that someone who is less exploitative of soil and water is able to capture the productivity benefits derived from having been so. Hardin's argument provides utilitarian reasons for placing soil and water under a regime of private property. It is an argument that is valid only when private property rights actually would produce the changes in conduct that take us from the upper-left corner of the Prisoner's Dilemma down to the socially optimal (and individually preferable) lower right.[6]

Hardin's utilitarian argument for private property coincides with a much older and quite different way of understanding the moral significance of soil and water as components of private property. Traditionally, ownership of property has been thought to be a natural right. Natural law philosophy posited a moral order as an existing reality, owing variously to God's grace to mankind, to principles of order evident to any rational being, or to the implicit terms of a social contract thought to undergird the foundations of civil society. Although the philosophical rationale for natural law has varied, the centrality of property is remarkably stable. In virtually any system of natural law, property rights govern the exchange and control of alienable goods. Property rights are alienable (as distinct from the inalienable rights Jefferson celebrated in the Declaration of Independence) because rights to use or exchange property can be alienated from one person and transferred to another. The notion that such transfers must hold some stability and permanence seems essential to the well-ordered society. If trades or exchanges could be abrogated arbitrarily (or even on unanticipated utilitarian grounds) social life would consist of little more than turmoil, dispute, and conflict. To this extent, then, recognizing private property rights amounts to common sense.

Rights come in at least two kinds, however. Property rights are usually thought of as *non-interference rights*. They protect the rightsholder's discretion or control over any and all uses (or non-uses) of the property in question. Property owners may do anything with the property they own but only as long as they do not violate the non-interference rights of others by harming their person, compromising their liberty, or infringing on the free exercise in the use of such property as *they* might own. No one (including government) may interfere in the exercise of property rights unless the property owner has already forfeited the right to this protection by interfering with someone else.

However, aside from a duty to desist from acts that harm or interfere, property rights do not form the basis for claiming that others should act on one's behalf. Non-interference rights protect goods that people already have—life, liberty, and property. They do not provide a basis for claiming that goods such as welfare payments or other benefits should be given to them. For example, people who claim a "right to education" are claiming that the government should provide schools. If so, someone must do the providing; someone must act on their behalf. A "right to education" or "right to health care" exemplifies the second kind of rights, discussed at some length below. A property right is *not* a "right to property" in a sense that parallels the right to education. In emphasizing the non-interference dimensions of property rights, we note that government must protect property owners from interference by others, but we do not say that people should simply be given property when they ask for it.

As a non-interference right, it is useful to think of property owners having a bundle of related rights to use their land and a bundle of restrictions based on the way their use of land might harm others. A property owner can decide who has access to the land, and who is to be excluded. Property owners have the right to decide how land is used. Property owners have the right to claim income or benefits that accrue from the use of land. Property owners also have the right to sell, trade, or give away any of these other rights for a limited time or in perpetuity. Interference in any of these rights constitutes a violation of the owner's property rights. However, owners have no right to use their property in ways that harm others. This feature of libertarian thinking could lead to policies that restrict farming practices that use chemicals or pose risk to others. For example, John Hospers (1971), one of the leading proponents of libertarian philosophy, wrote that activities that expose other people to pollution count as interference and should not be allowed.

People who believe that social order would be perfect if no rights other than those of non-interference are recognized are called *libertarians*. Libertarians believe that private property rights are absolute as long as they are not abused through interference in the life, liberty, or property of another. Most libertarians also believe that any property not used in performing essential state functions should be privately owned. In the libertarian view, soil and water should be used just as their owners' desire. A property owner who practices stewardship or uses soil and water to produce beneficial food and fiber commodities might deserve our praise or gratitude, but there is no waste or profligacy in the use of soil or water that could justify interference in the owner's property right. Ample evidence shows that many of the land-owning farmers and ranchers in millennial America have strong libertarian tendencies.

Rural property owners may be operating with a subtly different conception of property rights, however. For them, ownership of land secures both the opportunity right to food and their right to employment. As later text demonstrates , such rights are not typically thought of in terms of non-interference. Food and income are often thought of as goods that everyone deserves, and that someone—usually the society as a whole—is obligated to provide. Historically, however, a landowner's right to food and income is protected when others are prevented from interfering in the landowner's use of the land. So again, the argument seems to shift back to an argument for non-interference. Property rights are thus generally analyzed in libertarian terms. However, the link to sustenance provides a reason for landowners to think that property rights are of added moral importance and to be especially wary of proposals that would limit the uses that they make of their property.

This discussion leaves many important philosophical issues untouched. Where do property rights come from? How are initial claims on rights to land justified? Do any cases call for redistribution of property rights? For present purposes, however, what is crucial is simply that landowners feel morally justified in claiming a non-interference right to use the soil and water resources under their control. They feel particularly justified in uses that secure their livelihood. As one might think of soil and water in exclusively utilitarian or asset value terms, one might also think that the libertarian analysis says everything that is ethically significant about soil and water. A libertarian argument asserts that private property rights in land give owners the right to use soil and water in suboptimal (but not harmful) ways. Because the landowner's rights are all that matter, the libertarian view excludes consideration of the wider benefits society can derive from land-use decisions.

Clearly, however, these different moral perspectives can come into conflict with one another. The Prisoner's Dilemma analysis of the Dust Bowl shows how individual property owners making self-interested decisions about the use of soil and water resources can cause ecological and social disasters. Nor are these situations confined to bygone days and faraway places. Ranching of the Western rangelands in the United States results in an average annual soil loss equivalent to the thickness of a dime. Ranchers believe, with some plausibility, that they have a right to make a living from those rangelands, and a dime's loss of soil hardly seems enough to challenge that right. But topsoils in many parts of the West are only a roll of dimes thick (less in some places). That means that the soil essential to the plants and animals of the range ecology will be exhausted in only fifty years! Thus it is questionable whether the centuries-old tradition of rights can continue to guide our moral thinking on soil and water in the future.

Soil and Water in Producing Food: The Egalitarian View

Property can also be understood as an *opportunity right*—a right that *would* require giving property to the landless. No one has to act on behalf of the rightsholder in order to respect non-interference rights, but when opportunity rights are claimed, someone (usually the government) must act to ensure that the entitlement or opportunity protected by the right is actually available. As already noted, when people claim rights to education or to health care, they are claiming opportunities that must be provided for them. Providing such opportunities usually requires that those who have must provide for those who have not. Opportunity rights thus equalize or level the distribution of resources in a society. Those who believe that we should recognize opportunity rights as well as non-interference rights are sometimes called *egalitarians*.

It is important that the right to food, recognized in the International Declaration of Human Rights, is an opportunity right. When we say that the poor and the hungry have a "right to food," we mean that they should have a meaningful opportunity to acquire food. It may come through private charity or public programs supported by taxes. People in modern welfare states generally presume that the right to subsistence will be maintained by entitlement programs, such as food stamps. For John Locke, however, who crafted history's most influential discussion of property, the opportunity right to subsistence was guaranteed by a right to claim land as property, to farm it, and thereby to live.

Today, common sense tells us that the claims of the poor and hungry would be poorly satisfied if we approached the right to food in this traditional way. In industrial societies, the opportunity right to sustenance is secured mostly through gainful employment. People in urban settings derive income from industrial or service jobs or from operating their own businesses. Property rights are linked to sustenance in that a person's right to expend income in any legal manner is presumed to secure that person's opportunity right to sustenance: we buy our food from the grocery store. In this setting, the crucial opportunity right is the right to employment. Thus, in the United States, separating a person's right to food from the ownership of land is easy. Opportunity rights tend to drop out of the argument, and the libertarian view of property rights (which sees them strictly in terms of non-interference) comes into prominence.

Yet it is possible to argue that a landowner's property rights can be overridden by society's need for food and fiber. Every living human being needs food. Without food, we die. When food is scarce, humans become susceptible to disease and suffer from a variety of reduced capacities. Agricultural land use does not produce just any commodity—video games or cuddle toys—that people are free to buy or not. Everyone must have food, and for the present at least, meeting world food needs depends on agriculture.[7] By extension, then, producing food for human sustenance depends on soil and water.

One of the most obvious problems with utilitarian and libertarian analyses of soil and water is that it seems possible to rationalize the current mal-distribution of food. The death and disease associated with this grotesque situation be "outweighed" by benefits to others, or by private property rights? To be sure, utilitarians such as Peter Singer (1993) have argued against such a proposition. Singer claims that the needs of the poor outweigh the wants of the rich. Similarly, libertarians have argued that individuals have a voluntary moral duty of charity (though *requiring* them to aid the poor is an injustice). However, these arguments arrive at assisting the poor through the philosophical back door. These two philosophical traditions do not provide any way of stating outright that every human being should be entitled to a fair share of the resources needed to sustain life.

Gordon has heard the moral proposition stated like this: Everyone has a right to food. As already mentioned, the United Nations Declaration of Human Rights includes a right to food. The rights listed in the Declaration are intended to specify what global society owes to individual human beings. It establishes a basis for individuals to claim food needed for survival, to have that claim recognized on the basis of common morality as well as by international organizations. Because food is one of the most basic human needs, a right to food would override rights to higher-level goods such as medical care or private property. Only when all humanity's food needs have been met would it be permissible to shift resources to the production of luxury goods. It would be impossible to imagine a situation where the poor's claim on food could be "outweighed" by the wants of the rich. One point of stipulating a right to food is to state that individual needs have priority over any norm or goal derived by weighing costs and benefits.

The right to food is different from a property right in several important respects. As already noted, it is *basic* and *universal*. People with no food have little interest in higher-level rights to spend their money as they wish, and *everyone* needs this right. Most important, the right to food is an opportunity right, not a non-interference right. As already noted, opportunity rights are the basis for claiming that a just society owes its citizens more than simple protection from others. An egalitarian believes that society owes each person the basic needs that are necessary for having a decent life.

The egalitarian view is often developed as a reaction to libertarianism. Many people are attracted to libertarianism because it seems to give each individual the maximal amount of freedom and autonomy over his or her actions. Libertarian non-interference rights restrict a person from doing things that harm others, and they preclude *requiring* one to do anything on others' behalf as well. But this result is compatible with a very uneven distribution of wealth and opportunity. Indeed, some may have so little wealth that they cannot feed themselves, while libertarian philosophy protects the property of the rich.

Clearly, if individuals may make a valid claim on the food that they need to survive, someone in society (or society as a whole) will have the responsibility to deliver that food to the needy. Respecting a right to food requires that people give *their* food to the poor, or, more likely, that all of us give money to support a program of buying food for the poor. The right to food demands more than non-interference; it demands that society provide everyone with the opportunity to nourish themselves. The inclusion of opportunity rights, which demand positive action on the part of others, is what separates an egalitarian view from a libertarian one.

The right to food is also a *narrow* right, whereas the right to property is *broad*. Property rights are claimed for specific material goods (such as a bowl of beans), for land holdings (the case discussed previously), and for ideas and discoveries (such as a gene with a specific agronomic function). The advocates for private property often write as if *any* restriction on these broad claims threatens all the others. It is the institution of private property that they are defending as much as a property owner's claim in any particular case. Clearly, recognizing one hungry person's right to a bowl of beans may be instrumental to alleviating hunger but it need not be understood as a general challenge to the institution of property rights.

However persuasive the case for a right to food, it may seem that this discussion has strayed far from a discussion of soil and water. In most industrial countries, the right to food will be protected through public assistance programs (such as the U.S. food stamp program) that give people the money they need to purchase food.[8] Yet in times past (and still today in other parts of the world), the right to food was understood as a right to the resources needed to produce food. Traditional farming societies did not secure the right to food through markets. The right was secured either by an opportunity to produce one's own food (to farm) or through an entitlement to a share of local crops. Agrarian reform movements were launched by Gerrard Winstanley in sixteenth century England, by Bolsheviks in 1910s Russia, and by Central American revolutionaries in recent times. In each case, they called for a redistribution of land to secure every citizen's opportunity to provide for their own sustenance through farming or grazing. The link between a right to food and a claim on soil and water is somewhat muted in industrialized and bureaucratized societies. Yet the connection between land and subsistence rights was clearly seen in history and continues to be important for people living in less developed countries today.

Linking soil and water to the right to food requires a shift in how we view the ethics of soil and water use. Whether we satisfy the right to food through a payment of money or through a redistribution of land, a certain portion of the earth's soil and water resources must be dedicated to the task of feeding every individual human being. For an egalitarian, the benefit-harm trade-off reasoning of the utilitarian is justified only when every individual's right to food is secure. The egalitarian rejects a libertarian view of property because the opportunity to eat must override rights that are less essential to the basic problem of survival. In either case, a right to food is viewed as more fundamental than the main concepts (utility and non-interference) in which utilitarian and libertarian theories have been framed.

The ethics of soil and water are more obvious when we think not only of the world's current population but also of people yet to come. Soil and water resources are in decline while human population continues to increase. Intensive agriculture currently produces enough food to feed the world, but maintaining our current levels of production will require a preservation of soil and water resources. If future generations also have a right to food, then the only way to give them an opportunity to claim this right is to bequeath soil and water (renewable resources) that is at least of comparable quality and quantity to our own. If future generations have a right to food, then we have a powerful ethical argument to preserve soil fertility and water purity at current levels.

Gordon is not convinced that these egalitarian considerations bear directly on the Chatham River case. No one is arguing that the Chatham basin should be used to grow food for the poor. But he believes that it is important to see how the ideal of equal opportunity confronts both libertarian and utilitarian thinking. And one *might* frame the question of access to outdoor recreation in terms of an opportunity right. One might, for example, argue that the need to preserve soil and water provides a basis for taking the stewardship viewpoint as a way to respect the opportunity rights of future generations. An egalitarian might also join environmentalists in opposing both development and exploitative farming because he or she thinks that equal opportunity requires us to ensure that *everyone* has access to outdoor recreation—hunting, hiking, or bird watching. In any case, the argument for opportunity rights conflicts with the utilitarian way of thinking because the rights of a single individual can override the principle of the greatest good. Egalitarian philosophy conflicts with libertarianism because it recognizes both non-interference and opportunity rights. Gordon concludes that even if it is not obvious how opportunity rights should be applied in the present case, it is best to keep them in mind in order to avoid neglecting important moral considerations.

Soil, Water and the Ecology of Virtue

Gordon sees that the utilitarian, libertarian, and egalitarian philosophies provide three different ways to understand soil and water, but he also has heard people express views that don't resemble these arguments in any discernable way. Sometimes, for example, people say that we should be stewards of the land. They argue that land has *intrinsic value*, value totally apart from the value it gets in virtue of the uses that humans make of it. Others believe that acclimatization to soil and water needs provides the basis on which moral character or *virtue* is to be measured. In placing land before human use, such statements implicitly reject the view that what is morally important about soil and water can be described strictly in terms of human use (or right of use). What do such claims mean?

Recent environmental philosophers have tried to provide some sense to the view that nature has intrinsic value. Some views (such as James Lovelock's Gaia hypothesis) propose that the entire planet is like an organism. The entire planet can flourish or it can experience degradation and death. Understood as an organism, land itself may be said to have interests. People who hold this view believe that forests, lakes and watersheds have interests in the sense that each can flourish or die. Such views are often called *ecocentric* (centered on ecology) as opposed to utilitarian or rights philosophies that are *anthropocentric* (centered on human values). Another view, often called *deep ecology*, states that we cannot understand or appreciate the significance or beauty of nature until we view it as having value utterly apart from any of the uses—productive or recreational—that humans might make of it. Human community (including human values) is built on a foundation of biotic community. Deep ecologists believe that we have a better understanding of our moral community when we consider those biological foundations first. Here, human values must be derived from ecology in some sense.

Sometimes deep ecology, or the claim that nature or land has intrinsic value, is just a way of saying that we should respect nature and desist from spoiling ecosystems, irrespective of any uses that we contemplate either for posterity or ourselves. One can come very close to the same view without adopting the premises of deep ecology or intrinsic value. Philosopher Peter Singer has argued that animals have intrinsic value because (like humans) animals have interests. Animals are *sentient*: They experience pleasure and suffer pain. In Singer's view any sentient being has an intrinsic interest in seeking satisfaction and avoiding pain. Singer's argument provides a way to expand the utilitarians' concern with the well-being of affected parties, but land and ecosystems are not sentient. Singer denies that ecosystems can consistently be said to have interests. So Singer (1993) concludes that human and animal interests provide a strong basis for protecting the environment but that attributing intrinsic value to nature is a mistake.

Like Singer, Gordon sees problems with ecocentric philosophy. Most important, he does not see how to apply it to agriculture. Ecocentric and deep ecology views assign privilege to *natural* ecosystems. In these philosophies, natural ecosystems must be preserved because they are intrinsically valuable. But agro-ecosystems are, by definition, not natural. Agro-ecosystems exist where the cumulative effect of farming and ranching has had a profound effect on the species that proliferate in the region, on the way in which water moves through the ecosystem, and on the transport of nutrients and microorganisms that sustain life. The Chatham environmentalists may be interested in protecting nature *from* agriculture, but it is not easy to see what they would have to say about an ethic for farming practice.

Gordon has heard that Aldo Leopold's land ethic provides the best statement of the stewardship ethic: "A thing is right when it tends to preserve the integrity, stability, and beauty of the biotic community. It is wrong when it tends otherwise." Leopold clearly believed that this ethic applies to farming; he also believed that farmers would be more likely to have the moral character needed to live up to the land ethic. He wrote, "There are two spiritual dangers in not owning a farm. One is the danger of supposing that breakfast comes from the grocery, and the other that heat comes from a furnace." In other words, people derive the knowledge and moral character needed to become stewards by living in immediate and intimate dependence on ecosystems. In such circumstances, ecosystem and human interests merge.

Leopold's argument harks back to some very old philosophical ideas. The belief that soil, water, and climate shape human culture and moral character has been expressed since antiquity. Long before Greeks invented what we now call philosophy, people have believed that human beings adapt, over time and generations, to a particular place. Over time, people develop habits of observation, patterns of response, and social norms of collaboration that make them better able to cope with the special challenges of a particular landscape. In some landscape myths, the human body is itself transformed through a form of subspecies evolution. In other myths, a people is said to be of the land or even owned by the land. In most of these myths, the peculiar adaptation of a people to the land is thought to give them a special moral claim on the inhabitation and cultivation of the land.

The moral content of these ancient beliefs is often expressed in terms of the virtues and vices that living in one way (rather than another) are thought to inculcate. Virtues include character traits such as being courageous, reflective, soft- or harshly spoken, industrious, or adventurous. Vices include character traits such as cowardice, hot-headedness, avarice, or laziness. Virtues and vices are evident in repeated and resilient expressions of mentality, personality, and conduct. The term *moral integrity* conveys the idea that individuals (and to a weaker extent, social groups) tend to engage in conduct that maintains the coherence of the community in which they live.

Moral integrity reflects and results from the reinforcement of virtue and the discouragement of vice. A good or positive character produces a pattern of conduct that is, on balance, personally and socially functional. A virtuous person is capable of coping with adversity, is seldom self-destructive or antisocial, and evinces intentions and feelings of good will and beneficence toward others. Strong moral character is associated less with extraordinary facility in coping with singular challenges than with balance. One habit of personality (one virtue) intervenes when another threatens to carry over into vice. A person with excellent moral character is so rich in these self-correcting tendencies that our very conception of what is admirable in a person tends to be defined by excellent role models rather than by criteria for virtuous conduct in any general situation.

If this is what Leopold had in mind when he refers to "integrity, stability, and beauty," or to "spiritual dangers," he is advocating a philosophy that is similar to the ethics of Aristotle. The ethical life consists in finding the mean, the balance point, where virtues check each other and do not devolve into vice. There are two points that must be made in linking Aristotle and Leopold. First, those who stress the importance of land as forming moral character see nature as a crucial balancing force in shaping moral integrity. Aristotle himself may have the thought that society, the *polis*, was more important. Second, nature can play a role in shaping virtue or vice in at least two radically distinct ways. One is in the sense that we commonly distinguish "nature and nurture." Nature is "blood," or genetic endowment, whereas nurture is the family and community environment. After the genetic endowment is fixed, nature is done with its work.

In an *agrarian* philosophy, however, it is nature as *natural environment* that is thought to be formative. As the environment in which virtues and vices are reinforced or corrupted, nature is more like nurture than nature-as-genetic-endowment. That is, nature continues its work throughout the education and lifetime of the person. It would appear that Leopold is advocating an agrarian conception of virtue. The moral virtues that a person forms from interacting with nature (that is, through farming, fishing, or otherwise making a living from nature) are thought to be more durable, more functional, and more carefully balanced than could be produced by any environment structured by the incentives of commerce or even manufacturing.

Whether this is what Leopold had in mind or not, the idea that humans respond to nature's own structure of incentives and reinforcement is sufficient to illustrate why this might be thought of as a form of moral ecology. An ethic of virtue would differ from a morality that might arise from the complex incentives found in a more socially constructed lifeworld precisely because the integration of the agrarian life demands attentiveness to nature. Yet it is also important to stress again that nature's influence on personality and social life must be reproduced repeatedly for every generation in this model of agrarian virtue. There need be no suggestion that any traits or dispositions of moral personality are carried "in the blood."

The disclaimer of genetic inheritance is crucial, for it is just this claim that has been at the heart of the most egregious abuses of virtue ethics. The basic pattern of logic just described opens the way to practices that treat certain individuals or groups as flawed—incapable of receiving or retaining nature's imprint of virtue and moral character. Combined with the view that a certain ethnic or racial group is "of the land" and that others are not, genetic determinism and agrarian rhetoric have been the basis for exclusion, racial slavery, and genocide. Perhaps for this reason, if no other, virtue arguments have fallen out of favor in recent years and people have sought to express what is morally important primarily in the language of utility or rights.

A Philosophical Depth-Chart

Gordon has now done his work. He can see how political or economic interest lines up with ethical philosophies. He summarizes these relationships in figure 8.2. The top of the figure shows the basis for the dispute: Everyone is concerned about how soil and water, (that is, land) are to be used. At the next level down are the interests groups that are likely to be in contention: economic developers, farmers, advocates of the poor, and environmentalists. One more level down shows the legal or policy option that each group would prefer. Developers want regulation to ensure that soil and water resources are used efficiently, especially when Prisoner's Dilemma situations lead to the possibil-

Land	Land	Land	Land
Economic Developers	Farmers	Advocates of the Poor	Environmentalists
Regulated Market	Property Rights	Redistribution	Stewardship and Preservation
Asset Value	Ownership and Control	Means of Subsistence	Intrinsic Value and Virtue
Greatest good principle	Non-interference	Fairness/ Equal Opportunity	Ecological Integrity
Utilitarianism	Libertarianism	Egalitarianism	The Land Ethic
John Stuart Mill	John Hospers	John Rawls	Aldo Leopold

FIGURE 8.2
A Philosophical Depth Chart

ity that individual incentives do not line up with the best use. As property owners, farmers want to maintain traditional property rights that give them control over the land. Advocates of the poor want to ensure that everyone has a secure right to food *before* farmers or developers start their work. Property may need to be redistributed to the poor (either through land reform or social entitlements) in order for that to happen. Finally, environmentalists want rules that direct people to steward the land and to protect soil and water ecosystems from degradation as the result of human action.

Each group has arrived at its preferred policy because its members tend to associate specific values with the use of soil and water. These values, which define the interest that each group takes in land use, are summarized on the fourth line of the depth chart (fig. 8.2). Developers (and some farmers) see soil and water in terms of asset value: Is land more valuable for agriculture, for industry, or for residential development? Differences in location, landscape and fertility will produce differences in asset value. As property owners, farmers stress the way that soil and water fall under their control, irrespective of whether they want to use land in the most beneficial way. Any law or policy that challenges their property rights is seen as a threat to their freedom and their ability to make a living. Advocates of the poor might stay out of land-use disputes in the United States, where the right to food is more reliably guaranteed by employment or food stamps. But in countries where land serves as people's main access to food, they will argue that soil and water must be distributed so that the right to food is fulfilled. Finally, environmentalists believe that the ecosystems in which soil and water occur have intrinsic value, or that they shape the formation of a virtuous human character in a fundamental and irreducible way.

On what basis can these values and interests be ethically justified? If we skip to the bottom row of the depth chart, we find the names of four men who devoted themselves to articulating broad principles for understanding ethics. The next row up gives the philosophical viewpoints with which they are associated. Nineteenth century philosopher John Stuart Mill is recognized as a key figure in the development of utilitarian philosophy. In the twentieth century, libertarianism was advocated by John Hospers and egalitarian arguments were associated with John Rawls. Also in the twentieth century, Aldo Leopold articulated a new land ethic by combining traditional ethical views on virtue and moral character with new insights on the vulnerability of nature to human abuse. Many other people could have been listed on the bottom row. Each of these philosophies has many advocates, and other examples could easily have been listed.

The fifth row in figure 8.2 shows the general principle that each philosophy would endorse most strongly. Utilitarians argue that all decisions should produce the outcome that is most efficient, leading to the greatest good for the greatest number. Libertarians stress non-interference and limit law, policy, and government to the protection of non-interference rights. Egalitarians see skewed distribution of property and opportunity as unfair and advocate the recognition of opportunity rights that redress this injustice. Environmentalists stress the need to preserve the "integrity, stability, and beauty" of the biotic community.

Each of these four principles supports the value judgment in the row immediately above it. Asset values allow one to see soil and water as tools for bringing about the greatest good. Ownership and control of property is at the very core of libertarian non-interference rights. Egalitarians will tend to think that fairness requires the protection of a right to food before any other interests are allowed at the table. The principle of ecological integrity supports the idea that soil and water should be managed so that the key agro-ecosystem processes are preserved, including the feedback loops that link them to the formation and development of human moral character.

Gordon's depth chart allows him to understand how one can look at land-use issues at several levels. Near the top, they seem to be political and economic conflicts of interest. Near the bottom, they seem to issue out of incompatible life philosophies. The conflict, in short, may be one of interests *or* philosophy. Gordon appreciates the deep way in which these philosophies contradict one another. In a court of law, he can see himself as an advocate, like Mill, Hospers, Rawls, or Leopold. Outside the courtroom, he sees himself as mediating conflict instead of advocating any given philosophy. He tries to find legal solutions that allow each of these principled philosophies to survive, and that help interested parties live with one another. Admittedly, this task is not easy.

As Gordon shows his scheme to Doug, he cautions about a danger in "depth chart" thinking. Cynics may see it as confirmation that moral disputes are irresolvable. True-believers will call it "relativist" because it suggests that we should take diverse viewpoints and styles of thinking seriously. But neither of these reactions is warranted, says

Gordon. Moral inquiry *proceeds* by placing incompatible viewpoints in dialog with one another. Democracy *requires* a delicate balance of advocacy and mediation. For either to succeed requires a vigorous debate, along with a search for policies that allow us to agree to disagree. Charting the depths of law and policy disputes yields an understanding of ethical differences. Gordon believes that charts start the process of ethical reflection and debate instead of ending it.

WORKS CITED

1. Avery, Dennis. (1985) "U.S. Farm Dilemmas: The Global Bad News is Wrong," *Science* 230: 408–412.
2. Brown, Lester. (1997) *The Agricultural Link: How Environmental Deterioration Could Disrupt Economic Progress*, Worldwatch Paper 136 (Washington, DC: Worldwatch Institute).
3. Hardin, Garrett. (1968) "The Tragedy of the Commons," *Science* 162: 1243–1248.
4. Hospers, John. (1971) *Libertarianism: A Political Philosophy for Tomorrow* (Los Angeles: Nash Publishing).
5. Leopold, Aldo. (1948) *A Sand County Almanac, and Sketches Here and There* (Oxford: Oxford University Press).
6. Simon, Julian L. (1980) "Resources, Population and Environment: An Oversupply of False Bad News," *Science* 208: 1431–1437.
7. Singer, Peter. (1993) *Practical Ethics* 2nd Ed. (Cambridge: Cambridge University Press).
8. Thompson, Paul. (1996) "Pragmatism and Policy: The Case of Water," in *Environmental Pragmatism*, E. Katz and A. Light, Eds., (London and New York: Routledge Publishing Co.), 187–208.
9. Wilson, Katherine and George Morren. (1990) *Systems Approaches for Improving Agriculture and Natural Resource Management* (New York: Macmillan Publishing Co.).

NOTES

1. The Chatham River case described here is drawn from Wilson and Morren, 1990. The case and its relation to interest group conflicts in water policy is discussed in Thompson, 1996.
2. Farmers have definitely *not* left land as they found it as far as the diversity of plant and animal species is concerned. Drained wetlands and leveled contours have dramatically altered habitat. Agriculture is the dominant force on the landscape (and hence on plant and animal life) in most areas where human habitation occurs.
3. There are two economic theories for explaining the value of land. According to *location theory*, the price of land will reflect the cost of getting goods back and forth,

so land near large centers of trade and population tends to be worth more than land that is distant or inaccessible. Both asset and location theory value soil and water as resources that can be converted into usable and exchangeable goods.

4. Although these assumptions are implausible, they are not as far-fetched as they may seem. Unsustainable agricultural practices can return high levels of productivity for many years before any noticeable drop in productivity occurs, but when the point of depletion nears, productivity drop-off can be sudden and irreversible. It is also implausible, of course, to suppose that over the course of a hundred years there will be no inflation, no change in interest rates, or no change in farming practices themselves. Nevertheless, any of these factors are unlikely to vary in a fashion that would make the underlying logic of the example lose force, or that would make an investor following the logic of the example lose money.

5. Julian Simon (1980) is well known for such a view, and economists including Dennis Avery (1985) have followed its logic in the belief that technological inputs will replace the asset value of soil and water for the future.

6. Hardin's argument applies to overgrazing of a commons because private property rights give ranchers a way to keep others' livestock off their land. But it does not apply to the Dust Bowl, for which no system of property rights would have made good stewards invulnerable to the wind-borne dust created by their extravagant neighbors.

7. Fishing provides a large portion of the world's food supply, but ocean fisheries are declining and most projections indicate that humanity will become more, rather than less, dependent on agriculture. High technology replacement systems for conventional agriculture produce food in hydroponic systems in which nutrients are delivered along with carefully rationed water, and some speculate that biotechnology can also be developed to virtually replace conventional agriculture. But it seems unlikely that such capital-intensive ways of producing food are likely to improve conditions for the needy anytime soon.

8. In many European countries, truly indigent and hungry people may not be prosecuted for simply taking the food they need to survive, even though they may not have the means to pay for it. This policy suggests a more direct right to food than U.S. welfare programs.

BIOTECHNOLOGY

Fred Gifford

CASE STUDY: DR. KRISTA THE SCIENTIST

The next guest speaker in Dr. Wright's class is Bo Krista, a full professor of Molecular Biology on campus.

"I understand," she begins, "that you have talked about many issues this term, including animal rights, environmental ethics, and duties to the poor and hungry in the developing world. I'm here to tell you that we may have a solution to world hunger that respects animals and nature. That answer is agricultural biotechnology."

Emily sees Rich look up expectantly. Indeed, the entire class seems to be leaning forward. Doug, on the other hand seems unimpressed.

"Agricultural biotechnology may be able to produce cost-effective nutritious food in a way that does not exploit animals or farmers or the land. For example, consider the following scenario."

Dr. Krista looks around the room. "Imagine yourself fifty years from now standing in the middle of a huge antiseptic warehouse staring at rows of tan-colored objects that look something like footballs. Shiny stainless steel pipes descend from the ceiling and disappear into mouth-like orifices on top of each object. Black rubber tubes are attached by suction cups to the bottoms. The only attendant in the building tells you that the pipes bring water and rations to what he calls 'the birds' while the rubber tubes carry excrement and urine to a sewer beneath the floor. Every twelve hours each bird drops a no-cholesterol egg onto a conveyor belt. 'Regular as clockwork,' he adds with a wink."

Dr. Krista continues. "You are staring at thousands of living Egg Machines, transgenic animals genetically engineered to convert feed and water into eggs more efficiently than any of their evolutionary ancestors, layer hens. The science fiction objects I am asking you to imagine are biologically descended from the germplasm of many species unrelated in nature, including humans, turkeys, and today's chickens, so the worker is not speaking in mere metaphor when he calls the objects 'birds.' But unlike today's poultry varieties, which are only treated as machines, the brave new birds I have in mind really seem to be more machine than animal. For, in coming up with the new birds, poultry scientists have not only selected for the trait of efficient conversion of feed into eggs; they have also selected for lack of responsiveness to the environment."

"The result is not a bird that is dumb or stupid but rather an organism wholly lacking the ability to move or behave in dumb or stupid ways. Scientific research shows that the egg machine's complete lack of any externally observable behaviors is paralleled by its lack of physiological equipment necessary to support behavioral activity. The brain of the bird is adept at controlling the digestive and reproductive tracts but the areas of the brain required to receive and process sensory input and initiate muscular movement have been selected against, bred away. The new bird not only has no eyes, no ears, no nose, and no nerve endings in its skin; it has no ability to perceive or respond to any information it might receive if it had eyes, ears, or a nose."

Doug raises his hand.

"That seems pretty unlikely," he says. "And, speaking as a dairy farmer who loves to see calves chasing each other around in the pasture, pretty disgusting, too."

"You're right," answers Dr. Krista. "The organism I have just described is a philosopher's fantasy, inspired by a remark of Bernard Rollin's, and it has a big 'yuck' factor attached to it. I have never heard a poultry scientist or agbiotech enthusiast describe anything like it as a viable goal for which agricultural genetic engineers should aim. But why not? Are the moralists ahead of the gene-splicers here? Suppose that a team of poultry scientists sees possibilities in the idea. Should we find them some funds, set them up in a lab, and encourage them to get to work?"

Rich makes a fist and murmurs "Yes!"

Doug rolls his eyes.

Now turn to Appendix A and perform Exercise 9.A before continuing with this chapter.

DISCUSSION OF ISSUES

Dr. Krista's futuristic farm vision of a warehouse full of bird-like machines (or machine-like birds) evokes in many of us an emotional, visceral reaction. Something seems morally askew here; something is not quite right. Yet pinpointing what is so objectionable is not easy.

CHAPTER 9: BIOTECHNOLOGY   193

These Egg Machines, or "football birds," are science fiction, but agricultural biotech-nology is not. Ag biotech is made up of a broad set of technologies and industries. At its heart are the techniques of recombinant DNA, or genetic engineering. These techniques enable the transfer of genetic material between organisms, whether microbes, plants or animals. Such transfers, in turn, make possible changes that are much more substantial and precise than those made with traditional breeding methods.

The kind of agricultural biotechnology products already developed include toma-toes with increased shelf-life, herbicide-resistant soybeans, insect-resistant corn, ice-minus bacteria to help prevent frost formation on crop plants, and a synthetic version of bovine growth hormone. Proponents claim that these and other products—crops that are resistant to pests and drought, or require less fertilizer—will aid in addressing a whole range of agricultural problems, thereby increasing agricultural productivity and helping to feed the world's hungry.

Critics object that this whole enterprise involves "tampering with nature" in a way that may have unanticipated consequences: Engineered microorganisms may escape into the environment, genes for herbicide resistance may get transferred to weedy rel-atives of the genetically modified crops, or there may be serious negative economic im-pacts on small farms or developing nations. Some criticize particular products as being motivated purely by commercial concerns rather than by the desire to enhance the public good. This is said, for instance, of such products as crops with sterile seeds, which make it impossible for farmers to collect the seeds from their crops for the fol-lowing year.[1] Some critics give more prominence to such notions as "tampering with nature," or to "in principle" arguments that genetic engineering is "inherently" wrong. For all these reasons, biotechnology, including its application to agriculture, often pro-vokes strong emotional reactions.

In what follows, I explore some of these criticisms and try to evaluate whether they provide us with good reasons. I begin by describing a traditional set of issues concern-ing the evaluation of the costs and benefits of ag biotech.

But the case of the Egg Machines described previously suggests another point of view, so a consideration of this leads me to consider various in principle arguments. Fi-nally, I examine some questions about ownership and patenting that in turn leads to further issues, including those of global justice.

TECHNOLOGY ASSESSMENT OF BIOTECHNOLOGY

Technologies are often assessed in terms of "cost/benefit analysis," which involves assessing alternative actions in terms of the overall positive and negative consequences that are expected to ensue from them. I can apply this analysis to ag biotech.

Because it is a general and powerful tool for making changes in agricultural organisms, biotechnology will have an impact on all of agriculture and everything that agriculture affects. Thus the ethical issues to be addressed cover a broad range. Sometimes ethical concern focuses on environmental effects. Sometimes it focuses instead on (or in addition to) economic impacts, such as those on small farms or on developing nations. Concerns are also sometimes raised over the safety of the food produced. In the case of animal biotechnology, questions are asked about whether the genetic changes could compromise the animals' health or cause them to suffer. There are many things to be said about each of these sorts of consequences and the ethical and policy issues they generate. Several of these issues have been addressed in earlier chapters.

The case of bovine somatotropin (BST, or bovine growth hormone) can be used to illustrate the broad range of consequences involved. It also illustrates how ag biotech includes different sorts of techniques. For this case differs from the others mentioned above in that it does not involve creating an organism that has had genetic material from some other type of organism inserted into its DNA. Rather, genetic engineering is used to create a synthetic version of a naturally occurring hormone, and this hormone is injected into cows in order to increase their milk yield. In the controversy that emerged as BST was being introduced, there was discussion of a broader set of consequences beyond increase in milk yield:[2] that the farm price of milk would decrease; that a number of dairy farmers will be forced into bankruptcy, having an effect on rural America generally, but also shifting the dairy industry to the Southwest; that the dairy industry that resulted might be more efficient; and that consumer milk prices would go down. Some claimed that there would be effects on the health of the cows (mastitis due to increased milk production) and further that this could have a detrimental effect on the quality of the milk (due to the possibility of increased amounts of antibiotics).

Analogous stories can be told in other cases. Consider a few of the possible consequences of herbicide-resistant crops.[3] One important effect is that on the amount of herbicide used, which has been controversial.[4] Some say it will increase use of those herbicides, with bad consequences for both worker safety and the environment. But others point out that we will be able to use *safer* herbicides as a result, hence the consequences in these areas will be positive. A different concern is that mentioned earlier: That the genes for herbicide resistance might get transferred to weedy relatives of the genetically modified crops as a result of naturally occurring gene transfer between plants.[5] Finally, farmers will have to buy as a package the herbicide and the variety of seed created specifically for that herbicide, constraining their choices about how to farm.

Note that the consequences that need to be evaluated include both *direct* ones on the product, more *indirect* ones on our lives as consumers, and *side effects* on the methods of production and the ability of certain groups of farmers to make a profit or stay in business. Clearly, a number of different consequences have to be weighed against each other. (For instance: Do gains in productivity or economic prosperity outweigh the fact that the possibility of ecological damage exists?) This prompts us to think about this in terms of a cost/benefit framework, a way of thinking that has a rationale in utilitarianism. Utilitarianism is a common starting point in discussions of ethical theory. Moral concerns surely have at least substantially to do with the *consequences* of our actions. According to utilitarianism (or, more broadly, consequentialism[6]), *all* moral considerations are solely a matter of the consequences of the action. For a utilitarian, we morally ought to do that which brings about the greatest good (or the greatest balance of good consequences over bad consequences) for the greatest number of people. Some utilitarians broaden this framework to include the welfare of other sentient animals in the calculation as well, because these creatures can also experience pleasure and pain. Other utilitarians expand the boundaries even further to include other inherently valuable states of the world, such as those in which there is a diversity of species and ecosystems. But in all cases, the core of utilitarianism is that all that matters in morality is maximizing good consequences.

This is not to say that this will give us a simple and straightforward algorithm for generating the answer to what we should do. There are serious difficulties—both practical and conceptual—with carrying out such an evaluation. Sometimes these difficulties are used to challenge cost/benefit analysis as an appropriate method for assessing what we ought, all things considered, to do. First, we need to be able to assess various factual claims: the potential outcomes (for example, that herbicide resistance will be transferred to weedy relatives, or that there will be a negative effect on human health) and the probabilities of each such outcome. These facts may be hard to assemble and even harder to assess objectively.

As a result, a central theme in discussions of biotechnology is scientific controversy and how properly to deal with this. Different scientific experts may not come to a consensus as to what the facts are. Even if most of those in the scientific community *do* come to such a consensus, there may remain skepticism by outsiders. For example, there was a consensus among essentially all of the scientific community that BST was safe[7] but this fact did not prevent continued public concern.

Second, we must be able to assess the values associated with each outcome—how good or bad that outcome would be, compared with other outcomes—and ultimately we must assess on a common scale such things as the extent of environmental damage, the increase in overall crop yield, and the change in product quality. Even if such assessment is possible in principle, as a practical matter, the tendency exists to focus only on those aspects of the consequences that are readily measurable; this can overrate the importance of such considerations as productivity and economic consequences.

DISTRIBUTIVE JUSTICE

Note that the consequences of the preceding examples included not only those for *overall* production and overall quality of life of consumers but also for *distribution* of risks and benefits. Should we say that it is acceptable that many small farmers will go bankrupt, on the grounds that overall production is maximized? This conclusion prompts many people to deny that we should in fact use utilitarianism as a moral standard, for this only concerns itself directly with *overall* good. They have the moral intuition that what matters morally is not just how much good is produced, but how that good is distributed; they think that there ought to be a more equal distribution of welfare. As a result, they say that, at the very least, we need to add another moral principle along side utilitarianism, such as a principle of equality, or a principle that tells us to reduce the gap between the rich and the poor as much as possible. Or they might adopt John Rawls' social contract theory, which says that the most just society is the one that treats its least well-off members as well as possible.[8] I return to the question of distributive justice at the end of the chapter.

Another ethical question to ask is the following: What is the appropriate kind and degree of *public involvement* in assessments about biotechnology? Do people have a right to consent, through some sort of democratic process, to actions that will have a profound effect on their lives?[9]

I have discussed three different general moral views or principles: (1) that utility or welfare should be maximized for the whole; (2) that goods should be distributed fairly (for example, that there should be a concern not to allow too great an inequality); and (3) that people should have some say over technologies that have a major impact on their lives.

Sometimes these three principles may conflict, so it might be thought that we can't make moral judgments until we have decided which is the correct principle. But it's worth noting that when people criticize a given biotechnology, they often give reasons to believe that *none* of these principles is satisfied. For instance, they may suggest that there may be harm to our common environment with profit benefiting a few in a way that neither maximizes overall utility nor allows a just distribution of welfare, and that the technology would not be chosen by the people affected if they were told the facts. Similarly, arguments *for* a biotechnology often claim that its introduction will bring about benefits in a broad-based way. Proponents often emphasize products or innovations that could prevent world hunger, or keep food prices low; this technology helps *a lot* of people and it helps the *least well-off* people.

Still, whichever of these principles we utilize, our moral assessment is likely to have to do with the assessment of consequences of the technology.

EGG MACHINES

"I agree with all this," says Emily. "You're making a lot of interesting points about how to think about whether developing or introducing a technology is a good idea, or how we might argue for or against it. I can think of a lot of cases in which this would help me think about that. But my reaction to these Egg Machines doesn't seem to have to do with any of these things."

We might describe Emily's idea here by saying that the case seems to be constructed in such a way that these concerns cannot be what is driving our intuitions. The Egg Machines don't appear to present a threat to the environment. Economic impact is likely but the extent is unknown. The main issue might *seem* to be the way the "animals" are treated but in fact no harm comes to the animals for they have no conscious experience of pain or stress. Indeed, the suggestion is that the use of Egg Machines is a great improvement over our present conduct in this regard. So just what could be wrong about it?

Yet we may find the prospect of these inert, unfeeling "Football Birds" quite disturbing, distasteful, repugnant, and "creepy." Our reaction may be even more intense if we fill out the thought experiment and imagine this practice on a *large scale*, if we imagine that we have transformed our egg production to be done almost exclusively in this manner, or perhaps even that something analogous occurs for *all* of our animal food production.

Some may decide, on the basis of such reflection, not to go down such a road. But because it is hard to say exactly why Dr. Krista's vision involves any sort of *moral* wrong, we are challenged to explain what could be the basis of our moral intuition here. One is challenged to answer the charge that our intuition is a mere emotional reaction, an irrational prejudice.

WHY EVEN CONSIDER THIS KIND OF CASE?

"But this case isn't real," says Doug. "This sort of thing isn't going to help to feed the world's hungry. There's nothing to worry about, either. It'll never happen."

Now, this scenario is indeed quite a strange one, and although the vivid image may make it seem interesting or even powerful, this might also be thought to be a problem—a symptom that this line of thought will simply get us off track in our attempt to think about ethics and biotechnology. In particular, it will be said that this example is simply too unrealistic or "sci fi," or too different from the actual cases of biotechnological innovations being introduced at present.

First, it will be said, we have no reason to believe that this particular technology will be developed by anyone. We don't even know at this point whether this would be technically feasible. Second, even if it is admitted that something like this might occur someday, the scenario is too unlike present reality for us to have clear intuitions about it, so our moral intuitions—our intuitive judgments about the moral acceptability of the practice, based simply on our confronting it in our own minds—will only be misleading. Finally, it will be said that focusing our attention on this sort of case distorts our view of biotechnology by making us think that this is the sort of thing biotechnology usually is, when in fact it is not. One consequence of this might be less attention paid to the more immediate and real challenges posed by biotechnology.

These are important points that must be kept in mind. Still, several reasons exist not to simply dismiss the case. First, it doesn't seem to be wise counsel to wait until a technology is upon us before considering whether to bring it about. It is often pointed out that in the case of biotechnology, we thankfully have the opportunity to reflect on these moral issues early on before we are too far along, unlike our predicament with the power of the atom. Although we should surely keep the speculative nature of our thinking in mind, it may nevertheless be a quite useful thought experiment to reflect on cases that are in some ways extreme. After all, if we only consider such cases as new crops with one or two altered genes designed to improve on one or two traits, such as shelf life or pest resistance, we might not adequately anticipate or comprehend the cumulative effect of many such products over time.

Finally, exploring our attitudes and reasoning in more extreme cases may be worthwhile for their vivid illustration of a more general phenomenon. Such exploration evokes broader concerns in discussions of biotechnology dealing with the patenting of life, or living things or animals, or human genes, or cloning or other artificial modes of reproduction, or just about altering genes at all. Polls over the years have shown many members of the general public to be uncomfortable about biotechnology.[10] Biotechnology as a whole tends to evoke deep concern or fear. We need to think about this carefully.

THINGS THAT COULD BE WRONG ABOUT THE EGG MACHINES AND SOURCES OF MORAL CONCERN

So, suppose that coming "face to face" with the possibility of a roomful of Football Birds arouses strong feelings and leaves us with the intuition that it is the wrong thing to do, that we should not go there. What can we identify as moral concerns? Here are some candidates:

- We are mixing genes from different species.
- We are mixing *human* genes with those of other species.
- We are creating a "species" that did not exist before.

- We are creating a living entity specifically to have a diminished capacity.
- We are blurring the line between animal and machine, or we are treating the animals as machines.

Although these concerns might all come to mind together, or in rapid succession, they are nevertheless distinct from one another and may have quite different sorts of concerns or rationales underlying them. Some may be more significant than others. Mixing genes from different species (creating transgenic organisms) is common in biotechnology; this is done in the products already brought to market. But to do this with *human* genes might raise the level of moral concern, and to create a "species" that did not exist before raises still further questions. The blurring of animal and machine perhaps takes us in yet another direction.

Are any of these sufficient reasons to object to the Football Birds from the moral point of view? What sort of argument or moral principle could be cited to support claims such as "It is wrong to mix human genes with those of other species" or "It is wrong to blur the distinction between animals and machines"?

Some might say that it's just obviously wrong; you don't need to say anything more. But others might claim that they don't find anything wrong here. So one has a responsibility to try to say more than this.

Here are some further possible rationales that might be cited as underlying our concern about such things:

- It is "playing God."
- It is *unnatural*; it interferes with Nature.
- It involves crossing species *boundaries*.
- It does not exhibit proper respect for life or, it is sometimes said, it does not respect the *telos* of the animals or the integrity of species.
- It "commodifies" life.
- It involves a reductionistic, mechanistic view of living things or of nature.

These purported moral considerations share a number of features.

First, on the face of it, they are distinct from "consequentialist" considerations, the sorts of considerations that were cited in relation to BST and herbicide-resistant crops earlier. Apparently the Egg Machine practice is being said to be *inherently* wrong, an assessment to be made *independent* of consequences. They can be called "intrinsic" concerns, as contrasted with "extrinsic" or consequentialist concerns.[11]

It might be thought that any inherent or "in principle" consideration cannot be reasonable, because morality is more complicated and nuanced than that. But note that to say that something is inherently wrong, wrong "independent of the consequences," need not mean that it cannot be done "*whatever* the consequences." It is to say that at least part of the reason the action is wrong is not due to the bad consequences that will ensue, but due to the very kind of act involved. For example, we sometimes say that lying is wrong *per se*—because of the kind of act that it is, not simply due to bad consequences that are likely to occur. Yet we might hold that if the consequences were weighty enough, they could override this.

So the person who is disturbed by and thus questions Egg Machines or mixing genes from humans and other species need not be saying that it should be forbidden even if it were important or necessary for creating a sufficient amount of nutritious food while avoiding the problems of harm to animals. Rather, they might only be saying that there are some real considerations against it that should be taken seriously, so that one should only do it if the potential gain is important and not well achievable in some other way.

A second feature of all these rationales is that some will be skeptical about whether they should really count as genuine moral concerns, or as a reasonable justification.

This doubt may arise in part due to the *kinds* of rationales or sources for the view. This might be said of a rationale that is religious (or quasi-religious), as in the case of playing God. A similar worry arises from the fact of being based on emotion or intuition.

Further skepticism arises from these principles being somewhat vague, metaphorical, and difficult to grasp completely or state precisely, or their being open to alternative interpretations. As we shall see, this makes them difficult to evaluate. Yet for all this, we may not be comfortable simply dismissing them as having no force at all.

Finally, note that another feature shared by each of these rationales is that they seem to apply to a broad range of biotechnologies—not just the Egg Machines.

I will say some things about each of these things as we go along, as I consider several of these rationales.

Unnatural

"Well, yes, I guess that one of the things that strikes me about these Egg Machines is that it's so unnatural. The natural thing is to have a bunch of chickens running around and laying eggs."

"This is definitely the sort of thing that would not occur in nature," agrees Doug.

Let us consider this claim of unnaturalness. This can be put as the claim that we are creating some product that is "unnatural" (the Football Birds) and that we shouldn't do this. Or sometimes it can be put as the claim about the *process*—that we should not *interfere* with Nature. Doing things naturally can seem like a good idea. Saying that the Egg Machines are unnatural may seem reasonable. But what exactly is being said? What is it we are doing when we are carrying out an unnatural intervention? One interpretation that makes some intuitive sense is that we are making some change in the world that could not have occurred without the intervention of humans. When humans do not interfere, the world goes on naturally.

But this makes building and using cars and airplanes and selective breeding count as unnatural as well. Yet we are not even tempted to say that there is anything wrong about these things; nor are we likely to call them unnatural. On the face of it, we have a counterexample to our principle, an implication of our principle that we cannot accept. If *this* is what *unnatural* means, then we simply cannot possibly avoid doing unnatural things, hence it can't be *wrong* to do them.

So, how can this be responded to? Why should these things (planes and selective breeding) *not* count as unnatural? Well, perhaps, because it's actually very much the nature of humans to invent new sorts of entities like this. Technology is natural to humans. It's statistically normal. It might even be said to be what distinguishes humans from other species. Further, it presumably has been adaptive for our species and appears to arise "naturally" out of our very human qualities of curiosity and intelligence.

But then, is there any reason to deny that more high-tech endeavors, such as genetic engineering and even Egg Machines, will be natural as well?

In response, one is still likely to say that *this* isn't what we meant by *natural* or *unnatural*. We had in mind some more specific sense of *natural* in which building cars and planes (and carrying out selective breeding) would be natural, whereas the Egg Machines would not. After all, surely there is something to the difference between small-scale sustainable organic farms and Egg Machines. If the above definition ignores this, then there must be some other way to make the distinction.

But it is very difficult to make this out. If we cannot give a principled distinction, then we will worry that this is mere prejudice, disguising a value judgment as a factual claim. The worry is that we first make a judgment that the Egg Machines are bad, and then, on the basis of this judgment, we label them *unnatural*. But doing so is unfair, because the assessment of its being unnatural was supposed to be the justification for the claim that it was bad. One wonders whether *natural* here just means the way we are *supposed* to farm (or *used* to farm). It is not at all clear that we can give a definition of natural that will show why Egg Machines stand out as unnatural.

In any case, note that most of what has been considered so far is whether a line can be drawn, between natural and unnatural, in a way that fits our intuitions. But even if we were able to give a clear definition of what counted as natural and what did not, it would not follow that we had picked out something good or bad. Why is the natural thing good? Indeed, *are* natural things necessarily good? Consider earthquakes, smallpox, or deer flies.

One answer might be that the unnatural thing had a greater likelihood of leading to bad consequences. This has certainly been a common theme concerning high-tech innovations. Of course, this is an extrinsic, or consequentialist, reason. The idea here was to uncover intrinsic considerations. But let's leave that aside for now. Perhaps what underlies the intuitive negative reaction to things that don't seem natural is in fact the fear that this is more likely to be dangerous. And if being unnatural were a good predictor of being harmful or dangerous, we might have succeeded in uncovering what is operating here. But in fact it doesn't seem to be that good a predictor, as lots of (intuitively) natural things can be dangerous (earthquakes, smallpox). Still, it might be argued that there is a *greater uncertainty* about unanticipated consequences when something is unnatural in the sense of new and untried.

In any case, insofar as the real issue is "significant likelihood of unanticipated consequences," then we should say that that is the issue and make our evaluations in those terms: Will Egg Machines have unanticipated consequences? Throwing in the term "natural" appears only to confuse things. Also present is a worry that *natural* simply gets used as a general statement of praise and *unnatural* as a general sort of condemnation.

So it is quite questionable whether we can use the term *unnatural* to criticize such things as the creation of new species or the mixing of human genes and genes from other species. For given the unclear meaning of *natural*, it doesn't seem that it can be used to make the distinction clearly and in the right place, and it is not clear what is bad about being unnatural.

Factories and Telos

Doug says, "What's wrong here, in my opinion, is that the Egg Machine system treats animals as factories." Emily agrees. "Yes, that's something more specific than whether it's natural or not. I think that maybe that's what's more unique to what we're doing in this case."

Now, to say that "We should not treat animals as factories (or perhaps as machines)" might seem to have a certain proper moral ring to it. It suggests that we should properly respect our fellow creatures. It seems to be a less all-encompassing critique than that it's *unnatural*, so perhaps it can avoid the problems of that view.

But can we hold such a principle and expect it to help us make judgments? Just as in the case of *natural*, we need to consider each of two challenges. First, can we accept the whole set of implications of this principle, or are there "counterexamples"? (For instance, are there cases for which we are clear that they are morally acceptable even though the principle forbids them?) Second, is it really clear what advice the principle provides, or does it end up being too vague or ambiguous?

There do indeed seem to be counterexamples to the "don't treat animals as factories" principle. For don't we already treat many animals as factories? Isn't this a correct description even of our use of dairy cows? Or, indeed, of all of our raising of animals for products for food and clothing? Is there any way to avoid treating animals as machines?

Emily says, "Yes, we use animals this way, but I'm not sure we *should* use animals this way. The arguments from the defenders of animals are starting to convince me. We can get all the food and other products that we need from plants."

Dr. Krista replies, "But isn't the reason that it's okay to treat plants this way that they lack consciousness, so that the process can't cause them suffering?"

"I guess I still don't think you should treat animals as plants," says Emily.

It might be objected that the sort of thing that goes on with raising milk cows, etc. is not necessarily wrong because farmers treat their animals with proper respect. If all farmers were like Doug and shared his attitudes toward animals, then farming would not necessarily involve viewing animals as machines, or as factories.

We are raising some important questions about character, about the kinds of attitudes we ought to take towards animals. This is important, but it might be said to be a separate issue from what particular technology to use. For instance, consider this question: Might it be possible for our tender of the Birds to take the right attitude toward *them*, and would this then solve the problem?

Or it might be argued that, with the move to large-scale agribusiness, we have for the most part already lost the ability to have this right attitude. Thus, it might be argued, this particular technology is not really going to make a significant difference in this regard.

Can we say what would be *wrong* with treating the Football Birds as factories? After all, the so-called Birds are unconscious. It is sometimes said that there are certain ways an animal ought to be treated—because of its nature or "telos"[12]. As an analogy, human beings, given their telos or nature, should (according to Kant) be treated as autonomous beings; it's wrong to treat them as mere means to an end and not at the same time as ends in themselves. The reasons one must do this are quite independent of the consequences of doing so. Other animals don't need to be treated with *that* kind of respect, but there is such a thing as treating them inappropriately given the kind of entity they are, so the claim goes.

Of course, one problem here is whether there is such a thing as a telos that really gives us a reason to treat them a certain way. We will not take this up here. But there is a special problem for using this rationale in thinking about the Egg Machines. Consider: Even in the case of "unmodified" species of animals (and we are stretching it to call domesticated animals unmodified), it is not clear how we are to show that a certain way of treating them is the right way (leaving aside pain and suffering). But there is yet another difficulty in applying this to a Football Bird. It would be one thing if we had anesthetized the organism or even made it unconscious. We might then say that it is not expressing *its* telos. But the situation here is different. For here it seems that we have *simply created a new kind of organism entirely*, and it is not clear why we should say it has the same telos. Perhaps it has *no* telos, perhaps it has a different telos, but if the latter, on what grounds can we say that it is not expressed here? So it is unclear how this helps us ground the claim that there is some particular way it ought to be treated.

One might conclude that the issue is not one of how we may treat certain animals, but of whether we may *create* certain kinds of entities, whether we may make "plants" (or mere unconscious factories) out of animals, create vegetative beings out of the raw materials of animals. Some may have the intuition that it is wrong, intrinsically, to change the telos or nature of the animal. But I am not sure we have yet been provided with a sound argument to justify this intuition.

THE ENVIRONMENTALIST ANALOGY (AND "STEWARDSHIP")

The general difficulties here bear some resemblance to some general dilemmas environmentalists face when they make claims about our moral obligations concerning the environment. In that context, we may have intuitions that such entities as plants and species and ecosystems (even the "Land"[13]) ought to be preserved (or respected, or held to be valuable) for reasons over and above their instrumental value to human beings. Yet these entities have no consciousness and thus cannot be made to experience harm. So perhaps our problem here could be amenable to solutions designed for the environmental case.

Many hold that we should in fact preserve the environment—and not just for instrumentalist reasons—even though it is somewhat unclear what the basis is for this obligation. Sometimes this obligation is seen as arising from "stewardship": we ought to act as a steward of the living world, of the species of our planet. But the general point can be made here even if we don't think about it in terms of stewardship. Just suppose we find some analogous reason or justification for an obligation toward the environment. If we find this reasonable, perhaps something analogous can be said concerning our question concerning Egg Machines.

But there are various problems with any such approach. First, and most generally, there is a problem of specifying *just what counts as* stewardship here, or just what our obligations are here. (How strong is our obligation to preserve other species when this is weighed against other things, such as other usages for a given area of land?) If we cannot do this, it is hard to see how we could show that this responsibility would apply to the Egg Machine case, or any other particular case. But second, there is a more specific problem for our use of this analogy: the "stewardship of our natural resources" approach is presumably a way of grounding *preservation*, yet this is not really what is at stake here. For again, we are talking about *creating* a *new* species, not destroying an already present one.

Of course, one might imagine that the agricultural species are only in existence because of their continued use for human consumption; we might well not continue them if we don't need them any more. So perhaps there *is* an issue of preservation. On the other hand, this might also be a consequence if we all became vegetarian. In any case, it is not necessary that we terminate those original species in order to create the Football Birds, so, ultimately, it's a separate question.

It appears that neither the stewardship concept nor the environmental analogy will apply to our case in a straightforward way.

Religious Concerns

"So maybe it really comes down to something religious," says another classmate. "Creating these new kinds of creatures does seem like 'Playing God.'"

Emily interrupts: "But so many things get called 'Playing God'."

"But does that mean it's not important?"

It was said above that several of these concerns seem to have their source in religious views. The most obvious is "Playing God," but this can be said of others as well. For example, if someone says that the making of Egg Machines doesn't properly respect life, this might be interpreted as there being something *sacred* about life. The species boundaries that we are not to cross might be seen as laid down by God.

It is worth mentioning that polls show that members of the public do indeed express religious concerns about biotechnology.[14] Of course, the religious basis might explain the depth of feeling involved here.

But moral objections or concerns based on religious views are often criticized for a variety of reasons. For some, of course, one reason for this would be that they don't believe that there *is* a God, or at least that they are not confident about this. But all must accept the following reason: we in fact live in a pluralistic society, with different people having different religions and some people having no religion at all. Therefore one cannot rely on such arguments to reason in a way that all will agree with. As a result, it seems that those who put forth such concerns may need to recast them as secular or perhaps consequentialist.

Playing God

Let us consider first the case of Playing God. This issue is sometimes understood in terms of the claim that we are trying to "improve on God's creation." One can read this in such a way that it sounds sacrilegious, thus the principle appears to be a profound one. Yet it might be objected that the pronouncement that we should not improve on God's creation will imply that we shouldn't engage in *any* technology. Surely this is not what is intended. So why is changing God's biological creation different? Why is selective breeding different? Is there any principled answer to this? For instance, could one find such an answer by examining scripture? At present, we don't have agreement on what would count as Playing God, or how to find out.

An important worry here is that appeal to this claim will function to cut off debate (not just in society, but also for oneself). For one thing, we are loathe to question publicly the religious beliefs of others. Further, a phrase such as "playing God" is metaphorical and ambiguous (as is the term *natural*). As a result, such claims tend not to get pushed further, thus some might be skeptical of such considerations, taking them not to be fully thought through, not subjected to critical reason, and difficult to elaborate further.

Still, one can be left with the feeling that there may be something very profound or important concerning this and other seemingly religious concerns and that perhaps they should not be dismissed. It may be worth noting that Bernard Rollin, in discussing this sort of issue, quotes John Dewey as saying that "putatively religious concerns may well be metaphorical ways of expressing social moral concerns for which no other ready language exists."[15] Perhaps over time, as a result of critical discussion, we might come to see that there *is* a rational and generally communicable basis for them. But others may say perhaps not.

Boundaries

Emily suggests a different tactic. "Okay. Let's not put our argument in terms of playing God. How about the idea that we should not "cross species boundaries"? That seems more specific, so maybe it will be more helpful. Is there anything wrong with *that* principle?"

Krista: "So our question, I take it, is whether we can appeal to this as a reasonable general moral principle. It will have to explain why the Egg Machine practice is wrong and it must not forbid activities that we are pretty sure are perfectly acceptable."

First, actually, we will have to clarify the nature of the claim: the nature of these boundaries, how we know they are there, and what their significance would be. Different people claiming that "we shouldn't cross species boundaries" might have different things in mind. We must be sure not to run them together.

One contrast is the following: There are two different kinds of things one can mean by "crossing species boundaries." First: taking genes from one species and putting it into the DNA of another (moving genes across species lines). Second: creating a new species in what might be called the "gaps in phenotypic space," creating a new type of individual that might share some properties of each of two or more species, but that is significantly different from any one of them. The former is commonplace in genetic engineering whereas the latter is not. The Egg Machines involve both.

It is important to keep this distinction in mind. For instance, suppose one succeeds in generating an argument that moving genes across species lines is intrinsically wrong, that boundaries should not be crossed in this sense. Then those who don't see transgenic animals *per se* as posing a problem, but do worry about the Egg Machines, will see the line as drawn in the wrong place.

The other important ambiguity about what is meant by crossing species boundaries is this: The objector might say that there are boundaries placed there by God. This would of course require an explicit religion-based view; this then inherits the problems of a religious-based view in that context, including that of there being no way to debate it in the society as a whole. On the other hand, one might instead offer a more scientific version of this, saying that there are boundaries established by evolution.[16]

As long as the issue is put in terms of a factual or scientific claim about what evolution has wrought, we can look more carefully at the facts involved. Actually, there are various natural processes whereby genes have always been transferring between very closely related species; this was the basis of the concern referred to earlier concerning herbicide resistant crops, where, due to hybridization, genes could possibly spread to the weedy relatives of the target crop.[17] Now, the question of whether evolution has erected a complete and absolute barrier to gene flow across species is complicated somewhat by disagreement about just where species lines are to be drawn. But on any definition of species, there is very little such transfer.

So some might argue that there is not an absolute, impermeable wall there, but that there is still in this sense an objective barrier out there in the world. Similarly, it's an objective fact that there are between most species a significant distance in phenotypic space; most species don't completely gradually fade into one another.

But what do these scientific facts tell us about what we should do? What is the argument that these are boundaries not to be crossed, rather than spaces to be filled in? It is not clear what this could be, except for the claim that as a matter of fact, it could be a dangerous thing to do, due to the consequences that might result, or perhaps the fact that we simply don't know what the consequences would be. As in the case of the argument from unnaturalness, it would be more honest to say that and try to evaluate it on those terms.

THE ROLE OF EMOTION

"So, I'm drawn back to this idea you mentioned at the beginning that our reaction to the case is a *mere* emotional reaction," Emily says. "What does this mean? I think we rely on emotion when we judge what's morally right. I'm not sure that's a bad thing."

Let us explore this. Suppose we were confronted with *conscious* organisms strapped to tubes and otherwise like the Egg Machines. Noting the sort of existence we would be subjecting these beings to, surely we would see this as a serious moral wrong, and no doubt this would be attended by an emotional reaction or revulsion.

But these reasons don't apply to the Football Birds. There's nobody home. If we nevertheless have an emotional reaction, if we still have moral intuitions against doing this sort of thing, this might be said to be simply a sort of "carry over": the reaction is due to the case being *similar* in various ways to the cases described previously that *are* morally problematic. But because the features that make those cases morally problematic are exactly what are missing from the Egg Machine case, the carry over is *merely* psychological and we should not give credence to it. It is a confusion in our reasoning, resting on a mistake. (Note that we can say that this is a mistake even if we think that it's a good and healthy thing that we have this emotional reaction.)

Now, it should be pointed out that, actually, to assess this as merely psychological begs the question. The Egg Machine opponent can simply say that, although the "animal welfare" features have been eliminated from the case (hence it is *less* disturbing), there remain other things wrong or worrisome about it. (The challenge, again, is to say what.)

Nevertheless, our skeptic might push further, arguing that emotion should be viewed as something that *gets in the way of* good moral reasoning by distorting our judgment. Now, there are certainly emotional and intuitive judgments that we need to discount. For example, some people have visceral reactions to people who are very different from them. We don't want to say that there must be something to their moral view that these other people should be treated badly or as inferior.

Still, it might be thought that we should listen to our feelings or impulses, other things being equal, for they appear often to be informative. If we have an emotional reaction to seeing starving people or tortured animals and we are moved to believe that something should be done about it, this seems perfectly appropriate and we shouldn't feel a need to squelch those feelings. Emotions play an important part in morality.[18]

Perhaps the appropriate view is that we should take an emotional or intuitive reaction only as a starting point, or as suggestive of something to explore further. We surely must be willing to overrule such a reaction if there are good reasons to, but it doesn't follow that it has no merit in general. Yes, we should feel obligated to look for some further rationale that can be shared in an open and rational discussion, but perhaps we should not give up too soon.

But there is another challenge to the use of our emotional reaction. Consider a different kind of "case." Suppose the proposal for a further way to cost-effectively produce the nutritious food was not Egg Machines, but instead the creation of food stuff via a process that was a much more radical shift away from the whole or flourishing animal—the creation of food stuff in a high-tech tissue culture (altering the nutritional content in whatever way we want). To make the case otherwise parallel, suppose we will again include in this tissue culture genetic material from a number of higher vertebrates, including humans.

Now, in neither this Tissue Culture case nor the Egg Machine case is there any consciousness to worry about. But in the Egg Machine case, perhaps because there presumably are physical reminders of what particular higher vertebrates are involved, there is more of a tendency to see a "distorted telos," whereas the Tissue Culture case involves processes so far removed from the normal cases (for example, chickens) that we don't worry about it or don't have the same emotional reaction.

Yet the presence or absence of feathers or recognizable body parts in the one case doesn't seem to be morally relevant. It thus looks like it's "mere" psychological difference that underlies the judgment, sort of like when people have more moral concerns about endangered species if they are cute and fuzzy. So, the argument goes, our reaction in the Egg Machine case should be discounted.

On the other hand, someone might look at all this from the other end. It might be said that the Egg Machine case simply "wears its morality on its sleeve" and is merely a more overt example of the same thing. It's not that the Tissue Culture case is clearly okay and the Egg Machine case differs only in a psychological way, so that we should dismiss the reaction. Rather, it's that the Egg Machines intuitions indicate what's really going on even in the Tissue Culture case, but there it is masked.

"I don't know," says Doug, "but let's go the Tissue Culture route, instead of making those icky Birds."

Krista: "Well, perhaps one would not be so unmoved by the Tissue Culture image if what was in fact brought to mind was a massive industrial complex of acre upon acre of this stuff, especially if one imagined it replacing one's image of an idyllic small town farming community."

"Hmmm. . . it *is* pretty distasteful. Okay. I vote for the idyllic small-town farming community."

"I'm not sure that's one of your options."

GENERAL STRATEGY: TRANSLATING INTO MORE SUBTLE CONSEQUENTIALIST ARGUMENTS

Now, suppose that we are still of the opinion that our negative reaction to the Football Birds is indeed to be taken seriously here. It doesn't seem to be about some harm to the environment, or diminishing some aspect of the quality of the food product, or some harm to sentient creatures. Appeals to notions of Playing God, naturalness, and boundaries seem unsound. We have a concern that our reaction is merely an emotional one, so we need to be able to give reasons that can be evaluated.

One general strategy at this point would be to see whether we can explain these concerns (that there is something disturbing about the Football Bird case) in some other way. Can we translate these potentially questionable concerns into consequentialist arguments that might otherwise have been overlooked, or note less obvious consequences that are not typically brought to mind in simple consequentialist analyses?

The alternative explanations I consider here are the *slippery slope argument*, the problem of how it might rub off on us, and the matter of the connection to *world views* and how we think about things. Some of these may be suggestive, but I do not put forward any of them as a definitive resolution to the problem. As with the earlier suggestions, they present serious difficulties. I hope that they can prompt constructive discussion.

The Slippery Slope to Human Genetic Engineering

One can imagine Emily saying, "But what I guess I'm concerned about is where this will lead."

Asking where something "will lead" is certainly one way of pointing to more distant or more subtle consequences that are likely not to be part of a straightforward consequentialist analysis. Perhaps nothing is wrong with this activity *per se*; in particular, it is not that the consequences of this particular practice will be bad (for the animals, for the consumers, for the economy), but if we take this step (engage in this activity, develop this technology), we will be led to take further steps, develop further technologies, and eventually we will be engaging in activities that are indeed clearly wrong (whether intrinsically or because of the harm they cause).

We are struck by the Egg Machine image and envisage it going much further. Perhaps this is part of what underlies our reaction, and perhaps this is something we should take seriously.

The argument of this sort that comes most readily to mind says that we will be led to apply such techniques to humans. Genetic knowledge is of course already being applied to humans. Attempts are made to address genetic diseases by genetic screening and genetic therapy, and these endeavors are being aided by the Human Genome Project, the attempt to map and sequence the entire human genome. Despite promise here, serious concerns are being debated about genetic privacy and discrimination, as well as about genetic enhancement and eugenics. Other aspects of artificial reproduction, including cloning, are sometimes raised here as well. So, we might worry about rigorous application of genetic engineering to agricultural contexts leading eventually to something like the wholesale design of our offspring to more and more exacting standards.

This general kind of argumentative strategy is called a *slippery slope* argument. Even supposing that nothing is wrong with the present action, if we take that step, each step will become easier and less noticeable, and we will eventually find ourselves in a situation uncontroversially judged to be totally unacceptable.

Presumably this *kind* of argument can sometimes be reasonable. If it really is the case that making Egg Machines would inevitably or very likely lead us to this undeniably bad outcome, then surely we have a good reason not to do it. But it is important to be clear about what is required for a convincing slippery slope argument and where it can go astray, or where it can seem more convincing than it really should be. Basically, we must be very careful to clarify exactly what the bad outcome is supposed to be and to clarify what is bad about it. Further, we must give reasons to believe it would really occur. Unfortunately, it is too easy to construct seemingly convincing scenarios without giving serious thought to each of these matters.

The image of biotechnology applied to the wholesale design of human beings to more and more exacting standards is indeed a scary one. Now, one might argue that there is an element of a "yuck factor" even here, and it may be worth pondering what exactly would be morally wrong with these human applications. But we can assume that such an outcome would indeed be a very bad one, to be avoided at all costs. Still, is there reason to believe that the use of genetic engineering to the hilt in agricultural contexts (or the production of Football Birds in particular) will increase the likelihood of the above *human* genetic engineering practices?

We can tell various stories: At one point, we will put a human gene in a pig. At another point, we will put genes from other species into chimps. Finally, we will put genes from other species into humans. But it's not enough to trace possible intermediate steps. The slippery slope argument requires a causal claim, and in this case it's a causal claim that is very hard to assess.

So, how much evidence should be required? This is a difficult question. Suppose that someone said: Because we don't at present have any specific evidence that this sort of thing would happen and we don't have any good way to test such speculative claims about very subtle effects on difficult-to-predict human actions, we should therefore not give any credence to this slippery slope objection. This is surely too strong; although we should not tolerate blind speculation, such blithe dismissal does not appear to be the right policy. After all, we're talking about events that have never before taken place, and we really don't have very clear evidence one way or another. Ignorance is not bliss. But clearly the proponent of the slippery slope argument needs to be able to say *something* about why we should take the outcome seriously.

So there remain unresolved questions about what will count as sufficient evidence to be a serious worry and where the burden of proof shall lie. But we can say the following: The slippery slope argument's prediction of inevitable slide is especially weak if it doesn't address the following question: Could we not have safeguards (perhaps regulations, perhaps public discussions) so that each further step won't be easier and less noticeable? If we had some such safeguards, then we really could stop before we got to the bottom of the slope. Thus the claim that the bottom of the slope is unacceptable might not be a strong enough argument for not taking the first step.

One reply would be to insist that the steps really will be very small and not easily noticeable, and there won't be sharp lines to draw, and there will always be strong economic forces driving us to move ahead. It also might be pointed out that there is a long distance between "completely inevitable" and being confident that it *won't* happen, so showing that it isn't completely inevitable is not enough. Another response would be to declare victory: Egg Machines raise serious moral questions in that they jar us into taking seriously that we must have these discussions and these safeguards.[19]

The Rub

Another sort of "subtle consequence" of Egg Machines might be the psychological effect on us.

Emily might ask, for instance, "Might 'how we treat' these Football Birds rub off on how we treat either other animals, or even other humans?"

In this view, again, it is not as though the nonconscious Football Birds themselves *matter*; it is not as though there is some way they ought to be treated, so that we should not treat them as machines. But if we do so, we increase the likelihood that we will treat in analogous ways entities that *do* matter: persons and sentient creatures.

Various thinkers (for example, Aquinas) have given an analogous argument for why it is wrong to be cruel to animals.[20] Aquinas worked within a framework whereby only beings with souls mattered intrinsically, and nonhuman animals didn't have souls. Hence nothing was wrong *per se* with harming or being cruel to animals. But the person who did so might be made more likely to mistreat other human beings.

Now, we might not think Aquinas' use of this argument adequate to ground the degree of obligations to sentient animals that we have (or intuitively think we have). Nevertheless, this is a legitimate *kind* of argument. It is, again, a consequentialist one, but one in terms of subtle consequences. So, we have no trouble understanding its moral force—there is no worry that it is "merely" a psychological feeling, a "carry over" that counts as a confusion of reasoning. This would also explain why it is hard to pinpoint what is wrong with the Egg Machine practice.

But is there any reason to believe that its factual premise will turn out to be true? That those engaged in this activity will come to act toward animals and persons in new and detrimental ways? This is another very complicated empirical matter that forces again the questions of burden of proof. But there also are some more specific reasons for skepticism, reasons for thinking that this sort of argument works *less* well in the Egg Machine case than in Aquinas' case concerning the treatment of animals.

First, note a certain lack of precision in the story:

The question arises not only of how this would take place but also who is going to be affected. Just who is it that is "engaged in this activity"? The scientists involved in the research and development? Those (like the Bird tender) who "interact" with the Birds daily? Unless the effect is rather severe, it might be pointed out that not many people really are affected. Is the idea in fact that *all* of us would be affected for taking part as consumers? How is this supposed to take place, especially given our rather distant relationship to the sources of our food? Second, note that in the Aquinas case, it was a matter of being *cruel* to the animals; the present case would seem to be a much more subtle action and hence less likely to "spill over" in the way suggested.

Again, we would need to ask why safeguards or countermeasures could not be effective in counteracting these effects.

EFFECT ON WORLD VIEW

Emily recalls something she has seen recently on TV announcing that "Biotechnology will transform our very lives! We will live in a different world." She remembers it having really cool graphics and a lot of emotional appeal. She is thinking big.

"Perhaps we should look more broadly," she says. "Perhaps what strikes us about the Egg Machine case is that its general world view is corrupt, especially in its general attitude toward nature."

Emily's idea here is that the creation of Egg Machines reflects or arises from a certain mind set, a mind set about how we see ourselves in relation to nature. We might note that "intrinsic" considerations are sometimes understood in terms of *symbolic* significance.[21] Such concerns may be hard to pinpoint, yet they remain of great potential import.

There are a number of different things that can count as a "world view" or that may be cited in this kind of suggestion. For example, the stewardship idea discussed previously can be seen in this way. But so can the following: having respect for nature; the tendency to believe that certain things (such as life, living things, ecosystems) have "intrinsic value"; what counts as *natural*; or on the other hand, a reductionistic or mechanical view of the world, perhaps even just the general tendency to use high-tech solutions to problems.

Perhaps the main point—what makes these "world views"—is that they are not *principles* that say "Always carry out actions of this sort" or "Never do X." Rather, we are talking about a general way of looking at the world, a view about the relationship of humankind to nature. So it's much more general than a particular principle.

REDUCTIONISM

One sort of world view often warned against is that of reductionism.[22] Like the other *isms*, it is not easy to say just what this is, but one might start by noting that it includes understanding everything in terms of its physical chemical parts and paying little attention to the organic wholes, or higher levels of organization, such as species, ecosystems, or even organisms.

So, the claim might go, genetic engineering in general—and the Egg Machine practice in particular—fits with the reductionist view in some way, and this is to be avoided.

But in order to use this as a tool for evaluating our present issue, we would have to spend a good deal of time teasing apart a number of issues. First, reductionism can mean several different things, from the metaphysical view that all that really exists in the world are certain kinds of entities (describable in terms of our theories of physics and chemistry) to the methodological view that we should try to do science in terms of such entities, to views (usually attributed by others) that don't fully respect life.

Unfortunately, these are often thrown together without clarification of which is central and how they are related.

Second, we would need to specify what we mean by some practice "fitting with" reductionism. Is this going to be understood as "is consistent with," "arises out of or is an expression of," or "will (or could) lead to"? Further, the last of these could be anything from "will increase the likelihood of" to "will inevitably lead to." We need to be clear about exactly what claim is being made, keeping in mind that a claim that is plausible may not be the claim that is morally significant.

Then we need to clarify what is *wrong* with those things. One difficulty is illustrated by the following: One sort of reductionistic view would involve not paying sufficient attention to the ecological relations between different variables in the course of such things as environmental safety analysis. Presumably we can understand why this would be a bad thing. But this is very different from being a reductionist in the sense of limiting one's attention to the problems' solutions that focus on changing one or a few genes in a given crop. For one could do this and yet be ecologically sophisticated in one's handling of environmental impact assessment. One might expect "genetic engineers" to be reductionists in the one sense, but they wouldn't need to be in the other.

"A Vast, Organic LEGO Kit"

Considering another world-view idea might be more fruitful. Mark Sagoff[23] utilizes a contrast between positions of Barry Commoner and Hans Bethe to describe a difference between two ways of viewing the natural world, ones that may be at play in different views of biotechnology and that might underlie why different people have different reactions to the Egg Machine case. The contrast is that between viewing the natural world as *raw materials* (to be used for manufacture) or viewing it as *natural resources* (to be managed or conserved). The natural resources side is described as asserting that "the history of evolution has made natural species and ecosystems what they are and has given them forms or essences we ought to respect. We engage in manipulation at our own peril." The other side tends to view "nature as a collection of materials that humankind manipulates to serve its interests and ends."

There is some of the language here from our earlier discussions, with a number of components combined. We might therefore worry that this analysis may end up being unclear, metaphorical, unable to give helpful guidance, and difficult to evaluate. Also, like any world-view analysis, it will have some vagueness due to the fact that we are attributing a certain mind set or general outlook, and not merely a set of specific beliefs or rules, to the person. It is crucial that we bear in mind that this is speculative.

Still, perhaps it will be useful to say that the Egg Machines arise from (and therefore indicate) a tendency to treat the entire natural world as raw material for commercial manufacture, things to be owned and built with. Sagoff cites Yoxen in putting this idea as viewing nature as "a vast, organic LEGO kit inviting combination, hybridization, and continual rebuilding."[24]

This description is quite suggestive. Note, for one thing, that it might make sense of the connection between our problem and the environmentalist's conundrum, yet it will avoid the difficulty we saw in the preceding discussion concerning preservation and creation of species.

What else can we say about exactly what it is to view the natural world as raw material for manufacture? Well, it's always to be at the ready to make things out of bits of living nature, or to view nature always and only in terms of *commercial* potential. It's to view this in the same way that timber, coal, and copper are viewed. Insofar as it's a matter of having in mind constructing things out of bits of nature, it might be said to be reductionist in spirit. We might say that the Raw Materials view also indicates a tendency to choose to modify the organism rather than the environment; we would change the organism so that it can deal with increased pollution rather than clean up the pollution. Relatedly, when confronted with animal suffering in agricultural contexts, we would change the animals so that they don't feel the suffering rather than change the conditions so that the animals aren't under stress.[25]

Note the following important advantage of this way of understanding what underlies our moral intuitions here. Recall that when we considered such things as "treating animals as factories" as moral *principles*, they seemed unacceptable on grounds that there were cases of such treatment of animals that could not plausibly be ruled out completely as always morally wrong. Are we in any better shape here? Maybe. Consider this: Perhaps the moral principle should not have been seen as "Never treat an animal as a factory" (analogous to Kant's dictum that we never treat persons as means only, and not at the same time ends in themselves), or "Never treat nature as raw material for manufacture." After all, how could we possibly follow these principles?

Rather, we should imagine dicta such as "Don't *always* treat nature as raw material for manufacture" or "Be careful about the extent to which you treat nature as raw material." It is at most a requirement that we *tend* not to do it—that we put limits on the degree to which we do this. It is worth emphasizing that these things *come in degrees*. It's acceptable to treat animals as machines to an extent, but not to *this* extent.

If we understand the Raw Materials view in this way, then the fact that certain perfectly legitimate activities are cases of acting on a bit of nature as if it is "raw material for manufacture" does not require us to reject the legitimacy of the rationale. Of course, we also cannot use this rationale to say that the Egg Machine practice is necessarily wrong. Our treating animals as machines or as raw material in this case (and to this degree) does not entail that we always do so. The moral assessment depends on what else we do, what we do in other cases, how far we push this way of looking at things. So if we were looking for a way to say that the Egg Machine case was simply morally wrong, that it ought to be forbidden straight out, then this will not do it. But this is probably as it should be.

The idea here is not so much that the Egg Machine practice will *cause* certain bad things, even long term, or even as part of a slippery slope argument. Rather, it's seen as a *symptom* of the Raw Materials view and *triggers* our reflection on the world view. The scenario pushes in our face an idea of how we are treating these other creatures as machines rather than as fellow residents sharing the planet, which is cause for moral reflection: Should we take this attitude—to this degree? Such reflection may show us an image of ourselves that we don't like, or that, if taken as a deep and extensive foundation of how we view the world, would be of some concern.

This discussion began with one particular application of ag biotech (the Egg Machines) and broadened at various points to biotechnology generally. What have we learned? One kind of learning involves getting clear about what questions to ask. Perhaps the most useful question is not "Should we engage in agricultural biotechnology or not?" but rather "Are there *ways* of doing biotechnology (what degree? which products or kinds of projects? under whose control?) that fit more than others with appropriate world views—for example, the Natural Resource rather than the Raw Material view? In other words, supposing we think there is something to this idea of the superiority of the Natural Resources world view, are there ways to use this to shape the biotechnological revolution?

PATENTING, OWNERSHIP, AND DEVELOPING NATIONS

"Okay, my head hurts, but there's another thing I've been wondering about all this time," says Emily finally. "The idea that somebody *patented* these things. I guess it's related to the idea that they somehow *own* them."

"Wait," Doug interjects, "people have been owning animals for a long time."

"Sure. But they own individual animals: this brown cow or that white pig. No, what I mean is that now the courts have said that companies can own the very *type* of thing, all the animals of a certain kind—they own the idea of it. They have a patent on this form of life. So I'm thinking that there's another issue here on top of the matter of our creating and modifying living things in this way. And, now that I think about it, it seems as though this might relate to the question we've been discussing of the attitude we take toward the living world."

Of course, there is nothing "extreme" or "futuristic" about this aspect of the case; patenting is a normal practice with genetically altered animals and crops. But this practice (and the attendant issues of ownership and commercialization) is disturbing to some and can be seen to raise various moral questions. In the remainder of this chapter, I explore briefly a few of these.

Patenting is a mechanism that protects intellectual property. It allows an inventor to guard against others freely using his or her work; patents give one the right to license to others for a fee. This encourages useful innovation as well as investment in the industry.[26]

The system of patenting that we have is in fact has its source in several different things: Supreme Court decisions, legislation, analogous practices in other nations, and agreements between nations. This system of patenting might be seen to have various moral rationales. Primary is the utilitarian one that the encouragement of innovation brings useful products to society and promotes economic activity. But it also might simply be judged to be *fair*, because those who have put labor into an invention *deserve* to be able to profit from it.

The rules for what counts as patentable include that it be novel, nonobvious, and useful.[27]

But up until 1980 it was unclear that living things would be considered patentable. In that year, the U.S. Supreme Court determined that a microorganism, an *E. coli* bacterium modified so as to be able to degrade oil, could be patented. Patents have since been given for multicellular organisms; for both plants and animals, including mammals; and for genes.

Emily brought up the issue in terms of owning types of animals, but concerns about patenting range over a number of questions, from whether there is something worrisome or inappropriate about any patenting concerning life, to what *kinds* of things ought to be allowed to be patented (an organism or only a process of manufacture? a mammal? a human gene?), to what sorts of impacts this will have on commercialization or the direction of research.

Note that these various questions about patenting can be understood at different levels: they can be evaluated in terms of what the law says (as in Supreme Court decisions), but they can also be evaluated in moral terms. The moral considerations raised are often usefully categorized along lines we have already discussed, some fitting the consequentialist mode, but others involving more intrinsic considerations.[28]

For instance, on the one hand, it's said that patenting in the realm of plants and animals will lead to the same in the human area—a sort of slippery slope argument. On the other hand, it's argued that patenting seems to involve treating the item patented as a mechanical object and an item of manufacture, for these are the sorts of things that have traditionally been patented; we are assimilating living things to that model. It may be threatening or offensive to some to act in ways that ignore the difference between living beings and mechanical objects or machines. It also may be thought unwise to treat animals in this way. We've come back to the idea of treating animals as machines, though it's in a somewhat different way.

Sometimes the concern is not so much patenting *per se*, but commercialization, with which patenting may be seen as intertwined and may be said to encourage. Perhaps Emily's concerns are tied to this. Some argue that there is something intrinsically wrong with commercialization of life or living beings (especially as that life is closer to humans). Of course, it may be hard to square this with the fact of commercial agricultural practices that have gone on for a long time before biotechnology.

But concerns about commercialization more often have to do with the distorting effects of the profit motive. These can concern how economic forces affect, in disturbing ways, *which products* get created; those products that promise the greatest profit won't necessarily promise the greatest social benefit. But they can also concern the impact on basic science, which might suffer if people are drawn instead to commercial endeavors. This broadens the moral questions to be asked to the following: Who should own the products of genetic engineering? Who should have a say in the direction of research? Might it be unfair and unwise to allow some people to own and have control over the biological or genetic resources of the world?

In this context, it will be useful to examine an important debate concerning genetic intellectual property—in particular, the debate over ownership of germplasm (the world's plant genetic resources that form the basis for the crops grown all over the world). This discussion illustrates certain consequentialist sorts of concerns, but there are also some other issues that don't reduce to these. But these are not really the "in principle" types of concerns that I dealt with earlier. Rather, they are concerns involving distributive justice and other rights claims, as well as questions about the nature of property.

Such issues as the ownership of parts of nature might suggest questions about our relationship to nature, thus raising certain "world view" issues discussed earlier. Of course, viewing *plants* as mechanical objects or raw materials will have less emotional valence than this same thing concerning *animals*. What is most significant about the issue, however, is the way it connects the issues of ownership to a conflict of interests between the developed and developing nations and raises questions of global justice. At stake are both who should own the products of genetic engineering and who should own the world's germplasm, which constitutes the raw materials for the former. This question also allows us to address Krista's claim that her high-tech solution is a solution to world hunger.[29]

The basic dilemmas over patenting and the world's plant genetic resources arise as follows[30]: Traditionally, seeds found in the "centers of diversity," mostly in the developing nations, have been freely collected by scientists from developed nations. These seeds have then been used to develop sophisticated, high-yielding cultivars (in part by use of genetic engineering). These products, viewed as the property of seed companies, have been available to others (including those in the developing nations from which the initial seeds were taken) only for commercial purchase, allowing the seed companies to make substantial profits. The seed companies hold that they should be able to patent and profit from the products that they develop in this way, but that the original germplasm should be viewed instead as "common heritage." Critics of this position can take one of two stances. They can say that *both* sorts of plant genetic resources—original germplasm from developing nations and the cultivars developed from them—should be viewed as "common heritage." Or else they can say that the original germplasm should be viewed as the property of the developing nations, so those who utilize it in any form should have to pay a fee for it.

You might think it unfair to have it "both ways." But there is a rationale for the seed companies' position, that the original germplasm from developing nations should be viewed as no one's property in particular, whereas the lines developed by entrepreneurs should in fact be viewed as their private property. First, it will be said that viewing the initial germplasm as the property of developing nations is completely impractical. The usefulness of germplasm is not clear until some time down the road, and at that point a given cultivar may have its source in several nations. There is no effective mechanism for *pricing* the raw genetic material; at least, the market doesn't generate that price for us.

Second, some further points about the nature of the basis of ownership and patenting can be brought to bear: The germplasm only becomes *valuable* after time and money have been invested in its improvement and it is incorporated into a commercial cultivar. So, no real rationale exists for viewing the raw germplasm as patentable.

One response to this is to argue that the germplasm is not actually "raw" but is instead a product of accumulated labor of indigenous farmers over generations, and it is unfair that this goes uncompensated. Besides, raw resources aren't typically viewed as common heritage (consider oil and coal). In response to this, it will be said that when one collects seed, the resource isn't "taken away" in the same way as these other resources; the first country still has the resource. Hence it doesn't deprive them of anything, or leave them worse off.

Still, there is a further response: Those in the developing nations do indeed lose something—namely, earning power; thus they are not as well off. After all, in what other way is an oil-rich nation made worse off if, say, one tenth of its oil resources are taken away? This debate is not easily resolved. These questions about the kind of thing that can be owned as intellectual property are complicated.

Finally, one can also argue for the seed company position via a consequentialist argument: It's important to encourage the development of new products, which requires that patents be given to those companies that develop the elite lines of seed. Indeed, the world's people—including, it might be argued, those in the developing world—are better off having the development of all these biotechnological products. But this consequentialist rationale doesn't hold for the developing nations case. Even keeping in mind the labor that went into the development of the germplasm over generations, the motivation that the lure of patent protection provides was not needed in order to get this work done. Hence, the argument goes, there really is an asymmetry.

I cannot resolve this debate here. But I want to step back and ask a broader question. We have some different possible starting points for moral arguments here. On the one hand are some general arguments about what constitutes property. On the other are some specific moral intuitions about the plight of the poor. Which of these is more important, or a more appropriate starting point? Should we start from a view of what constitutes property (and what the rules and rationale for patenting have been) and then let the distribution of welfare fall where it may? Or should we start with the facts about global inequality, or about the uncompensated work that went into the gradual development of the "raw" germplasm in developing nations? Some might argue that it is important not to violate reasonable or established rules concerning ownership and patenting. But others might argue that we should start with some intuitive moral judgments about the justice of the situation and then mold our rules or conventions about ownership and patenting to bring about a more fair distribution of welfare.

Someone who chose the latter route might say this: Developed nations and their seed companies ought to pay such and such a sum to the developing nations from which they obtain raw germplasm, but not because that follows from our present rules of ownership and patenting—it may not. Rather, this should be done because it will lead to a more just distribution of food and welfare. Indeed, some might challenge the view that patentable property is the right way to conceptualize these materials and resources.

Of course, on the other side it will be said that we cannot be all that confident about (or may not agree about) our intuitions about what counts as a fair distribution of welfare in the world. Better to follow rules of property seen independently as reasonable.

All this requires that we be able to reason about what counts as a fair or just distribution. So I return briefly to some general positions about this question. Recall that on a utilitarian theory of justice, we should do whatever maximizes the welfare of the whole community; say, for example, of the human community. The distribution *per se* does not matter. If an unequal (even very unequal) distribution maximizes overall utility, then so be it. On the other hand, there might well be utilitarian reasons for opting for a more equal distribution. First, because there are *so many* people in the developing nations; their welfare has a great impact on the calculation of *overall* welfare. Second, it can be argued on utilitarian grounds that we should focus on the welfare of the less well-off people due to diminishing marginal utilities of an increase in welfare to those already well-off. It might be argued further that great inequality is to be avoided on the grounds that it threatens global stability.

Other theories of justice require more directly that we close the gap between rich and poor as much as we can. As stated earlier, this could be accomplished either by having a principle of equality, or by saying, as Rawls' social contract theory does, that the most just society is the one that treats its least well-off members as well as possible. Keep in mind, however, that the world is not a single society, thus it is less clear how all these theories of social justice should apply to justice between nations.

As Paul Thompson's chapter indicates, a libertarian view of justice does not stress welfare; it stresses freedom and rights, including property rights. It urges us to consider only questions of procedure, not to look directly at what the pattern of distribution is. We should apply appropriate views of what counts as legitimate acquisition of property and "legitimate transactions," such as voluntary market transactions. One would probably expect this to be used to argue for positions favoring the seed companies of the developed world. But one could also argue for a different position, taking more seriously the above point that the original germplasm is actually the result of the accumulated labor of indigenous peoples. Those in the developing world are responsible for a large part of the value of many of the products, which have been taken from them unfairly. So they should be compensated for it.

Of course, to fully address these issues, we would also need to resolve a number of factual issues, such as what economic impacts various policies would have. Further, it should be pointed out that there are in fact a number of policy options besides the two basic opposing positions outlined here. For instance, other ways can be created for compensating the developing nations. Thus a full analysis of what to do would have to take account of this wider range of options, as well as geopolitical forces of a variety of sorts. But my goal here is simply to show that we need to make a judgment about the appropriate conception of distributive justice. It is also to show how conceptual issues concerning the nature of property and the role of patenting become part of the debate.

This brief discussion of the "germplasm problem" indicates another way in which biotechnological innovations have profound effects on our world—ones that force us to ask questions about whether biotechnology will benefit members of developing nations and what our obligations to them are. It also provides a different sort of challenge to the view that we should be utilitarians and simply add up the consequences. This challenge does not pose questions about the relevance of various "in principle" considerations, such as naturalness, but rather about rights and distributive justice. It looks from another angle at the issues concerning our general view of the world and our relation to it, for it concerns ownership or property relations applied to the living world.

CONCLUSION

Emily might take a lesson from the first part of our discussion to be that we have to broaden our ways of thinking about the ethical evaluation of a technology. She might, for example, say that some moral objections to biotech need to be seen as intrinsic ones, ones that go beyond concerns about consequences. She may continue to hold that even if all of the consequences of the Egg Machine system of agriculture are positive, there might still be something intrinsically wrong with it. Others might disagree with her, holding that such a view is based *merely* on emotional reaction, indicating a soft, "unscientific" way of thinking. These people will call for us to discount our initial reactions and take a more hard-nosed consequentialist approach. On the other hand, our last topic, concerning germplasm, requires Emily to think about yet a different set of issues, from property to global justice. Overall, she sees that the Egg Machine thought experiment not only jars her mind into thinking hard about what role emotions and intuitions play in one's reasoning and whether morality is solely a matter of consequences, but also how there are connections to such things as people's relationship to the natural world and the developed nations' relationship to the developing nations.

NOTES

1. Service, R., "Seed-Sterilizing 'Terminator Technology' Saw Discord," *Science* 282, 30 October, 1998, 850–851.
2. Shulman, "Bovine Growth Hormone: Who Wins? Who Loses? What's at Stake?" in *Agricultural Bioethics: Implications of Agricultural Biotechnology*, Gendel, S., et al., eds. (Ames, Iowa: Iowa State University Press, 1989), 111–129; and Comstock, G., "The Case Against bGH," in *Agricultural Bioethics: Implications of Agricultural Biotechnology*, Gendel, S., et al., eds. (Ames, Iowa: Iowa State University Press, 1989), 309–339, reprinted in Comstock, *Vexing Nature? On the Ethical Case Against Agricultural Biotechnology* (Boston: Kluwer, 2001), 13–33.
3. Reiss, R and Straughan, M., *Improving Nature? The Science and Ethics of Genetic Engineering* (Cambridge: Cambridge University Press, 1996), 139–144.
4. Krimsky, S. and Wrubel, R., "Engineering Crops for Herbicide Resistance," *GeneWatch* 11, nos. 1–2 (April 1998), 7–9.
5. Reiss, R and Straughan, M., 142, 147.
6. Utilitarianism gets defined in different ways, sometimes including a particular view of what good is to be maximized. The main point for us in this chapter concerns consequentialism, the claim that right and wrong actions are to be defined in terms of the consequences of those actions.
7. "NIH Technology Assessment Conference Statement on Bovine Somatotropin," *Journal of the American Medical Association* 265, n. 11 (Mar. 20, 1991), 1423–5.
8. Rawls, J., *A Theory of Justice* (Cambridge: Harvard University Press, 1971).
9. Kline, D., "Introduction: Agricultural Bioethics and the Control of Science," in Gendel, Steven et al., eds., *Agricultural Bioethics: Implications of Agricultural Biotechnology* (Ames, Iowa: Iowa State University Press, 1989), xi–xxi.
10. Hoban, T.J. and Kendall, P.A., *Consumer Attitudes About the Use of Biotechnology in Agriculture and Food Production*, Report to Extension Service, USDA, 1992. Lee, T.R.

et al., *Consumer Attitudes towards Technological Innovations in Food Processing* (Guilford, UK: University of Surrey, 1985).

11. Reiss, R and Straughan, M., chapter 3.

12. Rollin, B., *The Frankenstein Syndrome: Ethical and Social Issues in the Genetic Engineering of Animals*, 1995, chapter 3. Fox, Michael, "Transgenic Animals: Ethical and Animal Welfare Concerns," in Wheale and McNally, eds., *The Bio-Revolution: Cornucopia or Pandora's Box?* (London: Pluto Press, 1990).

13. Leopold, *A Sand County Almanac* (New York: Oxford University Press, 1949).

14. Hoban, T.J. and Kendall, P.A., *Consumer Attitudes About the Use of Biotechnology in Agriculture and Food Production*, Report to Extension Service, USDA, 1992.

15. Rollin, B., *The Frankenstein Syndrome: Ethical and Social Issues in the Genetic Engineering of Animals* (1995), 24. (Dewey, J., *A Common Faith* [New Haven: Yale University Press, 1934].)

16. Sinsheimer, R, "Two Lectures on Recombinant DNA Research," in Stich, *The Recombinant DNA Debate* (Englewood Cliffs, NJ: Prentice Hall, 1979), 85–98.

17. Reiss and Straughan, 151.

18. It is sometimes said that all there is to morality is expression of emotion. This is not at all what is being suggested here. (Cf. Rachels, J., *Elements of Ethics*, 2nd edition [New York: McGraw-Hill, 1993], chapter 3 on subjectivism and emotivism.)

19. Others might use the slippery slope argument to argue for stopping much sooner—on the grounds that we might end up making Egg Machines! Would *this* be reasonable?

20. Rachels, J., *Created From Animals: The Moral Implications of Darwinism* (Oxford: Oxford University Press, 1990), 209.

21. Wachbroit , R., "Eight Worries about Patenting Animals," in *Values and Public Policy*, Claudia Mills, ed. (New York: Harcourt Brace Javonovich, 1992), 66.

22. Rifkin, J., *Algeny* (New York: Penguin Books), 1983.

23. Sagoff, M., "The Biotechnology Controversy," in *Values and Public Policy*, Claudia Mills, ed. (New York: Harcourt Brace Jovanovich, 1992), 43–47.

24. *Ibid*, 45.

25. Rollin, B., "Genetic Engineering of Animals for Confinement Agriculture," *Ag-Bioethics Forum*, v. 8, n. 1, June 1996.

26. Krimsky, S., *Biotechnics and Society: The Rise of Industrial Genetics* (New York: Praeger, 1991).

27. Office of Technology Assessment, *New Developments in Biotechnology: Patenting Life* (Washington, D.C.: U.S. Government Printing Office, 1989), 37–40.

28. Wachbroit , R., "Eight Worries about Patenting Animals," in *Values and Public Policy*, Claudia Mills, ed. (New York: Harcourt Brace Javonovich, 1992), 66–71.

29. I discuss this in relation to plant genetic resources from developing nations, but analogous things could be said concerning animals such as chickens or any of the animals of which the Egg Machines are "descendants."

30. Kloppenberg and Kleinman, "The Plant Germplasm Controversy," *Bioscience*, vol. 37, no. 3, 190–198, 1987. Juma, C., *The Gene Hunters: Biotechnology and the Scramble for Seeds* (Princeton, NJ: Princeton University Press, 1989).

10

FARMS

CHARLES TALIAFERRO

CASE STUDY: ROY THE DAIRY FARMER

The semester has drawn to a close, and Emily is reminiscing with Doug about the class.

"Remember how this all started? How I almost got into trouble cheating on Wright's stupid quiz the first day?" she laughs.

Doug grins at her. Then his face turns sober. He picks up his cup of coffee.

"I need to tell you something."

"What?"

"My mom and dad are thinking about selling the farm."

"Why?"

"Well, you've seen the place. It's an open-lot dairy with just over two hundred cows. The farm is about five hundred acres total with most of the acreage, four hundred acres more or less, in hay and grain. We use the forage to feed the cows. The cows are fenced in a ten-acre area. My folk's house, the barn, and various buildings and holding pens for calves take up roughly another eight acres, and there is a two-acre lagoon in which run-off wastewater from the animals is stored. It's a small place. My folks think they can't continue to make it here unless we dramatically change what we're doing."

Doug speaks almost reverently about the place. Then he looks at his watch and bends over to grab his backpack.

"Sheesh, I got class," he says. "Why don't you come back to the farm this weekend?" he yells over his shoulder.

The next Saturday, Emily helps Doug with milking in the morning, and they spend the afternoon talking with Doug's father, Roy.

Roy explains the problem to Emily. "You know that Doug has an older sister, Karen. Both she and Doug would like to become partners in the farm. To bring in two more families successfully, however, would mean that the farm would have to triple its profits. I think we can do this only by tripling the size of the operation. The dairy farm, that is, must either become exclusively a business proposition, competing on the basis of high-volume production of a low-cost product. Or . . ."

Roy does not finish his sentence, but Emily urges him on.

"Or?" asks Emily.

"Or, we must sell."

"The problem with expanding," Roy continues, "is that milking six hundred or more cows would place our farm in a new class. State environmental regulations dictate that a dairy of that size have many more acres over which to spread manure and wastewater. Consequently, we would have to take on sizable debt loads in order to buy not only cows but also land."

"Are there no other options?" asks Emily.

"Well, actually, there is one. It's the one Karen and her husband prefer, actually. There are two small abandoned farm homes nearby, one just across the road and the other about a half-mile away. We could buy those houses, one for Doug and one for Karen, and then split the current herd into three small herds of sixty-five cows each. I'd keep one herd, Doug would have one, and so would Karen. Each of us would place our cows in a rotational grazing system on land I now have in grain production. Rotational grazing of cows is a more labor-intensive form of farming that does not rely on heavy machinery and purchased inputs to move feed and manure; it relies on the cows to find their own feed in the pasture and to deposit their own manure there. The problem is that the profit margins to be expected from the alternative system would not be as high as the expected profits from the traditional purchased inputs system."

"Am I missing something?" Emily asks. "Why would anyone want to adopt a style of farming that would not only be less profitable but also seems likely to involve a lot more work?"

Roy turns to Doug. "Do you want to explain?"

Doug tells Emily that Karen and her husband are financial managers at a major corporation in an urban area about a hundred miles away. They are tired of commuting to work, tired of seeing little of each other, and tired of not having time to spend with their three young children. They have also been reading books about farm life, some of them written by Wendell Berry, a poet and farmer from Kentucky who praises the virtues of rural life and family farms. They are well educated and well informed and, after years of discussion and study, they have decided they would like to invest their considerable savings into an alternative system because they see it as a superior lifestyle option from several points of view.

"They often say it's a 'morally superior' way of life," winks Roy. "I thought you'd enjoy that remark, Emily, because you're taking that Ag Ethics course."

"Yes. But what do they mean by it?"

"Karen tells us that she and her husband think family farms are morally superior in four ways: they allow parents to spend more time with their children because everyone is working together on the farm; they promote closer relationships with neighbors, who for city dwellers are often unknown; they allow people to spend more time out of doors and being around animals; and they promote sustainable use of land, something that is good both for the environment and for future generations of humans."

"Yeah," chimes in Doug, "Karen's husband says three little farms supporting three families are three times better than one big farm supporting one family."

"Well," concludes Emily, "after thinking about agricultural ethics all term, I certainly resonate with that sentiment. But I also see the difficulties you face."

Roy is looking out the window. Doug is looking at his shoes.

Emily doesn't know what to say. At last she whispers, "I'm just glad I don't have to make the decision."

See the appendix to perform Exercise 10.A; then return here to continue reading this chapter.

DISCUSSION OF ISSUES

At first, Roy's decision may seem an entirely personal or private matter; it is a decision he is free to make in whichever direction he likes. If he decides to try to foster a small- or medium-sized family farm, well and good, but he should not be required to do so by us or by his family. In ancient Greek ethics, a decision that is not morally binding would be called *adiaphora* (from the Greek meaning *indifferent*). Today we may describe the adiaphora as the morally permissible. Even if it is granted that family farming has some moral superiority to other ways of living, such farming may still not be morally obligatory. Arguably, some actions are morally good—acts of charity or courage, for example—but are not morally required. An act of charity is charity in part because it is *above and beyond the call of duty*; those who courageously save the innocent when it is not their duty to do so (for example, when the rescue operation imperils their lives) are rightly considered to be moral heroes precisely because they do something not required of them. So, initially one may well conclude that family farming between consenting adults is not morally forbidden, and if farm families think they are living morally superior lives, let them. Although ethical problems arise in family farming (such as environmental degradation and contamination, the health of farm animals, world hunger and food policy), these are problems faced in agriculture at large and do not by themselves make it obvious that "family farming" is a distinct category of ethical concern.

But perhaps the decision is not such a private matter after all. The status of family farming may command the attention of the public and a case might be made that Roy *should* keep his farm small. What would this case look like?

The family farm has had a central place in North American social, economic, religious, philosophical, and political history. It has been a vital part of American heritage and, at present, is endangered. Family farms have decreased radically, and we may witness the virtual collapse of this way of farming. The decline has not always been because of voluntary migration to cities but rather because of bankruptcy. (Here we might return to the story and ponder whether the unoccupied farms that Roy's family might purchase are on the market because they have been abandoned because of bank foreclosure.) From 1954 to 1992 the number of farms in the United States declined by sixty percent. Now the farm population is less than two percent, whereas in 1840 it was forty percent (Bureau of the Census, 1992, 1994). The attrition has also hit African-American farmers especially hard; thirteen percent of the farm population was African-American in 1990 whereas it is now only one percent (Demissie, 1990). Marty Strange is right in his observation that "Hardly anyone in the United States knows a real farmer" (Strange, 1988, 15). So, we do well to reflect on family farming insofar as this is essential to reflection on our own identity as a nation and culture, and also so that we may discover whether we have any responsibility to preserve this aspect of our identity. The obligation to assess the nature and value of our national identity applies to us as citizens in a democracy, but it may also press home to us in certain specific ways. For example, if you are a practicing life scientist, then you have some responsibility to reflect on what parts of society are benefited or disadvantaged by your research. Is your biotechnology serving industrial, corporate agriculture over a sustainable, alternative agriculture, the family farm? If so, are you contributing to the loss of something of great value?

Another rationale for public debate emerges near the end of the conversation, when Roy reports on Karen's view, that small farms highlight a range of values that call for public exploration. Claims about several goods are at play, goods that involve family, neighborliness, health, and sustainability. These represent important civic goods and virtues. Wendell Berry, perhaps the leading advocate today of family farming, considers family farming an art that enshrines and fosters a deeply humane way of living that is threatened by industrial agriculture. Family farming offers an integrated way of life that actually prohibits or at least discourages the commonplace, market-driven division between the producer of a product and the product. In the end, claims Berry, it is the integration of a farmer's livelihood and life that both demands that the farmer be a craftsperson or artist in his or her trade and also secures a commitment to the quality of the food produced by farms:

> I am more and more convinced that the only guarantee of quality in practice lies in the subsistence principle—that is, in the use of the product by the producer—a principle depreciated virtually out of existence by industrial agriculture. . . . People who use their own products will be as concerned for quality as for quantity, whereas people who produce exclusively for the market will be mainly interested in quantity. . . . Industrial agriculture has tended to look on the farmer as a 'worker'. . . . We have neglected the truth that a *good* farmer is a craftsman of the highest order, a kind of artist. It is the good farmer—nothing else—that assures a sufficiency of food over the long term. (Berry, 1991, 85)

If Berry is right, then a rich, neighbor-centered agrarianism can highlight values that deserve the honor and allegiance of society at large. And if these values in quality and skill are publicly evident, then perhaps family farming merits the protection and assistance of society. Given some credibility to Berry's thesis, then, at a minimum there may well be grounds for a social commitment not to deploy protection and assistance to large-scale, industrial agriculture when this unfairly places family farming at a disadvantage. In the end, even if we conclude that such a public stand is not ethically required or politically wise and that a family's decision to foster and protect family farming is entirely private, it is still desirable to encourage public deliberation on the values involved, if only to enable persons to make intelligent, informed decisions about their own vocations. The decision to participate in the practice of family farming can be made more responsibly to the extent that rural *and* urban education makes available the facts about what Berry calls a craft "of the highest order."

Just as there are what may be called public reasons for society at large to engage in the family farm debate, there are reasons why family farmers—or those aspiring to become family farmers—should consider the public dimension of this practice. Some reason exists to think that a decision to family farm is not an entirely private one from a conceptual point of view. The concept of "family" is a public, socially informed one. Of course, the term "family" is used to describe nonhuman, biological life, but the concept of "family" in human social contexts has a profoundly public, political meaning. (The concept of "family" has a role in other areas of bioethics in accounts of medical health and responsibility.) Moreover, the very rationale Roy offers involves "neighborliness," which further secures the public dimension of any decision to family farm. The very concept of *owning a farm* involves a nest of public relations. According to one popular theory of property, ownership is a social and political construction, secured by social contracts and backed by credible enforcement. So, even framing the question to oneself about whether to engage in family farming involves public concepts and relations. Furthermore, if one can successfully make the claim that family farming is indeed a vital part of American heritage or that it offers American culture moral insight, then family farming may constitute an important influence on farm policy, agricultural research, and development.

PRACTICES AND POLICIES

The values that are cited by Roy in the story can be split up and examined in isolation, but in doing so, we risk missing the overall picture of what is at stake. Broken up, the case for family farming is tenuous. Take the components separately and they do not immediately seem very persuasive. For example, you do not need a farm to have a rewarding family life. There are many ways in which families can enhance their time together. Arguably, large-scale, corporate agriculture may yield conditions in which family values are cherished. Neighborliness can be achieved in a variety of ways and it is not obvious why owner-operated farming is the optimal route to securing it. Exercising outdoors can be accomplished through any number of routes, and sustainable agriculture can be carried out on large, corporate farms. If soil erosion is what bothers us, won't it be more effective to establish and verify good soil stewardship with several corporations, rather than rely on the more scattered, decentralized framework of many family farms?

But this breakdown of the rationale for family farming taken from the case study does not capture the allure of family farming or what its advocates argue is its overriding value. According to Wendell Berry and other advocates of family farming, it is a practice or a way of life. The goods of family, neighborliness, health, and sustainability are folded into a way of life. They are not simply patched together like a quilt, nor does the family farming way of life represent a kind of container in which these goods are placed in a haphazard, sentimentalized fashion. Rather, these goods are complementary, internally related, and mutually supportive. Ideally, the goods are constituents in a greater whole that, according to some of its defenders, involves virtue and human flourishing.

Like Berry, Marty Strange describes family farming as a way of life in which various values are at work, quite distinct from market-driven employment.

> Above all, family farming carries with it a commitment to certain values, entirely independent of the pettiness of economics. The agrarian tradition, of which family farming is a part, calls for people to be neighborly, to care for future generations, to work hard and to believe in the dignity of work, to be frugal, modest, honest, and responsible for and to the community. Family farming may be a business, but it is not just a business. It

is a way of life as well. The farms in a family farming system operate in a social milieu which constrains the business behavior of farmers. Perhaps the best test of whether a farm is a family farm is this: Does the farmer feel more pain at the loss of a neighbor than joy at the opportunity to acquire that neighbor's land? (Strange, 1988, 35)

If Strange is right, then a moral and psychological component exists at the core of family farming.

I use the term *practice* as well as *way of life* to describe family farming. It may seem odd to think of it as a practice. Certainly it is not a particular practice, narrowly defined as a way of harvesting, but the term *practice* has often been employed by philosophers to refer to comprehensive and sometimes highly complex, principled ways of living. The chief end is to delimit such ways of living from highly abstract, purely theoretical projects. I am also using the term *practice* to explicitly juxtapose the way of life in family farming with policy making. Family farming, as a practice, is influenced by government policy. At the most general level, government policy enables family farming to continue to exist (whether through aid to all forms of agriculture, special treatment, establishing trade policies), but family farming is not itself constituted by nor founded by policy making. As a practice or way of life, it is influenced by various forces (religious and economic institutions, and others), but it is not a creature of law in the same way that a corporation is. Corporations are legal entities, created by contracts and the institutions that define and enforce them. The notion of a family farm is not the notion of a body of carefully crafted, explicit rules of responsibility and power, but rather a way of life in which the rules are embedded in the way some people live.

The portrait of family farming as a virtue-laden, fulfilling way of life or practice fits well with Aristotle's treatment of excellence and happiness. According to Aristotle (384–322 BC), human flourishing or happiness consists in the virtuous exercise of human powers. Virtue is discovered through the exercise of practical wisdom or *phronesis*. Practical wisdom enables one to find the right balance of desire and appetites, pleasure and pain. In Aristotle's ethics, happiness or flourishing (he used the Greek term *eudaimonia*) involves more than pleasure or the satisfaction of desire. In his view, you may have all your preferences satisfied and yet, because your preferences are disordered or confused, you do not flourish and any happiness you claim is a sham. This understanding of happiness and flourishing is in close accord with what we shall see in the next section on family agrarianism with its emphasis on virtuous fulfillment through active collaboration with others.

The next section explores the case for family farming by taking seriously the interwoven nature of the various reasons marshaled by Roy, Doug, Karen, and those like them. Various areas of philosophy show a growing appreciation of how a theory or world view rarely stands or falls on the basis of a single line of reasoning. More common now is to appreciate how a position may be bolstered by a series of arguments or reasons. But before proceeding on this tack, some further reflection needs to be devoted to the project of defining a family farm.

Different definitions of family farming exist. Wendell Berry offers the following characterization of the family farm:

> What I shall mean by the term "family farm" is a farm small enough to be farmed by a family, and one that is farmed by a family—perhaps with a small amount of hired help. I shall not mean a farm that is owned by a family and worked by other people. The family farm is both the home and the workplace of the family that owns it. . . . Furthermore, the term "family farm" implies longevity in the connection between family and farm. A family farm is not a farm that a family has bought on speculation and is only occupying and using until it can be profitably sold. (Berry, 1987, 347, 348)

In *Family Farming: A New Economic Vision* Marty Strange offers this general definition of a family farm, which is a farming system that either is or tends to be the following:

- Owner-operated
- Entrepreneurial
- Dispersed
- Diversified
- At equal advantage in open markets
- Family centered
- Technologically progressive
- Production is in harmony with nature
- Resource conserving

These features are not advanced by Strange as individually necessary conditions for family farming. Perhaps a farm may legitimately be considered a family farm if it satisfies some but not all of the conditions. Strange's goal is to delimit family farming from mainstream, industrial agribusiness. He delimits the latter by some of the following characteristics:

- Industrially organized
- Financed from growth
- Management centered
- Capital intensive

- At an advantage in controlled markets
- Standardized in their production processes
- Resource consumptive
- Farmed as a business

This way of defining terms works with what the twentieth century philosopher Ludwig Wittgenstein called *family resemblances*. Wittgenstein despaired of finding strict analytic definitions of terms and noted, famously, that even the term *game* does not admit of an exact definition free of vagueness and borderline cases. But just as we might be able to note certain resemblances among family members, we may also be able to recognize the resemblance and right grouping together of different cases of games. We may be in a similar fix when it comes to the concept of *family farm*.

I believe we should not be any more precise at this stage than in using Berry's description and Strange's lists as guides rather than rigorously delimited, tightly bound concepts. After all, there may well be cases of farms that seem to trade in both corporate and family camps, on the format proposed by Wendell Berry and Marty Strange. Corporations can own farms that are run by families in a traditional fashion. Multinational corporations may own small and medium-sized farms that foster an Aristotelian-Marty Strange list of virtues. The descriptions proposed by Berry and Strange should thus be used simply to identify a prototype or ideal case. Later I consider the prospects of more precise definitions and what to do with cases that aim at such ideals but fail miserably. Right now, as Gary Comstock points out, "the experts are at odds not only about what size farm qualifies (to be a family farm), but about whether size ought to be the deciding criterion" (Comstock, 1987, xxv; see also Sontag and Bubolz, 1996, and Headlee, 1991).

I suggest that we also begin with a fairly flexible interpretation of the term *family*. The family farm in American history refers to one or more heterosexual couples with children (parents, grandparents). But of course the constitution of heterosexual families raises many ethical issues (the status of adopted, not jointly conceived, children; child development; health care; children's rights; the scope of parental authority; blended families; maternal surrogacy; and so on) and there is now a significant movement to legally recognize homosexual couples to ensure that they have the rights, privileges, and responsibilities of heterosexual couples. For now, however, I let the term *family* stand so that it covers the traditional definition and allow that, if the case for legally recognizing single-sex domestic partnerships as families is successful, then the term *family farm* includes this broader spectrum. I personally believe that this broader spectrum is warranted, though I do not argue for this point here.

One of the most important factors in the Berry-Strange case for family farming that affects the debate over the traditional definition of the family is that family farming is usually advanced as being intergenerational. If nontraditional families can meet this condition, securing a stewardship of land over generations, then the case for expanding the description of family farming will be strengthened.

FAMILY AGRARIANISM AND STEWARDSHIP

Certain concepts seem to have criteria of valuation built into them. Thus, *being a parent* seems to involve or entail judgments about what makes a good or bad parent. Should the parent abuse his or her child, we may well say the person has ceased acting as the child's parent. Maternal or paternal love is not simply love of a child by the biological parent, but a certain kind of nurturing care aimed at fostering the well being of the child. According to some advocates of family farming, being a family farmer is in this respect like being a parent. A family farmer is one who cares for the family, the community, and the land. Wendell Berry locates the concept of "family farming" in the midst of such a network of intelligent and wise care for others:

> If family farming and good farming are as nearly synonymous as I suspect they are, that is because of a law that is well understood, still, by most farmers but that has been ignored in the colleges and offices and corporations of agriculture for thirty-five or forty years. The law reads something like this: land that is in human use must be lovingly used; it requires intimate knowledge, attention, and care. (Berry, 1987)

Berry contrasts the notion of a farmer with that of a strip miner:

> I conceive the strip-miner to be a model exploiter, and as a model nurturer I take the old fashioned idea of ideal of a farmer. The standard of the exploiter is efficiency; the standard of the nurturer is care. The exploiter's goal is money, profit; the nurturer's goal is health—his land's health, his own, his community's, his country's. . . . The exploiter wishes to earn as much as possible with as little work as possible; the nurturer. . . to work as well as possible. (Berry, 1977, 7–8)

The farmer is ideal when acting as a nurturer and is not true to him or herself when acting as an exploiter of the land, family, and community.

Some agricultural policies are driven by terms that fail to capture this nurturing dimension of being a farmer and neglect, too, the way in which farming needs to be understood in a broadly ecological context. Berry places his philosophy of farming into a broader philosophy of interdependence:

> Obvious distinctions can be made between body and soul, one body and other bodies, body and world, etc. but these things that appear to be distinct are nevertheless caught up in a network of mutual dependence and

influence that is the substantiation of their unity. Body, soul (or mind or spirit), community, and world are all susceptible to each other's influence, and they are all conductors of each other's influence. (Berry, 1977, 110)

In light of this broad understanding of interdependence, Berry envisions the family farmer as making a vital contribution to healthy human culture:

A healthy culture is a communal order of memory, insight, value, work, conviviality, reverence, aspiration. It reveals the human necessities and the human limits. It clarifies our inescapable bonds to the earth and to each other. It assures that the necessary restraints are observed, that the necessary work is done, and that it is done well. A healthy farm culture can be based only upon familiarity and can grow only among a people soundly established upon the land; it nourishes and safeguards a human intelligence of the earth that no amount of technology can satisfactorily replace. (Berry, 1977, 43)

This interconnected social context is one that defines the farmer's identity.

Berry contends that, at best, the farm is radically different from an industrial plant for producing commodities. Its function (Aristotle would have referred to its *telos*, its end or purpose) is to foster care for others, future generations, and the authentic (not merely artificial or contrived) integration of work and play, family and community. Living in an overly mechanized, urban culture, we too often lose sight of the origin of the objects around us. Aldo Leopold claimed that there were "spiritual dangers in not owning a farm. One is the danger of supposing that breakfast comes from the grocery. . . " (Leopold, 1949, 6). Leopold and Berry stress that this is no mere intellectual failing. An intellectual failure can often be corrected in formal education, but the point is that there is an affective failure—a failure to appreciate the nature of plants and their harvest, animals and their care—in our losing touch with family agrarian culture. To live responsibly from the land is to invert the values of industrial, mechanical culture.

When one undertakes to live fully on and from the land the prevailing values are inverted: one's home becomes an occupation, a center of interest, not just a place to stay when there is no other place to go; work becomes a pleasure; the most menial task is dignified by its relation to a plan and a desire; one is less dependent on artificial pleasures, less eager to participate in the sterile nervous excitement of movement for its own sake; the elemental realities of seasons and weather affect one directly, and become a source of interest in themselves; the relation of one's life to the life of the world is no longer taken for granted or ignored, but becomes an immediate and complex concern. In other words, one begins to stay at home for the same reasons that most people now go away. (Berry, 1969, 88, 89)

236 PART II: LIFE SCIENCE ETHICS

This collection of home, labor, and land is articulated with an eye on the greater ecological context of our lives.

This bid for a richer appreciation of natural origins and our environment calls for a richer self-understanding and development of integrated skills. Wes Jackson, along with Berry, stresses how industrialized agriculture can tend not to hone the human intellect but rather to further atrophy the drive to develop ingenious, native ways of solving environmental problems. Jackson advances "regenerative agriculture":

> If someone were to ask me to define regenerative agriculture, my answer would be that regenerative agriculture is a state of mind that will cause us to constantly shift our focus from solving our problems through industrialization to solving our problems through the land. (Jackson, 1982, 23)

Berry notes the fragmentation that occurs in our overly mechanistic approaches to agriculture and other enterprises.

> What happens under the rule of specialization is that, though society becomes more and more intricate, it has less and less structure. . . . The community disintegrates because it loses the necessary understanding, forms, and enactments of the relations among materials and processes, principles and actions, ideals and realities, past and present, present and future, men and women, body and spirit, city and country, civilization and wilderness, growth and decay, life and death—just as the individual character loses the sense of responsible involvement in these relations. (Berry, 1977, 21)

Berry argues that alternative agriculture fosters a greater coherence and integration of goods.

For Berry, Jackson, and others, family farming is an agrarian way of life that is informed by a comprehensive notion of the good. *The good* here includes both human and nonhuman well-being. Common to many advocates of family farming is the project of living with and in nature, not over and against natural processes. In *Small is Beautiful*, E.F. Schumacher commends a wide view of the tasks of agriculture.

> A wider view sees agriculture as having to fulfill at least three tasks:
>
> - to keep man in touch with living nature, of which he remains a highly vulnerable part;
> - to humanize and enable man's wider habitat; and
> - to bring forth the foodstuffs and other materials which are needed for a becoming life.
>
> I do not believe that a civilization which recognizes only the third of these tasks, and which pursues it with such ruthlessness and violence that the

other two tasks are not merely neglected but systematically counteracted, has any chance of long-term survival. (Schumacher, 1973, 113)

Schumacher locates the good of farming in the greater context of the general good of civilization and nature.

The kind of farming that Schumacher, Berry, Wes Jackson, Jim Hightower and others commend is one that emphasizes the ways in which the goals of profitability can be melded with natural processes. In working with nature, by employing integrated pest management as opposed to intensive use of chemicals, for example, farming can be both ecologically healthy and economically rewarding. The agricultural industry has been facing increasing attacks from an array of sources to exercise greater ecological care, and this attack has been just what contemporary family farming advocates have sought to exploit in making their case for small and medium-sized, sustainable farms. We are now at a point where almost all parties to the debate realize the importance of an overall ecological ethic. As Paul Thompson observes, "Agriculture cannot continue without an environmental ethic, or at least it cannot continue happily" (Thompson, 1995, 15). Some environmentalists look to this broader, ecological context "to see if an argument for preserving the small farm can be found" (Hill, 1992, 278). Marty Strange in *Family Farming: A New Economic Vision* stresses how smaller farms are more likely to safeguard the soil and seek out long-term goods. Strange is well aware of cases when abuses have been perpetrated by family farmers, such as in soil erosion, groundwater contamination, and the like. But he and others have focused on ways in which small or medium-sized farms lend themselves well to crop rotation, terracing, and such and how larger farms have difficulty accommodating more ecologically sensitive practices (see Strange, 1988, also Hightower, 1973 and 1975, and Kreb's 1992 detailed discussion).

Two additional positions need to be brought to the fore in the family farm debate: agrarian democracy and religious stewardship. Both have been cast as favoring family farming.

AGRARIAN DEMOCRACY

Early American colonial life was comprised of family farming, which was appreciated by some early founders of the United States, the most famous being Thomas Jefferson. In his *Notes on the State of Virginia*, Jefferson described the family farmer in these high terms:

> Those who labor in the earth are the chosen people of God, if ever he had a chosen people, whose breasts he has made his peculiar deposit for substantial and genuine virtue. . . . Corruption of morals in the mass of cultivators is a phenomenon of which no age nor nation has furnished an example. (Jefferson, 1984, 290)

Jefferson's reasons were largely pragmatic. He thought that if one owned one's land and earned one's living from it, then it was highly likely that one would exercise great prudence and care in government. Jefferson had deep suspicion of industrial culture in which persons may be treated as replaceable parts, highly mobile, and thus easily directed to pursue merely economic as opposed to richer political and cultural ends.

Jefferson's position has been echoed in some family farm literature. But it has also been supplemented by attaching it to points brought up earlier about family farming's promotion of virtue. Although the end of the twentieth century ushered in a growth of nationalism and tribal identity as well as powerful radical religious groups, it also bore the flowering of a more cosmopolitan ethic that highlights global or universal citizenship. Any political theory that stretches our commitment to democracy in our own republic to a broader opposition to despotism and promotion of global republicanism must be built on a compelling view of the virtues of individuals. It may be argued that the kind of civic virtues that are promoted in family agrarianism—environmental and personal accountability, responsible stewardship for the community and future generations—bolsters a broader republicanism that promotes human flourishing across national boundaries. As idealistic as Berry and others may at times sound, it may well be that nothing else will do if we are to adopt a global republicanism.

Some advocates of family farming have argued that the practice does more to ensure liberty than corporate agribusiness. The latter risks the development of monopolies that can stifle free competition between relatively independent parties (Shrader-Frechette, 1991; see also the contributions to Thompson and Stout, 1991).

RELIGIOUS STEWARDSHIP

The preceding arguments may be developed in the framework of humanistic naturalism, that is, a view of nature that does not include any God. One can rephrase Jefferson's claim about "the chosen people of God" and simply refer to the people whom human evolution has favored; perhaps it is "nature" that has made "the peculiar deposit for substantial and genuine virtue." But much of American family agrarianism has been theistic. According to theism, there is an omniscient, omnipotent, all good, God who has created and conserves the cosmos. As revealed in Judaism and Christianity, this God conserves a cosmos that is fundamentally good, though it is marked also by profound evil, some of which is due to human action and irresponsibility. Christianity has fostered at least two environmental ethics, often referred to as a dominion ethic and a stewardship ethic. On the dominion model, human beings are given primacy over other creatures and, within limits, allowed to use them for human welfare. Responsible use is promoted over waste and overuse of natural resources that leave little or nothing to future generations.

On the stewardship model, human beings may have some primacy over other creatures, but we also have the privilege and duty to be good stewards, caring for other life forms and living humbly among other, nonhuman goods.

Christian theists have adopted an array of environmental philosophies (Paddock et al., 1986; Comstock, 1996 and 1997). But both a dominion and a stewardship ethic can be used to bolster family agrarianism. A dominion model can appeal to family farming's promotion of values and commitment to the welfare of future generations. A stewardship model can be joined with the earlier ecological case for family farming. In these respects, theism would serve not to add an entirely new argument for family agrarianism but rather to offer additional strength to the positions just outlined. It would intensify the case for family agrarianism (Taliaferro, 1992a and 1992b). I have reviewed many interrelated reasons to back family agrarianism and to not regard the loss of family farms as a matter of mere *adiaphora* or indifference. One way to characterize these reasons is by taking note of the root meaning of the words *obligation* and *duty*. Today, we do not distinguish these in terms of meaning, but they were once quite distinct. The concept of *duty* (like its Latin equivalent *officium*) was used to designate particular actions one should do. One may have a duty to care for the vulnerable, for example. But *obligation* (like its Latin equivalent *obligatio*) refers to the bond or relationship in virtue of which one has particular duties. So one's duty to care for the vulnerable may be in virtue of one's obligation as a fellow citizen, a fellow human being, or (for the theist) being a creature of God. The reasons for promoting family agrarianism rest largely on the grounds of the greater picture of one's obligations to oneself, family, neighborhood, civic society, and perhaps even as one vocation among others to the glory of God.

IMPARTIALITY AND PARTICULARITY

The preceding case for family agrarianism and stewardship draws on a variety of sources, from the ecological to the testimonial. It can be assessed within a broad, impartialist ethical framework, but it also invites reflection that is at odds with impartialism. Impartiality appears to be the benchmark of moral reflection. That is, it seems as though an essential condition for ethical reflection is to seek to minimize the influence of particular, personal commitments. We do not permit a judge to settle a case in which her own family is involved. The impartiality that is at work in ethics is evident in Kantianism, utilitarianism, virtue theory, and so on. For example, the British philosopher who founded modern utilitarianism, Jeremy Bentham, steadfastly opposed giving moral authority to the claims of friendship and family unless these could be vindicated by an appeal to the greatest good. No single individual or community can have a value that is independent of what would be detected from an impartial, abstract point of view.

But if we focus on the whole and construct what amounts to a kind of ethical aerial perspective, what are we to do with the testimony from the ground, the specific commitments that we each have and the testimonies of friendship, family, and community? Gary Comstock writes about the lived experience of family farming and the difficulty of capturing such experience in terms of economics, the sciences, or in purely intellectual contexts:

> Being a family farmer means caring for one's land. Such love cannot be taught in agricultural colleges; it is a practice that one learns at the feet of a master. It is knowledge of the heart, not the head, and it is best passed from generation to generation, not from agribusiness expert to agricultural student. This does not mean that newcomers cannot love the land; only that their doing so requires that they learn right emotions and intentions, not just right equations and ratios. This sort of care comes from lived experience and tradition—from memories, from the past. This provides a clear moral justification for giving preferential treatment to those farms that have long histories of having been family undertakings. (Comstock, 1987, 416)

Comstock is backed up by other critics of an abstract impartialism.

Samuel Scheffler, for example, has noted how we are deeply invested in our particular projects, which conflicts with the universalizing tendency of utilitarianism (Scheffler, 1982). If we allow moral reflection to take place only on an abstract level that is "unencumbered" by our individual projects and forms of life, then we risk cutting off moral reflection from what gives our lives meaning.

I briefly describe four contemporary movements in ethics and one in the philosophy of language and meaning that seem to give prominence to what may be considered the personal and particular. If any of these have merit, then they provide some reason to believe that the impartialism of an ethical point of view needs to be hedged or at least informed by specific personal or particular claims.

COMMUNITARIANISM

Communitarians have proposed that we are shaped by our particular traditions in a fashion that anchors us and gives us a moral balance (A. MacIntyre, C. Taylor, M. Walzer, D. Bell). One conclusion of this movement is that the sources of our moral identity consist of specific communities and social conditions and not abstract, ahistorical moral impartialism. Daniel Bell writes: "If you ask yourself what matters most in your life. . . the answer will involve a commitment to the good of the communities out of which your identity has been constituted" (Bell, 1993, 94).

COVENANTAL ETHICS

This school of ethics is most common in Christian religious traditions; it is like communitarianism but in a specific religious form. Both Catholics and Protestants look to ancient narratives of a covenant that involves God and the people of God, securing a particular identity and subsequent view of social and ecological responsibility. The Biblical background is located in Exodus 19–24 (see also Exodus 34:5 and Joshua 24:1–13). Ed Langerak offers the following picture of the covenantal community:

> Covenanting puts people in moral community with each other, a community in which both the common good and the good of each individual member are sought. Thus individuals' identities are shaped by their communities—they *are* their caring relationships—and communities' identities are shaped by the individuals the communities encompass. . . . Covenants, by their identity-shaping privileges and responsibilities, tend to endure over time and are influenced by new developments in unspecifiable and open-ended ways. (Langerak, 1989, 84)

This idea of a covenant diverges from relations that are defined by explicit contracts. A covenantal ethic diverges from an ahistorical impartialism and can be used to bolster an ethic that is defined and developed within specific moral traditions. Such an ethic seems to inform the 1986 National Conference of Bishops and their stewardship ethic (reprinted in Comstock, 1987).

FEMINISM

Feminism today has been largely fueled by the conviction that an abstract, strict impartialism is not gender-free but has tended to advance a male agenda. Over an ethic of justice as a set of rules, some feminists promote an ethic of care (Carol Gilligan and Nell Noddings), maternal thinking (Sarah Ruddick), and the loving or personal gaze (Maria Lugones). Whether or not one adheres to a form of feminism, the testimony of these philosophers is that an overriding impartialism that does not take specific relations and particularized emotions seriously is incomplete.

CONTEMPORARY JUDAISM

Martin Buber (1878–1965) and Emmanuel Levinas (1906–1995) have introduced fascinating work on the importance of personal as opposed to impersonal relations, the I-You as opposed to I-It relation, in Buber's terminology. Levinas has focused his attention on the importance of encountering the face of the other person. The resulting picture of the ethical life is profoundly personal and specific. Also in this school, Lawrence Blum has made great strides in highlighting the importance of particular moral perceptions, especially as these are represented in literature (Blum, 1994).

This more particular, personal outlook seems to allow for just the kind of reflection that family agrarians need to advance their position. The family agrarian position may also be strengthened by some recent accounts of language and meaning. Stanley Cavell insists that our language and social life are not grounded in codified rule-following. We are, rather, shaped by specific "forms of life." This appreciation for the practical context of our forms of life provides further reason to take seriously the accounts of farming and other practices from the people themselves, and an engaged investigation into people's stories of their lives as opposed to simply examining the case for and against family farming in the abstract.

> We learn and teach words in certain contexts, and then we are expected, and expect others, to be able to project them into further contexts. Nothing insures that this projection will take place (in particular, not the grasping of universals nor the grasping of books of rules), just as nothing insures that we will make, and understand, the same projections. That on the whole we do is a matter of our sharing routes of interest and feeling, modes of response, senses of humor and of significance and fulfillment, of what is outrageous, of what is similar to what else, what a rebuke, what forgiveness, of when an utterance is an assertion, when an appeal, when an explanation—all the whirl of organism Wittgenstein calls 'forms of life.' Human speech and activity, sanity and community, rest upon nothing more, but nothing less, than this. (Cavell, 1976, 52)

A preoccupation with impartialist rules will not give one a credible view of language and basic human activity. If Cavell is right, then if family agrarianism rests on a form of life as opposed to impartialist, rule-governed reflection, it still may be no worse for that.

Constructing a picture of a form of life will involve different skills than abstract, utilitarian calculation. Field experience may be crucial. This tendency to incorporate field experience philosophically and ethically seems to be a vital point among some environmentalists at large, such as Holmes Rolston's appeals to the experience of the "wild" as an irreplaceable resource for environmental philosophy (Rolston, 1986). The case for a family farming form of life may also involve literature and poetry. Thomas Auxter has made strides in showing how poetry can shape our self-awareness in agricultural settings (Auxter, 1985). The poetry of Wendell Berry is especially fitting here. For older literature, Xenophon's *Oeconomicus* is highly recommended for its extolling farm virtues in the 4th century BC. The book consists of a dialog in which the character Socrates articulates and praises agricultural virtues, a dialog not entirely different from our exchange between Doug and Emily. It may be that we need this broader back-drop of experience and literature; we cannot fall back on the way business ethics is typically taught (debate over the ethics of bargaining, deceit, bribery, advertising, whistle-blowing, trade secrets, and so on). A form of life such as family agrarianism requires a broad form of inquiry. In the end, stories by farmers themselves may be the key to our inquiry (for example, Hart, 1991; Gery, 1993; Rosenblatt, 1990).

Berry underscores the need to take seriously the particularity of our positions, as we also come to appreciate the greater framework in which we find ourselves:

> Harmony is one phase, the good phase, of the inescapable dialogue be-tween culture and nature. In this phase, humans consciously and consci-entiously ask of their work: Is this good for us? Is this good for our place? And the questioning and answering in this phrase is minutely particular: It can only occur with reference to particular artifacts, events, places, ecosystems and neighborhoods. When the cultural side of the dialogue becomes too theoretical or abstract, the other phase, the bad one, begins. Then the conscious, responsible questions are not asked; acts begin to be committed and things to be made on their own terms for their own sakes, culture deteriorates, and nature retaliates. (Berry, 1987, 143)

In this fashion, Berry thinks that the notion of a human economy should be hooked up into a bigger economy, the economy of nature:

> It is only when we think of the little human economy in relation to the Great Economy that we begin to understand our errors for what they are and to see the qualitative meanings of our quantitative measures. If we see the industrial economy in terms of the Great Economy then we begin to see industrial wastes and losses not as "trade-offs" or "necessary risks" but as costs that, like all costs, are chargeable to somebody, sometime. (Berry, 1987, 70)

Here it might well be noted that *economy* and *ecology* both come from *home*. By *the family farm*, we mean a home within a bigger home.

This argument for family farms may be less than rigorous but is perhaps no worse than many arguments in ecology. In ecology, various sciences come into play in form-ing comprehensive descriptions and accounts of natural phenomena in ways that com-pel one to expand beyond the limits of one's individual scientific practice.

OBJECTIONS, REPLIES, AND REFINEMENTS

Many objections exist to the preceding case for family agrarianism. Here, I consider four. Neither the objections nor the replies are presented as definitive points on either side but rather as points and counterpoints that need to be taken seriously in the debate. A final section of this chapter raises some further points to weigh in the case for family farming.

Objection 1:

Family farmers have been responsible for enormous ecological damage

This was part of the thesis of Rachel Carson's classic, *The Silent Spring*. Jim Hill's succinct observation about small farms is telling against Schumacher. "Small, qua small, is not necessarily beautiful" (Hill, 1992, 277).

A family farm may be the model of sustainability, exercising exemplary water and soil conservation, but it may also use dangerous levels of pesticides. The family farm is just as capable as agribusiness of going for short-term profits at the expense of serious soil erosion, the overuse of chemical fertilizers, and so on. It would be better simply to promote through incentives (or through penalties for failing to achieve) stable, ecologically sound forms of irrigation, crop rotation, animal welfare, and such. If ecological integrity can be achieved competitively through family farms, well and good, but doing so through agribusiness is fine as well. If we want to make the goal ecological health, why not simply outline and achieve that goal through legislative regulation?

Agribusiness may also be better able to meet the standards set by society. Turning to Roy's dairy farm, imagine that we are concerned with decreased genetic diversity among dairy cattle, or we want to promote the general health care of the livestock and cull ill cattle. Wouldn't attending to several large farm institutions be easier than to a series of farms?

Reply

The data here seem mixed. Some evidence showing that owner-operated, in-residence farmers are more highly motivated to care for their land and to promote environmental protection, though we can find many counterexamples. Moreover, refuting the charge that agribusiness can meet sustainable goals may not be easy (Boehlje, 1987). In my view, the best case for family farming that is found in Berry's writings, Jackson's and many others, is part of the case for sustainable agriculture. If this is right, then the best strategy is to directly challenge the objection and to argue that the smaller, sustainable farms have been significantly more successful than agribusiness in terms of avoiding environmental degradation. As Carol Hodne puts it: "I feel that perhaps our greatest moral obligation to save family farms stems from the moral imperative of preserving our soil and water resources" (Hodne, 1987, 54). Future work in alternative agriculture may bolster her suggestion. And it could be that some techniques, certain methods of harvesting for example, are more difficult when performed on a large scale. Wes Jackson and others have advanced this position (Jackson, 1987 and 1988). For a detailed case study that argues for the link between family farming and sustainable, ecologically beneficial agriculture see M. Suzanne Sontag and Margaret Bubloz's *Families on Small Farms* (see also Rogers, 1985).

An alternative defense is to follow Strange and Berry all the way and to build into the concept of "family farm" the goods they describe of sustainability, wise stewardship, neighborliness, and so on. In this view, to suppose that a family farm fosters ecological damage would be akin to claiming that parental love can compel one to exploit one's children. This strategy attempts to cut off at the roots the objection that family farming is ecologically hazardous. The advantages and disadvantages of this reply are explored in the next objection and reply.

OBJECTION 2:

The case for family farming is built on a highly romanticized, perhaps even sentimental portrait of farm-bred virtues

The family farm in the United States has often housed sexism, racism, and isolationism. The virtues that are at the heart of family farming have been shaped by a traditional patriarchy. From Luther Tweeten:

> The personality of the farm sector is basically healthy and has many of the favorable attributes embodied in the image of the family farmer as self-reliant and independent; and as committed to fair play, due process, and democratic ideals. But a darker side of the farm personality emerged in the course of American history and is characterized by scapegoating, armed confrontation, violence, and commitment to ideologies that would bring fewer gains to farmers than losses to consumers, taxpayers, and society as a whole. (Tweeten, 1987, 73)

This darker side is especially sinister when we note the implied perfectionism in some agrarian writing. A perfectionist account of property holds that ownership is tied to moral virtue; the vicious are not as clearly entitled to their property. This notion can be dangerous when we lack a clear-cut account of virtues in our pluralistic culture.

Harkening back to the claims about agrarian democracy, we do well to consider Gary Comstock's comment: "If Jefferson is right, how is it that we still have a democracy when less than two percent of us live on family farms?" (Comstock, 1987, 412) We may grant that cosmopolitan republicanism would be bolstered by the kinds of virtues outlined by Berry, Jackson, and others, and yet we can counter that these virtues may be acquired in nonfarming ways of life.

Objection 2 may be furthered by also pressing this point: If the concept of "family farming" is constructed in such a way that values and goodness are already built into it so that "bad family farming" is a virtual contradiction in terms, then the defenders of family farming have simply begged the question. That is, they have assumed at the outset the very thing that requires argument. A proponent of large-scale agribusiness could adopt a similar strategy and simply define such an enterprise as inherently valuable. Defining one's position into moral respectability accomplishes very little.

Reply

Granted, the case for family farming cannot be made through stipulative definition. Clearly Berry, Jackson, et al. do not make this move. The point that Berry and others wish to make is that what they identify as family farming stands for a way of life or practice that is governed by a rich, responsible understanding of the good—good community and good land stewardship. In their view, there is an ideal and dignity to *being a family farmer*; some small owner-operated farms are worthy of this title, some are not. Just as a biological family can become so dysfunctional that it is no longer a family in any culturally and ethically respectable sense, so can dysfunctional family farms lose their right to being called family farms. Consider an analogy in the theory of law. One strand of the natural law tradition holds that the very concept of "law" has a dignity and worth such that the concept of an "unjust law" would be like the concept of a square circle, a contradiction. *Lex injusta non est lex*, or an "unjust law is not a law," is the traditional claim. In this view, a view adopted with great conviction by Martin Luther King Jr., the enactment of white supremacist laws should be considered "laws" only in sneer quotation marks. They are, in fact, not *bona fide* laws but simply rules enforced to tyrannize people of color. The tactic Berry adopts of building into the concept of family farming a host of virtues and goods is no more a matter of question begging than a natural lawyer's view of the dignity of law.

Of course, a critic can still charge that there simply are few, if any, such family farms in this enriched, value-laden sense. Empirical scientific and sociological studies, testimony of agricultural communities, and the like are the place for such an objection and reply to be forged. (For an examination of some of the "darker side" of small American farms, see *Agriculture and Human Values* 2:1.). Perhaps, then, family agrarianism represents an ideal to be worked for. Comstock's question cited in the preceding objection is appropriate, but we may well pause to consider what kind of democracy we have, and the ways in which our democracy would be enhanced if the land wisdom available in family farming were more integral to our culture.

OBJECTION 3:

The Freedom objection

This objection does not take aim at the good of family farming but rather highlights the perceived good of freedom. Given a liberal, free market democracy, shouldn't any proposal to protect the family farm by legislature (either penalizing agribusiness or employing public monies to fix prices and incomes for family farmers) face an enormous burden of proof? That is, in a context in which freedom is a perceived right, shouldn't the fate of family farms be settled by a free and open market? If family farms falter, this could be due to a Darwinian weeding out of the weak. Perhaps regrettable but not unfair.

Reply

Two replies may be explored.

(A) One is to argue that there are many goods that we currently protect from free market exchange because of their value. If family agrarians can make the case that family farming constitutes and promotes virtues or that it has a vital standing in our heritage and is worthy of pursuit, then they may be on the way to an effective rejoinder. The heritage argument will have to be hedged, though, for clearly family agrarians do not wish to promote the equivalent of a series of museums but rather a viable practice. There are many cases in which our society does protect goods that are not given market value (educational institutions, inner cities), and family farm goods may be sufficient to merit protection.

(B) A second rejoinder is to contend that it is agribusiness, not family farming, that poses a threat to human freedom. It has been argued that agribusiness threatens the individual because of its tendency to create monopolies. It does not compromise individual civil rights (the right to vote, to be free from arbitrary arrest, and the like) but it does restrict individual liberties in terms of economic activity, the freedom to enter into fair competition (Shrader-Frechette, 1991). Another reply that seeks to overturn the freedom objection falls back on the charge that agribusiness tends to be less able to offer food with good nutrition and low environmental costs (Campbell, 1979). If this is right, then individual freedom to have access to good nutrition may bolster family agrarianism.

OBJECTION 4:

The use of a religious stewardship ethic suffers from two problems

First, theism seems to lend itself equally well to environmental responsibility and to recklessness. Second, it imports religious considerations into public debate; this is not fair in a democracy in which the state is supposed to be religiously neutral.

Reply

Two very brief replies. First, there is some reason to believe that Christian theism is generating a growing consensus on the importance of land stewardship. Although Christian language has been employed in the past in justifying the exploitation of natural resources, it is increasingly apparent that such appeal to Christianity was politically expedient and not the outcome of deep, authentic Christian convictions. Belief in a loving Creator who upholds a creation that is fundamentally good cannot be easily yoked with ecological exploitation (Comstock, 1997; Attfield, 2001). Second, even if the appeal to religion should not have a direct role in policy making, it does not follow that it should have no role in public discourse, public philosophy, and culture. Insofar as it does, and it does enhance a case for stewardship-based family farming, then the appeal to religion will be significant in shaping the politically relevant value of family farms. Thus, a liberal secular state (France) may have political reasons to protect certain religious institutions because of their overall contribution to culture.

CONSIDERATIONS FOR FURTHER REFLECTION ON FARM POLICIES

This chapter has focused on the structure of the debate about family farming. It has concentrated on the philosophies at stake and the different ways in which testimony and ecological findings can be marshaled. In closing, consider three important factors that will require attention in furthering more detailed argumentation on either side.

First, the conduct and aim of the debate will depend on the prevailing political and economic framework. A debate on the federal level will aim at uniform national standards, whereas state-by-state debate may permit great regional differences and employ different accounts of what comprises family farming. One's economic commitments will also play a great role in how to read the case for the family farm. If you are a staunch defender of the free market system with only minimal government intervention, it is likely that you will protect family farming only if you oppose subsidies to agribusiness that disadvantages family farming. My point here is that background political and economic theories will inevitably play a large role in the debate.

Second, if family agrarianism is imperiled today, it will be important to consider who has the responsibility for such a plight. If family farmers themselves bear such responsibility, then the duty to preserve family farming may be diminished. Marty Strange supports aid to the family farm but notes that the farmers themselves seem to have brought about these difficulties:

> Farmers . . . seem to have welcomed the very economic policies that have placed them in such jeopardy. Were they not among the most supportive of President Reagan when he sought reelection in 1984, even as the farm crisis deepened? Did not 70 percent or more of the farmers vote for him in that election when he pledged to reduce the budget deficit by cutting social spending? How could they be surprised and outraged when his first veto in 1985 was of an emergency farm-credit bill that would have added to the deficit? Weren't they being a little hypocritical to think he would cut all social spending *except* agriculture? (Strange, 1988, 16)

But if they did not know what they were agreeing to, one might well charge that family farm action was not fully voluntary.

Third, if family agrarianism represents a great good, that is a reason to support it, though perhaps not a decisive one. There are many goods in the world and it may be that a greater good requires our allegiance. (Imagine, for example, that world famine is best addressed through large-scale agriculture.) The loss of family farming may then be rightly deemed regrettable but not, under the circumstances, reversible.

One need not believe that family farming is an unqualified good in order to appreciate this regret. To believe that family farming is an unqualified good would appear to be a form of what Luther Tweeten calls "farm fundamentalism":

> Farm fundamentalism is the belief that farming is not only a superior way of life but also represents the highest ideals of the nation. Farm fundamentalism holds that the nation's political and social system cannot survive without the type of person the farm way of life produces. (Tweeten, 1987, 67)

Without being a farm fundamentalist, one may still hold that family farming represents an enormous good that we should either promote or, if the way of life that Berry and others celebrate is to be lost, deeply regret its passing. At the end of the day, whatever one's conclusion and qualifications, it is hard not to admire and take seriously a way of life that, at its best, incorporates stewardship, a commitment to long-term productivity, cooperation between neighbors and generations, the cultivation of civic virtues, and wisdom.

Turn now to the appendix to consider the questions in Exercise 10.B.

REFERENCES

Attfield R. (2001) "Christianity," in *A Companion to Environmental Philosophy*, D. Jamieson, ed. (Oxford: Blackwell).

Auxter, T. (1985) "Poetry and Self-Knowledge in Rural Life," *Agriculture and Human Values*, II: 2, Spring 1985.

Bell, D. (1993) *Communitarianism and its Critics* (Oxford: Oxford University Press).

Berry, W. (1991) *Standing on Earth* (Ipswich: Golgonooza Press)

Berry, W. (1987) "A Defense of the Family Farm," in *Is there a Moral Obligation to Save the Family Farm?* G. Comstock, ed. (Ames: Iowa State University Press)

Berry, W. (1987) *Home Economics* (San Francisco: North Point Press).

Berry, W. (1985) *Remembering* (San Francisco: North Point Press).

Berry, W. (1977) *The Unsettling of America: Culture and Agriculture* (New York: Avon Books).

Berry, W. (1969) *The Long-Legged House* (New York: Harcourt, Brace and World).

Blum, L. (1994) *Moral Perception and Particularity* (Cambridge: Cambridge University Press).

Boehlje, M. (1987) "Costs and Benefits of Family Farming" in Comstock, 1987.

Campbell, K. (1979) *Food for the Future* (Lincoln: University of Nebraska Press).

Comstock, G., ed. (1987) *Is there a Moral Obligation to Save the Family Farm?* (Ames: Iowa State University Press).

Comstock, G. (1996) "Toward an Evangelical Environmental Ethic," in *Christianity and the Environmental Ethos*, S. M. Postiglione and R. Brungs, eds. (St. Louis: ITEST Faith/Science Press).

Comstock, G. (1997) "Theism and Environmental Ethics," in *A Companion to Philosophy of Religion*, P. Quinn and C. Taliaferro, eds. (Oxford: Blackwell).

Demissie, E. (1990) *Small-Scale Agriculture in America; Race, Economics, and the Future* (Boulder: Westview Press).

Emerson, R.W. (1904) "Farming," *The Complete Works of Ralph Waldo Emerson*, Concord Edition, vol.7 (Boston: Houghton, Mifflin).

Gery, M. *Founding Farms; Portraits of Five Massachusetts Family Farms* (Amherst: University of Massachusetts Press).

Gilligan, C. (1982) *In a Different Voice* (Cambridge: Harvard University Press).

Hart, J. (1991) *Farming on the Edge; Saving Family Farms in Marin County* (Berkeley: University of California Press).

Headler, S. (1991) *The Political Economy of American Capitalism* (New York: Praeger).

Hill, J. (1992) Review of "Is There a Moral Obligation to Save the Family Farm?", *Environmental Ethics*, volume 14.

Hightower, J. (1973) *Hard Tomatoes, Hard Times* (Cambridge: Schenkman).

Hightower, J. (1975) "The Case for the Family Farmer," in *Food For People, Not for Profit*, C. Lerza and M. Jacobson, eds. (New York: Ballantine).

Hodne, C. (1987) "We Whose Future Has Been Stolen" in *Is There a Moral Obligation to Save the Family Farm?* G. Comstock, ed. (Ames: Iowa State University Press).

Hunter, D. (1990) Stories of the Minnesota Farm Advocates (Duluth: Holy Cow! Press).

Jackson, W. (1987) *Altars of Unhewn Stone: Science and the Earth* (San Francisco: North Point Press).

Jackson, W. (1988) "Ecosystem Agriculture" in Global perspectives on agroecology and sustainable agricultural systems. *Proceedings of the Sixth International Conference of the International Federation of Organic Agriculture Movements*, vol. 1, P.Allen and P. van Dusen, eds., 15–20. Santa Cruz: Agroecology Program, University. of California.

Jackson, W. (1982) "A New Vision for Farming's Future" in *The New Farm*, July/August.

Jefferson, T. (1984) *Writings* (New York: Literary Classics of the United States).

Kegan, S. (1989) *The Limits of Morality* (Oxford: Clarendon Press).

Krebs, A.V. (1992) *The Corporate Reapers: The Book of Agriculture* (Washington: Essential Books).

Langerak, E. *et al* (1989) *Christian Faith, Health, and Medical Practice* (Grand Rapids: William Eerdmans Publishing Company).

Legones, M. (1987) "Playfulness, 'World-Traveling,' and Loving Perception," *Hypatia* 2:2.

Leopold, A. (1949) *A Sand County Almanac, and other Sketches Here and There* (London: Oxford University Press).

Levinas, E. (1969) *Totality and Infinity* (Pittsburgh: Duquesne University Press).

MacIntyre, A. (1981) *After Virtue* (Notre Dame: University of Notre Dame Press).

McCorkle (1989) "Toward A Knowledge of Local Knowledge and its Importance for Agricultural RD &E." *Agriculture and Human Values*. VI: 3, Summer.

Merchant, C. (1989) *Ecological Revolutions* (Chapel Hill: University of North Carolina Press).

Nagel, T. (1980) "The Absurd" in *Mortal Questions* (Cambridge: Cambridge University Press).

Noddings, N. (1984) *Caring: A Feminist Approach to Ethics and Moral Education* (Berkeley: University of California Press).

Paddock, J., Paddock P., Bly, C. (1988) *Soil and Survival; Land Stewardship and the Future of American Agriculture* (San Francisco: Sierra Club Books).

Rogers, S.C. (1985) "Owners and Operators of Farmland: Structural Changes in U.S. Agriculture," *Human Organizations* 44 (3), 206–221

Rolston, H. (1986) *Philosophy Gone Wild* (Buffalo: Prometheus Books).

Rosenblatt, P. (1990) *Farming is in our Blood; Farm Families in Economic Crisis* (Ames: Iowa State University Press).

Ruddick, S. (1989) *Maternal Thinking* (Boston: Beacon Press).

Scheffler, S. (1982) *The Rejection of Consequentialism* (Oxford: Clarendon Press).

Schumacher, E.F. (1973) *Small is Beautiful* (New York: Harper and Row).

Shrader-Frechette, K. (1988) "Agriculture, Ethics, and Restrictions on Property Rights," *Journal of Agricultural Ethics* 1: 1.

Shrader-Frechette, K. (1991) "Property and Procedural Justice: The Plight of the Small Farmer" in *Ethics and Agriculture*, C. V. Blatz, ed. (Moscow: University of Idaho Press).

Smiley (1991) *A Thousand Acres* (New York: Knopf).

Smith, A. (1933) *The Wealth of Nations* (New York: E.P. Dutton).

Sontag, M.S. and Bubolz, M. (1996) *Families on Small Farms; Case Studies in Human Ecology* (East Lansing: Michigan State University Press).

Steinbeck, J. (1939) *The Grapes of Wrath* (New York: Modern Library).

Strange, M. (1988) *Family Farming: A New Economic Vision* (San Francisco: Institute for Food and Development Policy).

Taliaferro, C. (1992a) "Divine Agriculture," *Agriculture and Human Values*, Fall 1992, 71–80.

Taliaferro, C. (1992b) "The Intensity of Theism," Sophia 31: 3, 61–73.

Thompson, P. (1998) *Agricultural Ethics* (Ames: Iowa State University Press).

Thompson, P and Stout, B., eds., (1991) *Beyond the Large Farm* (Boulder: Westview Press).

Williams, B. (1973) "A Critique of Utilitarianism" in *Utilitarianism For and Against* (Cambridge: Cambridge University Press).

Wolf, A. (1987) "Saving the Small Farm: Agriculture in Roman Literature," *Agriculture and Human Values*, IV: 2, 3, Spring–Summer.

Xenophon (1994) *Oeconomicus* tran by S. Pomeroy (Oxford: Clarendon Press).

NOTE

I am grateful for helpful comments on earlier versions of this chapter to Gary Comstock, Ken Casey, Ed Langerak, and Gretchen Ross.

Part III

CASE STUDIES

Chapter 11

ENVIRONMENT

RARE PLANTS

Lynn G. Clark

SPECIES X: A CASE STUDY

The Atlantic forests of coastal Bahia, Brazil, harbor some of the greatest diversity of plant life on the planet. Within the last few decades, however, these formerly extensive forests have been reduced to approximately three percent of their original cover due to the cultivation of cacao and other crops. An extremely rare but evolutionarily significant species of an angiosperm family, referred to here as Species X, occurs in these forests. This species is known from only three populations along a six-kilometer stretch of road in the cacao-growing region of Bahia; at last count in 1994, a total of about 80 to 100 plants were found in the three populations, although a more recent count found even fewer plants. One of the populations grows at the edge of a cacao grove and none occurs within a protected area. It is possible that additional populations of the species occur in the area, although botanists have looked for it without success. Recent studies have shown that Species X is one of the few existing representatives of the earliest lineage of its family. These ancient forest plants (or plants very much like them) may have evolved in the late Cretaceous and coexisted with the dinosaurs, but the clade of which Species X is a member certainly had evolved by 55 million years ago.

Several botanists have visited the natural populations of Species X over the last twenty years, and a few live plants were removed for cultivation in Brazil and the

United States during that time. The plants were collected and taken out of Brazil with the proper authorization, although it is not clear whether documentation is available. Regulations in force today (including principles agreed upon at the Rio summit) would probably permit the collection of such plants for research purposes but would not allow for their commercial distribution without some form of compensation to the Brazilian government (assuming that the plants have any commercial value, which does not seem likely in this case).

Exact Geographic Positioning System (GPS) coordinates for the three populations of this species have been obtained but have not been released to the general public or scientific community. Species X is currently in cultivation in two places in Brazil, but a dozen or so plants are cultivated at various universities and botanical gardens in the United States. Although the species has some attractive qualities, it grows slowly and probably has little potential for development as an ornamental. It would be of interest to collectors mainly because of its rarity. Species X is very rare and extremely significant evolutionarily (effectively, it is a living fossil, showing us what the earliest members of its family may have looked like), a combination that would give it the highest priority according to some conservation biologists. By any criteria, Species X is a rare, endangered species but has not yet been listed formally as such.

DISCUSSION QUESTIONS

1. Should a coordinated attempt to preserve one or all of the natural populations be undertaken, even if the effort creates local hostility? Or is it sufficient to leave well enough alone, given that the species has survived this long, and hope that additional but as-yet-undiscovered populations are out there somewhere? Should (or can) international or national scientific interests supersede local politics and concerns (that is, what if the Brazilian government has an interest in protecting Species X but local residents don't support this)?

2. Scientific data, if gathered with funds from public sources (for example, the National Science Foundation), are considered to be in the public domain once published. Should the GPS coordinates for Species X be published, even if that exposes the species to unscrupulous collectors (amateur collectors have been known to extirpate species in the wild and, in a few instances, overzealous biologists have apparently collected species to extirpation/extinction)? How much influence should Brazilian authorities have over this decision?

3. If money to fund research and/or conservation efforts related to Species X could be raised by selling plants grown in captivity, should any of that money be returned to the Brazilian government?

4. International and federal regulations governing species officially listed as rare or endangered can inhibit research efforts by making it extremely difficult to legally im-

port material while at the same time calling attention to the rarity of the species. Those who smuggle rare plants and animals are often not caught anyway, but a legitimate researcher or grower cannot afford to ignore legal restrictions. Should Species X be listed formally as being rare or endangered (for example, with the Convention on the International Trade in Endangered Species – CITES)?

5. What if permit regulations for the collection and exportation of such plants for research purposes required that the material be destroyed upon completion of the research, unless the species were officially listed as rare and endangered?

MARINE MAMMAL PROTECTION AND MANAGEMENT

Donald J. Orth

RELEVANT FACTS ABOUT MARINE MAMMALS

Biologically, marine mammals are those members of the class Mammalia that are morphologically adapted to life in the ocean.[2] They include three taxonomic orders, the Cetacea (whales and porpoises), Pinnipedia (seals, walruses and sea lions), and Sirenia (manatees and dugongs). Other groups (sea otters and polar bears) are considered marine mammals in U.S. legislation. Many coastal cultures hunted whales and thrived on the meat, skins, and other products of whales, seals, and polar bears. Threats to these creatures have been well publicized. Some whales were hunted to near extinction; porpoises were killed during purse netting for Pacific tuna; contaminated sea lions aborted young; Northern fur seals were overharvested; and manatees were injured from motorboat collisions.

The Marine Mammal Protection Act, passed by the United States in 1972, is the most comprehensive protective mechanism for marine mammals. This act established a moratorium on hunting, capturing, or killing marine mammals in U.S. waters and by U.S. citizens on the high seas, and on importing marine mammals and marine mammal products into the United States. The MMPA also directs that commercial fishing operations reduce incidental kill or serious injury to marine mammals. Taking of marine mammals is permitted only after scientists determine that a population is at or above the Optimum Sustainable Population (OSP) level. Additionally, the U.S. Endangered Species Act (1973) protects sixteen marine mammals (as of August 1996) threatened with extinction. Internationally, thirty-five marine mammals receive indirect protection under the Convention on International Trade in Endangered Species of

TABLE 11.1. POPULATION ESTIMATES.

Species/stock	Pre-whaling	Current
Blue	160,000–240,000	9,000
Bowhead	52,000–60,000	8,200
Bryde's	unknown	66,000–86,000
Fin	300,000–650,000	123,000
Gray (E Pac.)	15,000–20,000	21,000
Gray (W Pac.)	1,500–10,000	100–200
Gray (Atl.)	unknown	extinct
Humpback	150,000	25,000
Minke	unknown	850,000
No. right	unknown	870–1,700
Sei	100,000	55,000
So. right	unknown	1,500

Wild Fauna and Flora (CITES), which regulates trade in threatened plants and animals. The International Whaling Commission (IWC), formed to serve the whaling industry concerns, now primarily addresses whale concerns.

CETACEAN BIOLOGY AND CLASSIFICATION

More than seventy-five species of whales and dolphins exist, arranged in nine families: Balaenidae (right whales), Balaenopteridae (rorquals, which include Brydes, blue, fin, Minke, and sei whales), Eschrichtiidae (gray whale), Physeteridae (sperm whales), Monodontidae (Narwhal and Beluga), Ziphiidae (beaked whales), Delphinidae (oceanic dolphins), Phocoenidae (true porpoises), and Platanistidae (river dolphins). The first three families are baleen whales, collectively known as Mysticeti (*mustache whales*), which feed by trapping prey in keratinous plates (baleen) that hang from the roof of their mouth. Baleen whales are usually found alone or in small groups during nonbreeding times. The other families are toothed whales, or Odontoceti, which are predators of squid and fish. The mysticetes have paired blowholes, whereas the odontocetes have a single orifice. The toothed whales are usually highly social and capable of rapid evasive action.

Several traits of whales make them particularly vulnerable to harvest. The population rate of increase is low (from 0.03 to 0.08 percent per year) due to delayed sexual maturity, long gestation periods (sixteen months in some whale species), a single offspring at each birth, and longevity up to ninety years or more (that is, K-selected). Their large size, air breathing, and social behavior have all further contributed to vulnerability.

Much has been written about intelligence of cetaceans. The widely held belief that a large brain implies a high level of intelligence has led to claims about intelligence of cetaceans. No objective definitions exist, however, of what constitutes "intelligence" and how to measure intelligence in animals (even human animals); whether cetaceans are more intelligent than pigs, for example, cannot be definitively answered. The neo-cortex (brain structure associated with advanced mental processes) in cetaceans is extensive, but that fact leaves many questions of cetacean intelligence unanswered. Cetaceans are sentient beings, capable of experiencing pleasure and pain.

PRODUCTS

Oil was historically the most economically important product from whales. Oil of baleen whales is similar to that found in plants and other animals, that is, triglycerides, consisting of one molecule of glycerine with three molecules of fatty acids. In the past, these oils were used for lighting, heating, foodstuffs, margarine, soaps, and lubricants. Oil of toothed whales is a wax used for candles, leather dressing soaps, and lubricants. One exceptionally valuable byproduct from sperm whales was ambergris, a gray, waxy substance formed as an impaction in whale intestines. Ambergris was incorporated in cosmetics, love potions, headache remedies, and perfume. Whale bones, or the keratin baleen plates, were used to make corsets. Bones were used to make furniture (vertebrae), fence pickets (ribs), and housing beams. Whale meat is still popular in Japan and Norway, where the lack of agricultural lands limits space for economical production of livestock.

THE HISTORY OF WHALING ("A WHALE SHIP WAS MY YALE COLLEGE AND MY HARVARD." ISHMAEL IN *MOBY-DICK*)

Early human-whale encounters (hundreds of years B.C.) were caused by whale stranding behavior, and today the mystery of whale beaching remains. Early descriptions of whales were based on beached specimens, and these encounters would eventually lead to whaling. Somewhere around 1000 B.C. the Basques began commercial whaling for right whales (so named because they had thick blubber, did not swim too fast for little boats, and floated when killed). Use of protected bays and inlets for breeding made some whales particularly vulnerable to whalers.

As stocks were depleted, the Basque, British, and Dutch whalers expanded the hunt to bowhead whales in arctic waters and eventually to sperm whales worldwide. Early whaling was a dangerous profession as men, armed with slim iron harpoons, attacked whales from rowboats and held on to wounded, thrashing sixty-ton whales. Despite the inefficient hunting techniques, bowhead, sperm, gray, right, and humpback whale populations were seriously depleted by the beginning of the twentieth century. Late in the nineteenth century, the Norwegians developed mechanized whaling (exploding grenade harpoons, bow-mounted cannons, and steam-catcher boats), which made the fast-swimming rorqual whales (blue, fin, sei, Brydes, Minke) vulnerable to capture.

Whales were a common-property resource, implying freedom of access. Consequently, local coastal stocks were quickly depleted, demonstrating the "tragedy of the commons." The response was for whalers to move farther in search of whales. Floating-factory ships with stern slipways allowed processing far from land and whale stocks in the Antarctic, and all oceans were targeted by various whaling nations. International agreements to regulate whaling were not made until the 1930s and 1940s. First regulations were intended more to stabilize the market, preventing overproduction of whale oil, and to increase the output of oil per whale. Harvest was restricted to after summer feeding had fattened the whales.

In 1946, the International Whaling Commission was formed to "provide for the proper conservation of whale stocks and thus make possible the orderly development of the whaling industry." The commission is open to whaling and nonwhaling nations. However, regulations of the International Whaling Commission are difficult to enforce; any nation may object to decisions of the IWC and thereby exempt itself from certain IWC rules. Because regulations are left to the national fisheries agencies to enforce, international pressure and trade sanctions are the only way to encourage compliance.

Scientific management was slow to be applied to whaling. In the mid 1950s, virtually no quantitative studies were being done on whale stocks and scientists attending IWC meetings had little, if any, quantitative expertise. As expertise in whale population analysis was applied in the 1960s, the world demand for whale products declined. Some American whaling persisted for pet food as recently as 1960s. However, by the mid 1970s, all eight great whales were widely regarded as endangered and the public acceptance of whaling was changing, especially in the United States.

Since 1970, U.S. society has treated whales as conservation symbols, and herculean efforts were made to "save the whales." For example, in 1988, $5.8 million was spent in an attempt to save three trapped gray whales (see Tom Rose, *Freeing the Whales*). During recent development of U.S. policy, the protectionist community has advanced a non-consumptive use philosophy, often supplanting scientific management with emotion.

Support for a moratorium on whaling shifted from a resource-management question to an ethical question. Former Chairman of the U.S. Marine Mammal Commission writes: "Whales are different. They live in families, they play in the moonlight, they talk to one another, and they care for one another in distress. They are awesome and mysterious. In their cold, wet, and forbidding world they are complete and successful. They deserve to be saved, not as potential meatballs but as a source of encouragement to mankind." Popular culture reinforces a mediagenic image of whales. In the movie *Star Trek IV: The Return Home*, humankind is saved because whales are brought back from the brink of extinction.

Management procedures followed by the IWC are based on the maximum sustained yield (MSY) concept. The MSY concept states that the surplus of recruits beyond natural mortality is greatest at some intermediate population level and the surplus can be harvested without depleting the population. Estimating the MSY and population levels for whale species proved very difficult, preventing agreement on management recommendations. Therefore, in 1982, the IWC imposed a moratorium on all commercial whaling because data on whale stocks and dynamics were deficient. The moratorium took effect in 1986 and was to last until 1990, by which time the IWC's Revised Management Procedure (RMP) would set scientifically defensible quotas. The moratorium remains in effect today, despite the unanimous recommendation by the IWC's scientists that the RMP quota setting is defensible. Japanese and Soviet whalers continued to take whales under scientific research permits; Japan has killed approximately three hundred Minke whales per year since 1987. The worldwide ban on commercial whaling cannot be justified as efforts to "save the whales" because, of the seventy-nine species of cetaceans, only nine (plus three species of river dolphins) are in fact endangered. However, it's important to realize that the demography of cetaceans is a highly imprecise science. Furthermore, the blue whale and humpback whale have been totally protected from commercial hunting since 1965, and right whales and gray whales since the 1930s.

LESSONS LEARNED

The history of whaling and attempts to regulate whaling highlight several important lessons for natural resource management:

- Sustainable use requires science-based management.
- Early intervention to limit access is needed to prevent overharvest.
- Accurate biological data are needed on each species and subspecies.
- Monitoring of users is needed to ensure compliance with regulations.
- Political agreement that your goal is desirable (for example, sustainable whaling) must be reached before you can do science-based management.
- Scientific advice is seldom neutral. It is generated in a cultural context, which influences the outcome. Consequently, interpretations of the same information by the whaling industry, environmentalists, and cetologists are usually conflicting.

CURRENT CONTROVERSY OVER WHALING

THE CASE FOR WHALING

After petroleum-based products replaced whale oil (circa 1900) and vegetable oils could be hydrogenated to make margarine (1960s), the justification for large-scale whaling ceased to exist. In the 1970s and 1980s, whaling continued to provide meat in those societies where it had historically been an important part of the diet. Meeting this need requires small-scale fisheries, not the industrial-scale factory ships that decimated the great whale stocks. Total value of whale products in 1972 was estimated at $100 million with potential value up to $500 million. Iceland, Japan, and Norway currently express interest in resuming commercial whaling, and most other nations have low consumer demand for whale meat. Therefore, the market demand for whale products is limited, alleviating fears of overharvest. Furthermore the Norwegian quota of Minke whales (425 in 1996) can be harvested from the northeast Atlantic population of 110,000 Minke whales with little or no risk to the population.

A ban on whaling does not safeguard ecosystem integrity. The three countries involved have limited land areas suited for modern agricultural meat production. Whaling is energy efficient and results in less environmental damage than land-based food production (which results in, for example, soil erosion, wildlife habitat loss, contaminants, greenhouse gases). Compared to coastal whaling, for which fossil-fuel energy input to protein-energy output ratios are 2:1, farm-raised chicken, pork, and feedlot beef production ratios are 22:1, 35:1, and 78:1, respectively. Small-scale whaling is also ten times more fuel efficient than major fisheries for finfish (cod, tuna, and others) and shrimp. Whale fisheries can be effectively regulated with quotas because the targeted whale can be identified by species and sex, and prohibitions against catching females with calves can be enforced.

The coastal communities of Japan, Norway, and Iceland have traditionally acquired most of their dietary protein from marine fisheries. Many whalers also derive income from fishing, thus whales and humans compete for seafood. Food consumption by sperm whales worldwide is 100 million tons per year; by comparison, the total world catch of seafood is approximately 100 million tons per year. Therefore, increasing whale stocks may threaten the livelihood of coastal communities dependent on fisheries.

The argument for whaling is not entirely based on use values. In Japanese coastal whaling villages, Minke whale meat and blubber are important for thirty different culturally significant events. Hunting, processing, distribution, consumption, and celebration phases of whale use are important components of the society's cultural identity. The promotion of whales as conservation symbols to be protected at all cost has ignored the cultural values of those communities that have historically harvested whales. Many Japanese view the protectionist attitudes of Western countries as ethnocentric or downright racist.

The Case against Whaling

Although historically the case against whaling has centered on the ethics of contributing to the extinction of whales, the rebound in whale populations has forced whale protectionists to develop an alternative position. Whales have intrinsic values apart from their human uses. This value can be protected only by recognizing cetacean rights and preventing inhumane treatment and killing.

The intrinsic values far exceed the economic value of whale products. Whales are unique in their intelligence level, playfulness, and grace. Their being sentient beings makes it morally wrong for humans to unnecessarily cause them pain and suffering. Furthermore, alternatives exist for most products derived from whales, and killing whales is not necessary to fulfill essential human needs. Other nonconsumptive uses of whales are more acceptable to our society and contribute to economies. For example, in 1991 more than four million people spent in excess of $300 million on whale-watching activities.

Oversimplifying the Case

Those opposed to whaling tend to talk about the whale in the singular, not the seventy-five or more species of cetaceans. Consequently, the image of the "super-whale" is created. The super-whale is the largest mammal on earth (blue whale), has a large brain-to-body-weight ratio (bottlenose dolphin), sings (humpback whale), has nurseries (some dolphins), is friendly (gray whale), and is endangered (blue whale, right whale). The super-whale is endowed with all the qualities we like to see in fellow humans: kindness, caring, playfulness. The super-whale is the image of a single whale possessing all these generalized traits. Such a creature does not exist.

Norwegian Whaling

Norway, Japan, and Iceland oppose the current IWC moratorium on commercial whaling. Norway has been most successful in preserving its whaling industry. Currently, Norwegian whalers operate out of small (fifty- to sixty-foot) family-financed boats. They do not see themselves as a threat to whale populations. Norway ceased commercial whaling in 1987, pending research into the status of the Minke whale population of the northeast Atlantic. They resumed harvest in 1993 with a quota of 293 Minke whales. Minke whales, at eight tons, are the smallest of the great whales. The quota in 1996 for thirty-one licensed boats was 425. An international Minke whale-sighting survey in 1995 produced an estimate of 110,000 whales (95 percent confidence intervals of 97,000 to 144,000). Whalers shoot them with a small harpoon; in the 1994 season 30 percent died instantly and the average time until death was three minutes. Whalers earn $13 a kilogram for the whale meat, which in shops cost four times that. What is your opinion on whaling by Norway? Is it wrong? Why or why not?

Aboriginal Subsistence Whaling

In 1982, the IWC distinguished between commercial and subsistence whaling. Aboriginal subsistence harvest means whaling for purposes of local aboriginal consumption, carried out by native peoples who share community, family, societal, or cultural ties related to traditional dependence on whaling or on the use of whales. The U.S. government requested an IWC permit for harvest of endangered bowhead whales by the Alaskan Inuits; the justification was to satisfy cultural and nutritional needs. The bowhead quota for Alaskan Inuits was a total of 141 for the three years 1992, 1993, and 1994. A maximum of fifty-four bowheads may be hit (by harpoons) every year, a maximum forty-seven may be landed every year (a number of wounded whales escape after being hit), and no mothers with calves may be hunted. The quota for 1994 through 1998 was 204 bowheads, and Russian and Canadian natives are now requesting quotas. The Alaskan Inuits continue to use seal-skin boats (umiak) but have adopted penthrite projectiles, which are small grenades designed to ensure a quick death when a whale is harpooned. The Bering-Chukchi stock in the Beaufort Sea (from which the Inuits hunt) is estimated to be 6,400 to 9,200 animals (the most likely number is 7,500). Scientists estimate that the replaceable yield of bowhead whales is 254 animals (most likely), or 92 animals (minimum) per year.

In 1996, the United States petitioned the IWC on behalf of the Makah Tribal Council to kill five gray whales off the coast of Washington. The Makah support resumption of the hunt for cultural reasons. Because they stopped hunting gray whales in the 1920s when gray whales were approaching extinction, the Makah cannot prove that they have a subsistence need. The gray whale came off the endangered list in 1994 and about 21,000 gray whales now exist. World Wildlife Fund, Sea Shepherd Conservation Society, Cetacean Society International, and the United States Congress pressured the U.S. delegation to drop the petition. How can the United States justify dropping the request of the Makah to hunt a recovered whale population while it supports Alaskan harvest of an endangered bowhead population?

ETHICAL DILEMMAS

The whale controversies involve ethical dilemmas in addition to scientific problems. Separating the two is important to make the rationale for one's position clear. Is killing whales (or other cetaceans) morally wrong? Should commercial whaling be banned? Should aboriginal whaling be banned? Whose rights take precedence—human rights to pursue traditions or animal rights? Think about the logical consequences of your arguments. What general moral principle did you use to support your argument?

Is it possible that antiwhaling forces are missing a larger threat? What is our ethical obligation to preserve ocean habitats for the whales, other marine life, and humans? Depleted fisheries, pollution from oil tankers and other sources, ozone depletion and phytoplankton productivity declines, coastal development, harassment of whales by en-thusiastic whale watchers, and other difficult dilemmas affect all forms of life dependent on the oceans.

EVALUATING MORAL ARGUMENTS

General moral principles guide our everyday decisions. For example, one principle is "one must respect human rights or be banned from society." Along with the rights come responsibilities, because rights granted to one individual may limit the freedom of an-other. Your moral argument regarding whaling should take the following form:

- Empirical premises
- General moral principle
- Conclusion

One example argument is the following:

(1) Whaling involves the infliction of unnecessary suffering and death to sentient beings.
(2) Causing unnecessary suffering, or unnecessary death, to sentient beings (whales) is wrong.

(3) Whaling is wrong.

Another possible example:

(1) Whale hunting is a part of the cultural tradition of certain societies.
(2) Whale hunting provides protein in coastal communities that have limited land for crop production.
(3) Whale killing can be done in ways to minimize pain and suffering.
(4) Whatever is a part of the cultural tradition of certain societies is sometimes per-missable, assuming it can be done in ways to minimize pain and suffering.

(5) Whale hunting is sometimes permissible.

To evaluate these and other arguments, you should evaluate three questions:

1. Are the empirical premises true?
2. Does the conclusion follow logically from the premises?
3. Is the general moral principle justifiable?

The best arguments will survive this scrutiny. Develop an alternative argument to support your position on whaling.

TAKE HOME MESSAGE

The controversies surrounding marine mammal protection and management are similar to many controversies in fisheries and wildlife. Scientists and decision makers are involved in making hard choices about dynamic world situations in the face of uncertainty. Some of the decisions are scientific, some moral, but all are difficult. To foster dialog and continued learning, we must alleviate tensions among conflicting interests and develop creative solutions. We also must learn to debate moral as well as scientific arguments and recognize the difference.

FOR MORE INFORMATION

A. A. Blix, L. Walloe, and O. Ulltang, eds., *Whales, seals, fish, and man* (NY: Elsevier Science, 1995).

J. Cherfas, 1992. "Whalers win the number game." *New Scientist*, 11 July, 12–13.

R. Ellis, *Men and whales* (NY: A. A. Knopf, 1991).

M. M. R. Freeman and U. P. Kreuter, eds., *Elephants and whales: resources for whom?* (Switzerland: Gordon and Breach Publ., 1994).

R. Gambell, "International management of whales and whaling: an historical review of the regulation of commercial and aboriginal subsistence whaling," *Arctic* 46 (1993) 7–107.

M. Klinowska, "Brains, behaviour and intelligence in cetaceans (whales, dolphins and porpoises)," in O. D. Jonsson, ed, *Whales and Ethics* Reykjavik Univ. Press.(1992) 23–37.

L. L. Lones, "The Marine Mammal Protection Act and international protection of Cetaceans: A unilateral attempt to effectuate transnational conservation." Vanderbilt J. Transnational Law 22 (1989) 997–1028.

R. Payne, *Among Whales* (NY: MacMillan, 1995).

M. Peterson, "Whalers, cetologists, environmentalists, and the international management of whaling." *International Organization* 46 (1992), 147–186.

D. Sarokin and J. Schulkin, "Environmental justice: co-evolution of environmental concerns and social justice," *The Environmentalist* 14 (1994) 121–129.

J. E Scarff, "Ethical issues in whale and small cetacean management," *Environmental Ethics* 2 (1980) 241–280.

M.P. Simmonds and J.D. Hutchinson, eds., *The conservation of whales and dolphins: science and practice* (NY: Wiley & Sons), 1996.

P. Stoett, "International politics and the protection of great whales," *Environmental Politics* 2 (1993) 277–302.

P. Stoett, *The international politics of whaling*, (Vancouver, BC: UBC Press,. 1997).

American Cetacean Society, PO Box 1391, San Pedro, CA 90733-1391

INFORMATION ON THE WORLD WIDE WEB

Cetacean Society International WWW site: http://elfnet1a.elfi.com/csihome.html

Tirpitz - Information on whales: http://tirpitz.ibg.uit.no/wwww/ss.html

Dancing Dolphin Institute: http://www.maui.net/%7Edolfyna/index.html

Dolphin Circle Homepage: http://www.premier1.net/~iamdavid/

Marine Mammals: Dolphins and whales: http://www.oregoncoast.com/Whales.htm

High North Alliance Web: http://www.highnorth.no/th-ec-pe.htm

Sea Shepard Conservation Society: http://www.envirolink.org/orgs/seashep/

NOTES

1. Reprinted from *Ag Bioethics Forum* Vol. 8, No. 1, June 1996.
2. Reprinted from *Ag Bioethics Forum* Vol. 9, No.2, Nov 1997.

Chapter 12

FOOD

INFANT DEATHS IN DEVELOPING COUNTRIES

Lois Banta, Jeffrey Beetham, Donald Draper, Nolan Hartwig, Marvin Klein, and Grace Marquis

You are sitting on an NIH committee that has been charged with the responsibility to evaluate two projects that have been submitted from university-based researchers. Only one project will be funded; monies come from U.S. taxes. You are responsible for deciding which project should be funded.

Project 1: This is a five-year, community-based trial of an oral, yeast-based, genetically engineered, rotavirus vaccine. Preliminary data show that the vaccine is 85 percent effective; the long-term risks have not been established but the safety of the vaccine has been verified by the FDA. The ethical committees at NIH, the researchers' university, and the Ministry of Health in the target country have given clearance.

The trial will take place in a low-income country in South America where infants zero to twelve months typically have nine episodes of diarrhea per child per year. Rotavirus causes approximately 10 percent of all diarrheas and is the cause of 80 percent of all hospitalizations related to infantile diarrhea. This virus is the primary cause of diarrhea-related mortality, resulting in approximately twenty thousand infant deaths each year (a 10 percent mortality rate).

The vaccine regimen is given at three, six, and nine months of age. The Ministry of Health already has a vaccine program for other diseases in place, into which this vaccine can easily be implemented. The full course of treatment costs $50 per child. Because we are going to inoculate all two million infants in the country, the total cost of the project will be $100 million.

Project 2: This is a five-year subsidy of an ongoing influenza vaccine distribution program of the Department of Health and Human Services, administered through the local health departments throughout the United States. This is a well-established vaccine that has been shown to be effective, and the risk of side effects is very low. The program provides influenza vaccinations for the elderly and immuno-suppressed individuals who are at increased risk of death from flu. The subsidy will allow the ten million Americans who could not otherwise afford the influenza vaccine to receive it at a final cost of $100 million. This vaccine is estimated to save twenty thousand lives.

QUESTIONS

1. Identify the factors that you consider the most important in helping you make this decision.
2. Which of these factors from question 1 are factual claims?
3. List the moral considerations that influence your decisions. (Values to consider: health, personal safety, security, human welfare, emotional well-being, compassion, altruism, empathy, equality.)
4. Which grant proposal would you fund?
5. What arguments can be made against your position?
6. Do any conditions (such as the political situation in the South American country in Project 1) exist that would lead you to a different decision? For example, what if this country is an emerging democracy, with its first elected leader after decades of repressive rule by a military junta? Should the U.S. government factor into its decision-making process a desire to ensure stable leadership and provide credibility for a precariously positioned neophyte president?

FURTHER DISCUSSION

1. Go through the same analysis assuming that Project 1 would save thirty-five thousand infants.
2. Go through the same analysis assuming that Project 1 would save five thousand infants.
3. Imagine a third alternative, in which the $100 million is used to develop the infrastructure and transfer the vaccine production technology to the target coun-

try so that that country is able to produce and distribute the vaccine. This implementation will take eight years.

EDIBLE ANTIBIOTICS IN FOOD CROPS

Mike Zeller, Terrance Riordan, Halina Zaleski, Dean Herzfeld, and Kathryn Orvis

Imagine that a large land-grant university has partnered with a major agricultural company to create a consortium to produce low-cost, high-quality phytopharmaceuticals. *Phytopharmaceuticals* are compounds that can be and are used as drugs, and can be natural products as well as genetically modified products derived from plants. In this case, corn was bioengineered to produce large quantities of a vital antibiotic: penicillin. The production of this crop containing the antibiotic in the seed will benefit mostly developing nations by providing a steady, reliable supply of cheap product that easily can be consumed orally. Ultimately, the cost of the drug will be 10 percent of the cost of producing penicillin using current production methods. Storage and transportation of antibiotic will be simplified by eliminating the need to refrigerate the drug. The use of needles and their associated risks will also be removed. In the United States, strict rules concerning genetically modified (GM) food crops exist and are routinely enforced. Presently, the USDA, FDA, and EPA have approved the modified maize for human consumption under prescription in the United States.

Opponents of the GM crop have raised the following issues. The potential for contamination of other, non-GM crops is very high when a GM crop such as corn expresses an allergenic compound. The reason is that corn is wind-pollinated. In addition to pollen drift, storage contamination and the potential for contamination through mixing of supplies raise serious risks for those allergic to antibiotics. Because of the seriousness of the consequences, it has been suggested that the risks be evaluated using the precautionary principle as opposed to risk assessment, the standard method currently relied on by regulatory agencies. Dosing and intake control have surfaced as major problems with consuming antibiotic in a food crop. Development of antibiotic resistance in infectious agents could pose serious risk. Potential environmental impacts include cross-contamination of neighboring maize fields with the GM crop pollen. Isolation and refugia (a "refuge" of GM crop among non-GM crop) of the genetically modified maize crop becomes undisputedly necessary.

An anti-GM activist group advances the claim that the consortium is not proposing the new crop as an altruistic action. Rather, the activist group claims, the consortium is proposing the new crop to make huge profits in the animal feed industry in the

United States. The idea is that the new crop would be grown primarily, on large acreages, in the United States. The major use of the new crop, in other words, would not really be for disease treatment in developing countries but rather for market animal growth promotion. In the United States, low levels of antibiotics are used in animal feed. These antibiotics modify the microorganisms in the gut of the animal, thereby improving the animal's weight gain and feed efficiency.

Genetically modified "traditional" pharmaceuticals are already in use and are widely accepted by consumers in the United States. These pharmaceuticals have been deemed safe by the relevant U.S. regulatory agencies. Recombinant insulin, for example, is widely used by diabetics. As a result of GM in the medical industries, insulin is now much cheaper and in greater supply.

WHAT ETHICAL ISSUES ARE AT STAKE HERE?

The notion of placing an antibiotic in a major food crop raises important questions on both sides of the issue. One side sings the praises of this breakthrough, while another side condemns the technology. Formulate a moral/ethical position for each of the following questions based upon moral values and factual analysis. Remember to consider the stakeholders in each question.

1. Consider each of these potential complicating factors: wind pollination; humans with allergies; underlying issues of giving away the product, yet acquiring large profits from animal uses in the United States; dosing of the "drug" and following up with taking the entire prescription; control of who eats and shares the food; regulatory issues; issues surrounding growing the crop in developing nations, including use of chemical and fertilizer inputs, intensive row cropping and weeding, to produce a sufficient quality and quantity of a crop for production to be profitable; resistance issues.
2. Should we be doing this?
3. How should it be regulated?
4. Will your agronomist become your pharmacist? Will your grocer become your pharmacist?
5. Should the GM maize be limited to human use? To animal use? How would such a limitation change the risks and benefits?
6. Is the opposition based on the actual risk implied or only on the alleged immorality of producing GM organisms?
7. Should the university receive benefits, financially or otherwise, from this product?
8. Should the consortium be allowed to patent and thus control the product?
9. If industry won't support this type of or this exact research, should the federal government subsidize the research? If this research is meant to help developing

countries then are we morally obligated to do it? Should government support depend on industry support?

10. Should the targeted users/audience have a say in the process? Should the process pass though international aid agencies or the governments of the developing countries?

11. Should U.S. agencies (USDA/FDA/EPA) or other agencies such as the WHO (World Health Organization) or FAO (Food and Agriculture Organization) regulate the product?

12. What might the effects of different cooking/culinary methods on the antibiotic imply for the consumer who is ill and needs the full benefit of the drug?

RESOURCES

W. H. R. Langridge, "Edible Vaccines," *Scientific American*, 2000.

WEB SITES

http://biotechknowledge.com
    Monsanto (industry) educational site
www.eurekalert.org
    Various articles on many scientific and technical topics, searchable content
http://scoped.educ.washington.edu/gmfood/
    Controversy Forum sponsored in part by the AAAS (*Science* magazine)
    Contains facts, e-mail list, discussion group, and an extensive resource/reference list
www.columban.com/gencon.htm
    A nice essay written from a religious perspective
http://216.129.146.198/Lauren's%20Lit%20Review
    A literature review written by a student on internship (Dietetic Intern)
www.anth.org/ifgene/proscons.htm
    A table of pros and cons of various aspects of genetic engineering
www.psrast.org/ecolrisk.htm
    Risks to the ecosystem of genetically engineered crops
www.newswise.com
    Searchable content of various news articles on science topics. See "Are Genetically Engineered Foods Natural?"
www.cast-science.org/
    Center for Agricultural Science and Technology: CAST is an excellent source for issue papers and reports, such as "Applications of Biotechnology to Crops: Benefits and Risks"
www.ers.usda.gov/publications/aib766/

Chapter **13**

ANIMALS

BEEF, MILK, AND EGGS

Gary Varner

ETHICAL VIEWS ABOUT ANIMALS

The literature on the moral standing of animals is complex and vast. To oversimplify, three major clusters of ethical theories exist: animal welfare, human dominion, and animal rights. Please read carefully the descriptions of these theories in Table 13.1, then answer the two questions that follow.

QUESTIONS

1. For those in the *human dominion* camp:

Dominionists deny that animals are conscious. How do you think they could defend this view on scientific grounds?

2. For those in the *animal welfare* and *animal rights* camps:

What do you think is the moral status of nonsentient animals for welfarists and rightists?

TABLE 13.1. REPRESENTATIVE VIEWS ON THE MORAL STATUS OF ANIMALS.

	Summary Characterization	Associated Attitudes	Typical Underlying Philosophical Basis
1. Animal Welfare	*We are stewards of animals. Their lives and experiences have intrinsic value but it is up to us to decide how to maximize value in the aggregate by using animals in various ways.*	Various traditional uses of animals are permitted as long as they serve nontrivial ends and are conducted in ways that eliminate unnecessary animal suffering. For example: • Medical research • Humane animal slaughter • Hunting, at least to prevent wildlife overpopulation	1. We have a moral obligation to balance benefits and harms. 2. If an animal can suffer pain, we have an obligation to balance this harm against the benefits of any human use of the animal. 3. So we should continue to use animals when the benefits to us outweigh the costs to them, but in doing so, we should eliminate unnecessary animal suffering.
2. Human Dominion	*We have dominion over animals. That is, they have value only as means to our ends.*	Everything under animal welfare is permitted, plus things such as: • Cockfighting, circuses, rodeos, and bullfights • Confined exotic animal hunting • Injuring animals for movies	1. Animals have no moral standing because they lack consciousness, including consciousness of pain. 2. So it doesn't matter, morally speaking, how we treat them; no treatment of animals can be judged immoral except by virtue of its indirect effects on humans.

TABLE 13.1. CONTINUED.

	Summary Characterization	Associated Attitudes	Typical Underlying Philosophical Basis
3. Animal Rights	*Animals have moral rights. And when individuals have moral rights, we cannot treat them as means to our ends.*	Many or most traditional uses of animals are opposed, including everything listed as permissible under either of the preceding views, plus such things as: • Consuming animal by-products (such as milk and eggs) • Captive breeding programs for endangered species • Keeping pets	1. If you have rights, we cannot justify harming you just because the benefits to us outweigh the harms to you. 2. Some nonhuman animals have mental lives similar to those of some humans (if only very small children). 3. So if we recognize rights for all humans (including very small children), we should recognize rights for those animals. 4. And so, for those animals, we cannot justify harming them just because the benefits to us outweigh the harms to them.

Note: The first draft of this summary was prepared by Gary Comstock of Iowa State University, based on Gary Varner's online lecture on animal rights and animal welfare philosophies, available at: http://www-phil.tamu.edu/~gary/awvar/lecture/index.html

Which kinds of animals are conscious of things such as pain, and how do you know?

ANIMAL AGRICULTURAL PRACTICES[1]

Read the following three cases. Using information from "Ethical Views about Animals," answer the questions that follow the text.

CASE #1: BEEF

Approximately thirty million cattle are slaughtered yearly in the United States. When it comes to the slaughter procedure itself, the large-scale, state-of-the-art facilities capable of slaughtering as many as four- to six hundred animals per hour are, perhaps contrary to popular belief, the most humane, at least if operated properly. The races approaching the stunning chute can be designed to look just like those through which cattle have passed previously for routine veterinary care; experienced handlers can move animals along without prodding; cattle do not "smell blood in the chutes"; and *stunning* is a misnomer for what happens in the kill chute, because a properly placed shot with a "stun gun" obliterates the animal's brain, making it impossible to regain consciousness.

Questions for Case #1

1. What would a person thinking from the animal welfare perspective say about this practice? Why?
2. What would a person thinking from the human dominion perspective say about this practice? Why?
3. What would a person thinking from the animal rights perspective say about this practice? Why?
4. In your own opinion, is this method of slaughter morally permissible? Should it be the legally required method?

CASE #2: MILK COWS

On average in the United States, milking cows spend between three and four years in production, after which they are slaughtered for relatively low-grade beef. Dairy farmers maintain high productivity by breeding cows to calve about yearly. The calves are removed from their mothers immediately or within days after birth, with most of the female calves becoming replacement milk cows and many of the male calves being raised for veal. Statistics indicate that about one seventh of the cattle slaughtered yearly in the United States come from dairy operations.

Questions for Case #2

1. What would a person thinking from the animal welfare perspective say about this practice? Why?
2. What would a person thinking from the human dominion perspective say about this practice? Why?
3. What would a person thinking from the animal rights perspective say about this practice? Why?
4. In your own opinion, is this method of slaughter morally permissible?

CASE #3: LAYING HENS

Today, more than 90 percent of laying hens in the United States live caged in intensive production facilities, which increased the average yield per hen from seventy in 1933 to around 275 today. In such facilities, birds cannot forage, flap their wings, dust-bathe, nest, establish dominance hierarchies, or even preen themselves in natural ways; culling of injured birds is economically inefficient; and the entire population of a battery operation is slaughtered and replaced periodically (every twelve to fifteen months in state-of-the-art operations).

Poultry are still exempt from federal humane slaughter legislation and, by comparison to state-of-the-art cattle slaughter facilities, poultry slaughter is still a relatively indelicate affair, with fully conscious birds hung from their legs on conveyor belts before being stunned and beheaded.

Questions for Case #3

1. What would a person thinking from the animal welfare perspective say about this practice? Why?
2. What would a person thinking from the human dominion perspective say about this practice? Why?
3. What would a person thinking from the animal rights perspective say about this practice? Why?
4. In your own opinion, is this method of slaughter morally permissible?

NOTES

1. The three cases originally appeared in *The Ag Bioethics Forum* 8 (December 1996): 4, 6, 9, and are based on information in Bernard Rollin's *Farm Animal*

Welfare: Social, Bioethical, and Research Issues (Ames: Iowa State University Press, 1995) and in Gary E. Varner, "What's Wrong with Animal By-Products?" *Journal of Agricultural and Environmental Ethics* 7 (1994): 7–17.

VETERINARY EUTHANASIA

BERNARD ROLLIN, JERROLD TANNENBAUM, COURTNEY CAMPBELL, KATHLEEN MOORE, AND GARY L. COMSTOCK[1]

ETHICS PRE-TEST

The purpose of this pre-test is to establish a baseline for the student's knowledge of ethics. After completing the entire exercise, the student will repeat the test to evaluate how much he or she has learned.

1. Read the following case:

   Ms. White brings in Tiger, a four-year-old, male, castrated, shorthaired cat. Tiger has recently begun "spraying" in the house, a behavior that began shortly after the birth of Ms. White's first child six months ago. She is at wit's end. In addition to the aggravation of the problem, she has already spent more than five hundred dollars on cleaning bills. She wants you to euthanize Tiger. You recommend several behavioral specialists, but she is too busy with the new baby to spend any more time or money on the animal. You became a veterinarian years ago with the intent of helping orphaned animals. You ask Ms. White whether she has tried giving the animal to a neighbor or family member. She says no. Should you euthanize Tiger?

2. Place a check mark next to all stakeholders, that is, individuals or groups who may be affected by your decision.

   A veterinary client _____

   Other veterinary clients _____

   All veterinary clients _____

   A veterinary patient _____

   Other veterinary patients _____

All veterinary patients _____

A cat _____

A dog _____

A cow _____

A horse _____

A pig _____

A domestic animal not named above _____

A wild animal _____

Cats _____

Dogs _____

Cows _____

Horses _____

Pigs _____

Domestic animals not named above _____

Wild animals _____

All cats _____

All dogs _____

All cows _____

All horses _____

All pigs _____

All domestic animals not named above _____

All wild animals _____

All pet owners _____

A farmer _____

All farmers _____

A rancher _____

All ranchers _____

Food manufacturing industry _____

Food consumers _____

People in developing countries _____

People in developed countries _____

Future humans _____

Future animals _____

Future human generations _____

Future animal generations _____

All living beings _____

All present and future living beings _____

3. Name the individuals and their interests that are in conflict in this case.

_____ v. _____

_____ v. _____

_____ v. _____

_____ v. _____

4. Philosophers and theologians develop ethical theories in part to help us to answer difficult questions, such as when, if ever, we are justified in killing. Which of the following theories discuss principles relevant to animal euthanasia? If you do not know, leave the space blank.

Utilitarianism _____

Social contract _____

Ecofeminism _____

Cartesianism _____

Talmudic law _____

Jainism _____

Christian dominionism _____

Animal rights _____

Buddhism _____

Christian stewardship _____

Islamic law _____

5. Which of the following theories is likely to object to euthanasia for Tiger? If you do not know, leave the space blank.

Utilitarianism _____

Social contract _____

Ecofeminism _____

Cartesianism _____

Talmudic law _____

Jainism _____

Christian dominionism _____

Animal rights _____

Buddhism _____

Christian stewardship _____

Islamic law _____

6. Which of the following theories is likely to approve euthanasia for Tiger? If you do not know, leave the space blank.

Utilitarianism _____

Social contract _____

Ecofeminism _____

Cartesianism _____

Talmudic law _____

Jainism _____

Christian dominionism _____

Animal rights _____

Buddhism _____

Christian stewardship _____

Islamic law _____

7. Which of the following ethical theorists would unambiguously endorse euthanasia for Tiger? If you do not know, leave the space blank.

René Descartes _____

John Stuart Mill _____

Peter Singer _____

Tom Regan _____

Peter Carruthers _____

Immanuel Kant _____

Total correct: _____

WRITE AND PASS

Ms. White brings in Tiger, a four-year-old, male, castrated, shorthaired cat. Tiger has recently begun "spraying" in the house, a behavior that began shortly after the birth of Ms. White's first child six months ago. She is at wit's end. In addition to the aggravation of the problem, she has already spent more than five hundred dollars on cleaning bills. She wants you to euthanize Tiger. You recommend several behavioral specialists, but she is too busy with the new baby to spend any more time or money on the animal. You are certain that Tiger would do well in a childless home with some modest behavioral assistance, and you have space in the back room to take him. On the other hand, you already have five stray cats back there, and each of them is, to the best of your knowledge, problem free. Each cat has been waiting for adoption for more than two weeks. Tiger will be more difficult to place and, of course, keeping the cats costs you time and resources. Ms. White asks you, again, to euthanize the animal. You remind yourself that one of your primary ethical values in becoming a veterinarian was to help orphaned animals. You ask Ms. White whether she has tried giving the animal to a neighbor or family member. She says no.

Question: Should you euthanize the cat?

Formulate your answer in a clear and complete claim and write it here:

(1) _____

_____

_____

_____

_____

_____

Pass this paper to the person on your right.

(2) Write down one good reason that supports the claim stated in (1).

_____

_____

_____

_____

_____

_____

Pass this paper to the person on your right.

Together, statements (1) and (2) form an argument. An argument is a claim and a reason that supports it. Every argument makes certain assumptions. (For example, from the claim that "All scientists are smart," it follows that "Lewontin is smart," but only if one assumes that Lewontin is a scientist.)

(3) List one assumption made in this argument.

_____

_____

_____

_____

_____

_____

Pass this paper back two persons to your left, to its original owner.

STEPS IN ETHICAL DECISION MAKING

You are able to convince Ms. White to keep her cat. She attends to his behavior problem and thanks you for your advice at his next check-up. A half dozen years pass. Tiger is in good health. Then, his blood work tests positive for the feline leukemia virus. The news comes as a shock to Ms. White because Tiger is her only cat and rarely leaves the house, although he has managed to get out on occasion to spend some time with a neighbor's cats. Other than testing positive for FeLV, Tiger is in excellent condition.

You explain to Ms. White that although feline leukemia has no cure, Tiger has no signs of the active disease and could live several years completely free of symptoms. Then, when Tiger does become sick, treating him with chemotherapy might be possible.

Ms. White responds very slowly and deliberately. She regrets to say it, but she has decided that she would like Tiger put to sleep now, before he becomes ill. "I don't want to see him get sick, and I don't want him to die a painful death, and that is all I have to say about the matter." You ask whether she would consider giving Tiger up for adoption. You tell her that your practice knows several people who adopt FeLV-positive cats. Ms. White refuses. She asks you, "Do you have objections to having him put to sleep?" "Well," you say, "yes, I believe it is unethical to put this animal to sleep. I know what you must be going through, but I would no sooner put Tiger to sleep because he will eventually become sick than I would euthanize a relative of mine with Alzheimer's disease."

Ms. White politely picks up Tiger and begins to leave your examining room. You intervene. "I'm sorry. Did I say something to upset you?" Ms. White looks at you with tears in her eyes. "My brother died six months ago from leukemia after considerable suffering. We were extremely close, and toward the end he asked me to help him put himself out of his suffering. But there was nothing I could do. I cannot bear to think about putting Tiger through similar suffering. So, when I ask for euthanasia for this little friend of mine, I do not ask for it without having thought long and hard about it. But I understand your position. I will find another veterinarian to assist us."

Question: Given the new information about the client's situation, and given your belief that she will easily be able to find another veterinarian to perform the procedure, do you think *you* should now euthanize Tiger?

1. *Identify the moral issue:* What ethical decision do you have to make in this case? Formulate the problem as a normative question. (Hint: Normative questions often contain the formulaic phrases "Should I do. . . ?" or "Ought I do. . . ?") Write your normative question here:

_____

_____

_____

_____

_____

_____

2. *Identify the stakeholders:* List every individual and group who may be affected by your decision. Indicate whether they will be positively or negatively affected, and briefly explain why.

_____

_____

_____

_____

_____

_____

3. *Identify the ethical principle most important to your decision:* Which of the following principles is most important to you in making this particular decision?

- **Love of wildlife, nature:** Promote and protect wildlife interests
- **Self-satisfaction:** Direct my own career, be allowed space for personal creativity
- **Love of large animals:** Promote and protect large animal interests
- **Obligation to client:** Promote and protect interests of my clients
- **Love of small animals:** Promote and protect small animal interests
- **Loyalty to profession:** Work cooperatively to advance vet profession
- **Religious faith:** Loving God, adhering to faith, following a religious call
- **Knowledge and science:** Discover and promote scientific understanding
- **Animal rights:** Act to benefit all animals, end animal pain and slaughter
- **Social beneficence:** Act to promote the good of all human beings
- **Fame or wealth:** Be financially independent, achieve recognition, honors

Most relevant principle:

_____

_____

_____

4. *Ethical theory:* Before deciding what to do, reviewing alternative approaches is us·
ful. At least three major theories exist in animal ethics. Review the brief descri·
tions of these three theories in Table 13.1. Then, using the exercise called "Ethi·
and Euthanasia" later in this section, explain whether euthanasia for Tiger wou·
be ethically justified according to each ethical theory.

5. *Moral closure:* Finally, taking into account the various alternative theories and o·
tions open to you, decide what you will do. Will you honor Ms. White's request t·
euthanize Tiger? Why or why not? Write your answer here and then defend it b·
writing at least two complete sentences:

_____

_____

_____

_____

_____

_____

ETHICS AND EUTHANASIA

Refer back to Table 13.1, "Representative Views on the Moral Status of Animals,·
earlier in this chapter. Using Varner's categories of Animal Welfare (AW), Human Do·
minion (HD), and Animal Rights (AR), explain whether euthanasia for Tiger is ethicall·
justified according to each ethical theory.

On the following lines, write *J, NJ,* or *D*

J = Ethically justified
NJ = Not ethically justified
D = Depends

Defend each of your answers in at least two complete sentences.

Euthanasia of otherwise healthy, FeLV-positive cat:

Animal Welfare : _____

_____

_____

Human Dominion: _____

_____

_____

Animal Rights :   _____

_____

_____

ETHICS POST-TEST

1. Read the following case:

   Ms. White brings in Tiger, a four-year-old, male, castrated, shorthaired cat. Tiger has recently begun "spraying" in the house, a behavior that began shortly after the birth of Ms. White's first child six months ago. She is at wit's end. In addition to the aggravation of the problem, she has already spent more than five hundred dollars on cleaning bills. She wants you to euthanize Tiger. You recommend several behavioral specialists, but she is too busy with the new baby to spend any more time or money on the animal. You became a veterinarian years ago with the intent of helping orphaned animals. You ask Ms. White whether she has tried giving the animal to a neighbor or family member. She says no. Should you euthanize Tiger?

2. Place a check mark next to all stakeholders, that is, individuals or groups that may be affected by your decision.

   A veterinary client _____

   Other veterinary clients _____

   All veterinary clients _____

   A veterinary patient _____

   Other veterinary patients _____

   All veterinary patients _____

   A cat _____

   A dog _____

   A cow _____

   A horse _____

   A pig _____

   A domestic animal not named above _____

   A wild animal _____

Cats _____

Dogs _____

Cows _____

Horses _____

Pigs _____

Domestic animals not named above _____

Wild animals _____

All cats _____

All dogs _____

All cows _____

All horses _____

All pigs _____

All domestic animals not named above _____

All wild animals _____

All pet owners _____

A farmer _____

All farmers _____

A rancher _____

All ranchers _____

Food manufacturing industry _____

Food consumers _____

People in developing countries _____

People in developed countries _____

Future humans _____

Future animals _____

Future human generations _____

Future animal generations _____

All living beings _____

All present and future living beings _____

3. Name the individuals and their interests who are in conflict with other interests in this case.

_____ v. _____

_____ v. _____

_____ v. _____

_____ v. _____

4. Philosophers and theologians develop ethical theories in part to help us to answer difficult questions such as when, if ever, we are justified in killing. Which of the following theories discuss principles relevant to animal euthanasia? If you do not know, leave the space blank.

Utilitarianism _____

Social contract _____

Ecofeminism _____

Cartesianism _____

Talmudic law _____

Jainism _____

Christian dominionism _____

Animal rights _____

Buddhism _____

Christian stewardship _____

Islamic law _____

5. Which of the following theories is likely to object to euthanasia for Tiger? If you do not know, leave the space blank.

Utilitarianism _____

Social contract _____

Ecofeminism _____

Cartesianism _____

Talmudic law _____

Jainism _____

Christian dominionism _____

Animal rights _____

Buddhism _____

Christian stewardship _____

Islamic law _____

6. Which of the following theories is likely to approve euthanasia for Tiger? If you do not know, leave the space blank.

Utilitarianism _____

Social contract _____

Ecofeminism _____

Cartesianism _____

Talmudic law _____

Jainism _____

Christian dominionism _____

Animal rights _____

Buddhism _____

Christian stewardship _____

Islamic law _____

7. Which of the following ethical theorists would unambiguously endorse euthanasia for Tiger? If you do not know, leave the space blank.

René Descartes _____

John Stuart Mill _____

Peter Singer _____

Tom Regan _____

Peter Carruthers _____

Immanuel Kant _____

Total correct: _____

NOTES

1. The case description in the "Write and Pass" exercise is based on the work of Bernard Rollin, *An Introduction to Veterinary Medical Ethics: Theory and Cases* (Ames: Iowa State University Press, 1999), Question 11, 119. The "Write and Pass" exercise framework upon which the present Write and Pass was designed was devised by Kathleen Moore, Philosophy Department, Oregon State University, and presented at the 1998 Bioethics Institute. The case description in the "Steps in Ethical Decision Making" is based on the work of Jerrold Tannenbaum, *Veterinary Ethics: Animal Welfare, Client Relations, Competition and Collegiality,* 2nd ed. (St. Louis, Mo.: Mosby, 1995), Case A-20, 588–9. The "Steps in Ethical Decision Making" exercise framework was designed by Courtney Campbell, Philosophy Department, Oregon State University; and presented at the 1998 Bioethics Institute.

14

LAND

HYBRID CORN

Jochum Wiersma, Deon Stuthman, David Fan, Donald Duvick, and Victor Konde

This case study consists of three parts. Please read the narrative and answer the questions following that section before continuing. This case is partially based on events in the past. It is very useful to limit yourself to the facts as they are presented and imagine the situation as though you were there. Several additional resources are listed in the "References and Background Information" section.

HYBRID CORN: PART I

In the 1920s and early 1930s, corn breeders in the United States developed a practical way to make hybrid corn. Seed of "double-cross hybrids" could be produced at a price that farmers could afford. Farmers therefore could take advantage of the benefits of inbreeding followed by directed cross-breeding of inbred corn lines that resulted in an increase in vigor and yield of the hybrid offspring. Donald Duvick describes the rise of hybrid corn in the article "Biotechnology in the 1930s: the development of hybrid maize" in the January 2001 issue of *Nature Reviews/Genetics*. He writes that the technology was introduced even though corn breeders and other scientists did not (and still do not) understand the genetic principles of hybrid vigor, one of the major reasons for increased yield of hybrid corn.

The introduction of hybrids also allowed, for the first time, a cost-effective protection of intellectual property in corn breeding. Farmers buying the seed could not maintain or recreate the hybrid themselves and thus needed to buy seed from the seed corn companies each year if they wanted to maintain the yield advantage the corn hybrids provided. This need gave rise to a viable plant-breeding industry. Numerous "seed corn companies" soon were responsible for much of the breeding and virtually all the production and sales of hybrid corn in the United States (see Figure 14.1). The farmers in the Corn Belt readily adopted this new technology, and the majority of the acreage was planted to corn hybrids in just a few years (Figure 14.2). Corn yields immediately started to rise and are still rising, a result in large part of annual improvements in hybrid genetics (Figure 14.3).

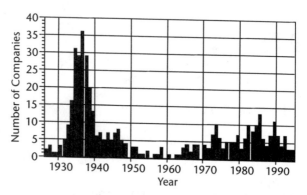

FIGURE 14.1. Year in which companies were formed or reorganized for business in hybrid maize. (From Duvick, D.N. 1998. Country case Studies: The United States, pp 103-211, *In* M.L. Morris, ed. Maize Seed Industries in Developing Countries. Lynne Rienner Publishers, Inc. and CIMMYT, Boulder, Colorado and Mexico, D.F.)

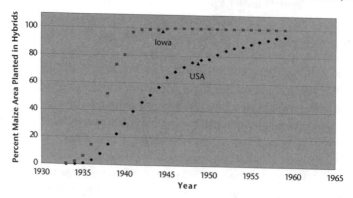

FIGURE 14.2. Hybrids as percent of total maize plantings, Iowa and U.S. (From Duvick, D.N. 2001. Biotechnology in the 1930s: the development of hybrid maize. *Nature Reviews Genetics* 2:69-74.)

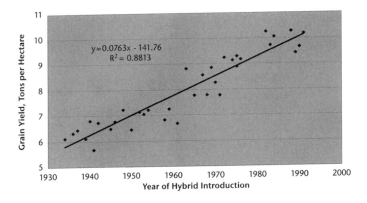

FIGURE 14.3. Grain yield of 36 popular hybrids introduced from 1934 to 1991. Tests conducted in central Iowa, 1991-1994. (From Duvick, D.N. 2001. Biotechnology in the 1930s: the development of hybrid maize. *Nature Reviews Genetics* 2:69-74.)

As with any new endeavor, the start-up companies did have their share of doubts and problems. Following are excerpts from two letters written by the president of one of those start-up hybrid corn-breeding companies. Both letters illustrate some of the dilemmas the president of the company was facing at the time.

Letter 1

July, 1934

[Our Company] is either something big or it is nothing at all. There is no halfway position. If it is nothing at all, we are wasting our time and effort and it makes little difference whether we waste it this way or that way. Our problems are imaginary, our solutions immaterial.

If [Our Company] is big, and we all believe it is, our problems are very real and the answers make a real difference to us and to the Iowa farmer. We have a selfish ambition to make money. We also have an unselfish desire to provide a better seed corn. Both ambitions are worthy and they are consistent with one another. We owe it to ourselves and to our neighbors to strive for proper solutions to our problems.

Our capital is $30,000.00. This is about 1% of what it should be. You know the task it has been to raise this 1%. The other 99% will not raise itself.

Letter 2

November 21, 1935

Dear ——,

I have always thought that [Our Company] had but one serious threat: The chance that we would someday be fooled by a cross which would pass our tests, put it out commercially and then discover it was not any good. . . . I feel we may have done this with [Hybrid X]*

I think it is a fine corn if conditions are just right [but I believe it is] too sensitive to minor adversities. If I am right it is not a proper seed corn for general sale.

We gave [Hybrid X] the benefit of the doubt last spring and sold it. We made a sales profit. I doubt if we made a real long-run profit. I am afraid of [Hybrid X]. I want to take it off the market. . . . If I jerk it off the market, [the Sales Department] will go wild for [they] can sell every grain of it. . . . [But] I have told the boys . . . to sell no more [Hybrid X] until further word.

———— ————, Pres.

*The hybrid had germinated poorly in 1935 and subsequently yielded less than expected.

QUESTIONS

1. What do you think were the reasons to introduce hybrid corn in the USA?
2. Given the information presented, can you think of any objections to the introduction of hybrid seed corn?
3. Given the pro and cons that you listed in questions 1 and 2, was it morally right to commercialize hybrid seed corn (defend your answer)?
4. On what grounds do you think the president of the company decided to halt the sale of the corn hybrid?
5. Was it morally wrong for the company to release the hybrid in the first place (defend your answer)?
6. Is it morally justifiable to protect and profit from the intellectual property and consequently create a dependency of the farmer on a seed source?

HYBRID CORN: PART II

Hybrid corn was, and still is, a scientific and commercial success. There were, however, unforeseen consequences of this technology. Existing (and genetically diverse) open-pollinated varieties throughout the Corn Belt quickly disappeared; consequently, uniformity in the cornfields greatly increased. This increased the potential for genetic vulnerability, as was demonstrated by the outbreak of Southern Corn Leaf Blight Race T, a fungal disease, in 1970. It was virulent to most hybrid cultivars at the time because of the genetic uniformity of the cytoplasm in those hybrids, due to use of a particular kind of Cytoplasmic Male Sterility (CMS) as an aid in production of the hybrids. The yield losses in the southern and central parts of the Corn Belt were disastrous. The average yield in the United States dropped considerably (see Figure 14.4). The seed companies responded quickly and unilaterally abandoned Cytoplasmic Male Sterility in favor of mechanical detasseling to produce the hybrid seed. By the next year hybrids were produced using inbred lines that were not Cytoplasm Male Sterile but rather male fertile and consequently not susceptible to the disease. Thus corn hybrids were no longer susceptible to Southern Corn Leaf Blight and individual farmers avoided any yield losses due to the disease.

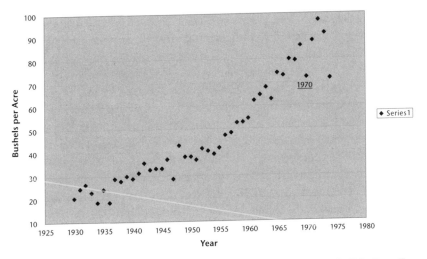

FIGURE 14.4. U.S. maize yields, 1930-1974. 1970: Southern Corn Leaf Blight Race T in southern and eastern Corn Belt. 1974: Abnormally cool summer, early frosts in northern Corn Belt. (Data from USDA/NASS: *Corn grain and silage—planted and harvested—area, yield and production—1930-1974.*)

QUESTION

1. Given the additional information provided in this section, would you change any of the answers you gave to questions 1 through 5 in part I?

HYBRID CORN: PART III

Now fast-forward to the present. Most of the opposition to genetic engineering and transgenic crops has focused on the potential risks of introducing genetically engineered plants into the environment and the purported unprincipled greed of multinational companies that, it is argued, will threaten food security and food safety. Some people are even willing to resort to violence to make their point, as is illustrated by the press release later in this part. In many of the debates, people use the precautionary principle to defend their stand on the issue. The Union of Concerned Scientists defines the precautionary principle as follows: "Potentially dangerous activities can be restricted or prohibited even before they can be scientifically proven to cause serious damage" (http://www.ucsusa.org/). The European Union has adopted the precautionary principle to be an integral part of policy making but does not provide an explicit definition of the precautionary principle (http://europa.eu.int/comm/dgs/health_consumer/library/pub/pub07_en.pdf).

According to Duvick, hybrid corn when first introduced encountered little opposition and produced little if any discussion about its potential risks. With the introduction of genetically engineered corn, the debate is very lively and discussions about the potential risks and benefits have been extensive. Pool and Esnarayra (2001) provided a nice overview of the potential risks and benefits of transgenic crops. At the present there are hundreds of genetically engineered corn hybrids on the market in the USA that are planted on a significant percent of the acreage in the USA.

As with the introduction of hybrid corn, unforeseen consequences with genetically engineered hybrid corn have occurred. In 2000, Aventis Crop Sciences was forced to remove the Starlink brand of corn hybrids from the market. The Starlink brand confers both resistance to the broad-spectrum herbicide Liberty as well as resistance to European corn borer and related pests (http://www.us.cropscience.aventis.com/AventisUS/CropScience/stage/starlink/starlinkcorn.htm). The insect resistance results from production by the plant of a compound that is toxic to the Leptidoptora class of insects. At the time, the Federal Drug and Food Administration had not yet approved the particular Bt toxin (Cry9A) incorporated in the hybrid corn for human consumption. However, traces of Starlink corn were found in several brands of corn tortillas, including Taco Bell, in the United States. Moreover, Japanese authorities found traces of Starlink corn in export shipments of corn to that country. These findings ultimately resulted in a voluntary withdrawal of any Starlink hybrid from the market by Aventis Crop Sciences on October 12, 2000.

Press Release

ANTI-GENETIC ENGINEERING GROUP SMASHES WINDOWS AT
WISCONSIN PIONEER SITE

GENETIX ALERT

NEWS RELEASE

FOR IMMEDIATE RELEASE

Date: October 29, 1999

Contact: Jeffrey Tufenkian 619-584-6462

An underground group opposed to genetic engineering (GE) claimed responsibility for breaking windows at the Eau Claire, Wisconsin Pioneer Hi-Bred facility on October 27th according to a communiqué released today. The group known as "Seeds of Resistance" charged Pioneer and other proponents of GE with deceiving the public and profiting off of growing GE crops. "Seeing their profits as a slap in the face of the earth and all its occupants, we took the liberty of paying them back," according to the communiqué. "We, Seeds of Resistance, smashed all the windows on one side of their disgusting building. Wisconsin is now another state that cannot hide from this growing resistance against GE culture." This action is the thirteenth known nonviolent destruction of GE crops or other property in the U.S. this year. Details of past anti-GE actions are available at www.tao.ca/~ban/ar.htm.

GenetiX Alert is an independent news center that works with other aboveground, anti-genetic engineering organizations. GA has no knowledge of the person(s) who carryout any underground actions. GA does not advocate illegal acts, but seeks to explain why people destroy genetically engineered crops and undertake other nonviolent actions aimed at resisting genetic engineering and increasing the difficulty for entities which seek to advance genetic engineering or its products. GA spokespeople are available for media interviews.

QUESTIONS

1. Are there differences between the introduction of hybrid corn and the introduction of transgenic crops? Why or why not?
2. Can you defend the precautionary principle on moral grounds?

REFERENCES AND BACKGROUND INFORMATION

Adventis Crop Sciences. 2000. Starlink Brand Corn (http://www.us.cropscience. aventis.com/AventisUS/CropScience/stage/starlink/starlinkcorn.htm)

American Corn Growers Association. Genetically modified crops: questions and answers (http://www.acga.org/gmobrochure/)

Duvick, D. "Biotechnology in the 1930s: the development of hybrid maize," *Nature Reviews/Genetics*, (January 2001).

European Commission, Comments of the use of the precautionary principle–(http://europa.eu.int/comm/dgs/health_consumer/library/pub/pub07_en.pdf), 2000.

Goklany, I. "Applying the precautionary principle to genetically modified crops," Policy Study 157, Center for the Study of American Business, (2000), St. Louis: Washington University,.

Pool, R. and Esnarayra, J. "Ecological monitoring of genetically modified crops," Workshop Summary, Board on Biology, Board on Agriculture and Natural Resources, National Research Council, (2001), Washington, D.C.

Union of Concerned Scientists (http://www.ucsusa.org/index.html)

TRAIT PROTECTION SYSTEM

THOMAS PETERSON AND BRYONY BONNING

DESCRIBING THE NEW TECHNOLOGY

Seeds for many high-yielding crop varieties are patented and farmers are legally obliged *not* to save and use seed from the crop that they grow, but rather to buy more seed from the company. A technique has been developed to prevent farmers from saving or re-using patented seed. This technology results in the F2 seed (offspring of the plants grown by farmers) being inviable. This technology is called the trait protection system and is covered under U.S. patent no. 5,723,765 (Oliver et al., 1998).

The technology works as follows: The plant genome is engineered so that it produces a protein that is toxic to the plant. The promoter that controls expression of the gene and subsequent production of the toxin is active only late in embryo development. To produce the transgenic F1 seed, a spacer is put between the promoter and the toxin gene so that the toxin gene is inactive. On either side of the spacer are sequences that are recognized by a "recombinase" enzyme that cuts out the spacer. This event brings the gene and promoter together so that the toxin is produced late in embryo development. A promoter that is activated by the chemical treatment of the seed sold to farmers controls the production of the recombinase. Thus, until the seed is chemically treated, the toxin gene remains inactive. The result of chemically treating seed purchased by farmers is that the farmer can buy viable seed and harvest the crop, but any seed collected from the crop will not grow. Opponents of the new technology refer to it as the *terminator technology*.

CURRENT CONDITIONS IN THE WORLD AGRONOMIC SEED INDUSTRY

The mode of reproduction of a crop plays a significant role in the seed industry. *Hybrid crops* such as corn, sunflower, and sorghum automatically require that farmers purchase new seed each year to maintain yield. If farmers kept their own seed of hybrids, 50 percent of the advantage of growing the hybrid would be lost in the next year. The incentive for farmers to buy new seed of a hybrid each year is quite large.

Self-pollinated crops, such as wheat, soybean, and rice, do not require that farmers purchase new seed each year. This is because the seed harvested from the crop is genetically identical to what was planted. It is common practice both in the United States and the rest of the world for growers of self-pollinated crops to keep and plant their own seed.

Brown Bagging

The practice of farmers saving their own seed, or *brown bagging*, as it is usually called, causes significant economic problems for commercial companies. Farmers may purchase seed of a new variety only once and in future years produce their own seed of this variety. This limits investment by the commercial sector in self-pollinated crop breeding because it creates limited potential return on investment. Much of the breeding of self-pollinated crops is done in the public sector, both in the United States and the rest of the world. Brown bagging has also limited the use of biotechnology in self-pollinated crops because controlling this practice is difficult. Hybrid crops do not have this problem and offer built-in protection for intellectual property.

Patenting crop varieties and requiring farmers to sign grower agreements has been one way of circumventing the brown bagging issue in the United States. Both the patents and the grower agreements prohibit farmers from saving seed to plant the following year. Obvious enforcement issues exist, but U.S. farmers have been prosecuted for brown bagging patented crop varieties.

In the Case of Corn

Even though hybrids offer a tremendous yield advantage in corn, hybrids are not grown by farmers worldwide. Hybrid seed production can be expensive and technically challenging to subsistence farmers. For example, with corn, nearly 100 percent of the U.S. acreage is planted with hybrids, and farmers purchase new seed each year. In contrast, 62 percent of the corn acreage in developing countries is planted with local germplasm or open-pollinated varieties, whereas only 38 percent is planted with hybrids. The local germplasm or open-pollinated varieties are grown from seed saved the previous year by the farmer, purchased from other farmers, purchased from a public or government agency, or in some cases purchased from a commercial company. Forty percent of the hybrid seed planted in developing countries is of public origin. The developing countries represent a potentially huge market for corn hybrids. The primary hindrance has been that farmers in developing countries do not have the capital to purchase hybrids; also, intellectual property laws are weak in developing countries.

In the Case of Wheat

The situation with a self-pollinated crop such as wheat is much different. Greater than 90 percent of the U.S. hard red winter wheat acreage is planted with publicly (usually from land grant institutions) developed varieties. The situation is very similar in developing countries. Biotechnology has not been introduced into wheat because companies have no way to protect investment in their intellectual property. Few commercial companies breed improved wheat varieties because the return on investment from breeding is very low.

CONTROVERSY OVER THIS NEW TECHNOLOGY

Following are two arguments presenting opposite sides of the controversial issues raised by this technology.

The Case for Introducing the New Technology: The Industry Perspective

The first point in industry's argument relates to the fact that the trait protection system is essentially a means to protect intellectual property. Seed companies invest a great deal of time and money to produce new improved varieties for farmers to plant. These substantial costs can be recovered only when the seed providers can be confident that their products will not be illegally reproduced and distributed. A similar situation exists in the illegal copying and subsequent sales of software, music CDs, and movies.

It is an accepted practice for these valuable electronic media to contain embedded "poison pills" that prevent their unauthorized reproduction. The trait protection system is a conceptually identical means of protection. Seed companies have a right to protect their intellectual property and recoup the costs they incur in developing improved varieties.

Second, the use of the trait protection system will increase availability of superior genetic stocks to third world farmers. Currently, seed companies are reluctant to sell their best germplasm in third world countries where a high probability exists that their lines would be illegally propagated and distributed. Because the trait protection system removes the potential for subsequent illegal propagation, seed companies will have no reason not to offer their best genetic materials to farmers in developing countries. The increased availability of superior genetic stocks to farmers worldwide will increase the choices available to farmers. Farmers will have the freedom to choose from either their traditional varieties or the best seed offered in the commercial sector.

Finally, seed protection technology will spur breeding efforts to improve genetic characteristics of many more crops than is done today. Currently, corn breeding is profitable because farmers purchase hybrid seed each year to obtain the superior performance of hybrids. Much less genetic research and varietal improvements have been achieved in other crops, partly because of the inability to recover the costs of producing improved varieties. Seed protection technology will protect investments in genetic improvement and will thereby promote such efforts in the private sector. Although such improved seed will no doubt cost more than varieties available today, higher yields will more than make up for this higher initial cost. Thus, the farmer will realize a greater income, and more food can be produced per acre, thereby lessening the overall environmental impacts of agriculture.

THE CASE AGAINST INTRODUCING THE NEW TECHNOLOGY: THE OPPONENT'S PERSPECTIVE

Opponents of the trait protection system believe that the use of terminator technology will result in a loss of biodiversity. First, the engineered seed will replace native seed. Second, relatively few varieties of any particular crop are suitable for engineering, so industry will be restricted in the varieties that can be engineered with the terminator technology. The result will be an increase in crop uniformity with potential increased vulnerability to pests and diseases.

The second concern is that the traditional role of farmers as plant breeders will be eliminated. Farmers who currently use traditional practices of plant breeding to produce varieties suitable for their local needs will cease to do so. In addition, the public sector that currently produces hybrids for distribution for profit in some countries will be forced to adopt the terminator technology to remain competitive.

Third, the cost of the engineered seed will be more than the cost of lower-yield seed, and farmers in underdeveloped countries will not be able to afford to buy seed every year.

The fourth point involves the risks associated with the use of this technology. First, a catastrophic risk is associated with dependence on terminator technology products. If seed production were interrupted or lost, farmers would be left with nothing to plant. Second, a risk exists that the killer gene would be transmitted to related species of plants via the pollen, which would have a deleterious impact on neighboring plants and farmers.

Finally, the use of this new technology could result in control of global food production by relatively few companies.

International dialogue is required to address the potential impact of this new technology on global agriculture and food production.

QUESTIONS

1. Should using biotechnology to sterilize second-generation seed for the purpose of preventing farmers from saving and replanting the seed be illegal?
2. Should industry be allowed to engineer seed for the purpose of protecting its intellectual property?
3. Should the U.S. government support this research, and if not, should this research be made illegal?

REFERENCES

López-Pereira, M.A. and Filippello, M.P. 1993/94 World Maize Facts and Trends: Maize Seed Industries Revisited: Emerging Roles of the Public and Private sectors" (Mexico, D.F.: International Maize and Wheat Improvement Center [CIMMYT]), 1995.

Crouch, M.L. "How the terminator terminates: an explanation for the non-scientist of a remarkable patent for killing second generation seeds of crop plants," An occasional paper of The Edmonds Institute, 20319-92nd Avenue West, Edmonds, WA 98020 (1998).

Oliver, M.J., Quisenberry, J.E., Trollinder, N.L.G., and Kelm, D.L. Control of plant gene expression, U.S. Patent 5,723,765, issued 3 March 1998.

Rural Advancement Foundation International Web site "The Terminator Technology" by Hope Shand, Programme Officer, RAFI-USA. (http://www.rafi.ca/communique/ 19982.html)

USDA FACT SHEET, Why USDA's technology protection system (aka 'Terminator') benefits agriculture.

Chapter 15

BIOTECHNOLOGY

GOLDEN RICE

Kristen Hessler, Ross Whetten, Carol Loopstra, Sharon Shriver, Karen Pesaresi Penner, Robert Zeigler, Jacqueline Fletcher, Melanie Torrie, and Gary L. Comstock

THE BACKGROUND

The World Health Organization (WHO) estimates that vitamin A deficiency affects 230 million children around the world, and at least one million children per year are dying of diseases related to this deficiency. Ingo Potrykus and his research group, with financial support from the Rockefeller Foundation, developed a variety of rice that contains beta-carotene, the plant pigment that is the precursor of Vitamin A. This rice supplies enough beta-carotene in a typical serving to supply 10 percent of the daily requirement for Vitamin A. Potrykus and Rockefeller have provided this variety of rice to the International Rice Research Institute (IRRI) in the Phillipines, which will breed improved rice varieties using its traditional rice breeding methods and make the seeds freely available to farmers in the developing world.

IRRI has been doing rice breeding for decades and has been on the front lines of the Green Revolution, developing and releasing new rice varieties with improved productivity (and increased dependence on fertilizers and pesticides). The institute's services are provided without charge to the farmers it serves and are supported by philanthropic foundations in the developed world (including the Rockefeller Foundation). Many people regard this development as an example of how biotechnology can be used to help developing nations, whereas others consider it a smokescreen to divert attention from the fact that biotechnology companies are trying to dominate the food supply.

Several questions surround golden rice, including when, if ever, it will be ready for commercial use and whether it might have unpredictable, untoward health effects on those who eat too much of it.

YOUR ASSIGNMENT

A charitable organization appeals to the WHO to stop the Rockefeller Foundation from releasing golden rice on the grounds that it isn't a good strategy for dealing with malnutrition. The WHO will soon convene a hearing (the WHO Panel of Arbitrators) to determine whether to block the development of golden rice. The organization has invited four different groups to advise it on this matter.

The four groups are as follows: (1) Friends of the Earth; (2) Philippine Partnership for Development Farmer-Research Scientists (MASIPAG); (3) People from Developing Nations; and (4) the International Rice Research Institute (IRRI). The first two groups argue that the technology should not be pursued because they think that golden rice is an expensive high-tech experiment, a gambit that is unlikely to solve the real causes of hunger in developing countries. The latter two groups see golden rice as a viable solution to some problems and argue strenuously for its development.

You will be assigned to serve on one of the four teams or on the WHO arbitration board.

GENERAL INSTRUCTIONS FOR ADVISORY GROUPS

Consider your group's position and generate reasons based on moral and scientific grounds supporting your position. Plan to include both factual statements ("Many children suffer from vitamin A deficiency") and moral principles ("We should provide aid"). Formulate a strategy for briefly presenting your position to the WHO panel in a persuasive manner, and be prepared to answer questions from the panel about your position. You will have only five minutes to present your position, so choose one representative to speak for your group.

Descriptions of the various groups, their positions on golden rice, and their specific tasks follow.

Panel: WHO Arbitrators

You will be asked to decide whether to block distribution of the golden rice. Use your time to decide what additional facts you need to make a good decision and what moral questions need to be answered. After you hear testimony from each interested group, you should ask the groups any questions that you think remain unanswered. After all the testimony is complete, you will have time to make your decision. Please select a representative to present your decision and support it with your moral reasons.

Witness Groups

Friends of the Earth

You represent an organization dedicated to protecting the environment and promoting sustainable development. You think that golden rice, as well as all genetically modified plants and animals, pose unknown threats to the environment and human health. In your view, the introduction of genetically modified organisms continues because of the power wielded by large agribusiness companies. You will argue that golden rice should not be released because of the environmental risks it poses, and you will urge the WHO to resist the influence of Monsanto and other large biotech firms.

Philippine Partnership for Development Farmer-Research Scientists (MASIPAG)

MASIPAG was formed in 1986 as a collaboration between farmers and agricultural researchers to improve rice farming practices. The organization hopes to make rice farmers independent of loans and chemicals through training in sustainable agriculture. You will argue that poor Asian farmers are not likely to benefit from golden rice. Instead, you see golden rice as a chance for the biotechnology industry to improve its image.

People from Developing Nations

Some (though not all) of you may be desperately poor and may have children exhibiting symptoms of vitamin A deficiency. No solution to this critical health problem is immediately available, and you will argue that golden rice is desperately needed because the effects of malnutrition are so severe.

International Rice Research Institute (IRRI) Scientists

IRRI is an organization that does research in plant breeding and development of new rice varieties for farmers in Asia. Products of IRRI development are freely distributed to farmers and local plant breeders. You will argue that distributing golden rice will benefit malnourished children.

For additional resources, see Appendix B.

ORGAN TRANSPLANTATION[1]

Christopher Baldwin, David Bristol, Emily Deaver, Bruce Hammerberg, Carole A. Heath, Surya Mallapragada, Gavin J. Naylor, Elaine Richardson, and Jim Wilson

CLONING HUMAN CELLS

Burn patients often require extensive skin grafts. Using current technology, physicians routinely take small skin samples from burn victims, clone the cells in tissue culture, and proceed to grow skin in sheets. These sheets are then transplanted to the burn site.

DISCUSSION QUESTIONS

1. What is your initial ethical reaction to cloning human cells in this manner? Is this a good or bad thing overall?
2. What facts support your response?
3. Does any underlying moral principle exist that supports your response?
4. Do any viable arguments against your position exist?

TRANSPLANTING ANIMAL ORGANS, PART A

Cats occasionally require renal transplants to survive. Currently, the owners of pet cats needing kidneys are required to obtain a donor cat. The life of the donor cat is usually sacrificed to save the life of the cat needing the kidney.[2] Recipient cats require chemotherapy to prevent kidney rejection.

DISCUSSION QUESTIONS

1. What is your initial ethical reaction to our current donor program? Is this a good or a bad thing overall?

2. What facts support your response?
3. Is there an underlying moral principle that supports your response?
4. Are there viable arguments against your position?
5. Is there a creative way to honor all conflicting moral obligations to some extent?
6. Break up into groups of five. Each member plays one of the following roles, and each member must justify their position using ethical principles and arguments:
   a. The veterinarian who wants to perform the procedure
   b. The loving cat owner who wants to save her cat's life
   c. The animal rights activist who opposes killing a healthy cat to save a sick one
   d. The employee of the Society for the Prevention of Cruelty to Animals (the source of many donor cats) who is willing to supply donor cats for the procedure rather than have to euthanize the animals
   e. The unhappy cat lover who has lost pets in the past and is concerned that a pet owner could unwittingly end up donating his or her pet.

TRANSPLANTING ANIMAL ORGANS, PART B

A combination of recent scientific breakthroughs now provides a unique opportunity for organ replacement. Researchers have recently used biotechnology to produce headless fetuses in mice. They produce headless mice by knocking out the gene responsible for the development of the head. By performing this action at an early stage in morphological development, the researchers can produce a mouse fetus that is born with its head missing. In other words, when the fetus is born, it has all its organs except the brain. Thus, it has no means of receiving or acting on any impulses generated by the peripheral nervous system. It is essentially a nonsentient being. Fetuses occurring naturally with this gene deficiency do not survive the early stages of pregnancy. However, with intervention, the mouse fetuses are maintained in a stable condition until birth, when they die.

Imagine using this technology to produce headless fetal cats. The cat fetuses would be an excellent source of tissue-type-matched organs. Kidneys could be produced to save the lives of sick cats. Thus, no healthy cats would have to be euthanized, and no chemotherapy would be necessary for the sick cats following this new transplant procedure.

Homework: Turn to Appendix A and fill out the form titled "Case Study Response: Individual" (Exercise 15.A). Bring it to the next class session.

Next class session: Break into the same groups of five that performed the role-playing exercise in Part A. Everyone receives a copy of the form "Ethics Case Study

Response: Group" (see Appendix A, Exercise 15.B). Drawing on the results of your individual homework assignment, work with your group members to produce a collective answer to each question. Hand in the results.

TRANSPLANTING HUMAN ORGANS, PART A

The same headless-fetus technology could be applied to people. That is, researchers could use biotechnology to clone your cells and produce a headless human fetus that matched you. If we went forward with the technology, the fetus would be born with all its organs present except the brain. Thus, having no means of receiving or acting on any impulses generated by its peripheral nervous system, it would essentially be non-sentient. The fetus would be an excellent source of tissue-type-matched organs for you. It would, for example, provide you with a kidney should you need it, and painful chemotherapy following the transplant operation would be unnecessary.

DISCUSSION QUESTIONS

1. What is your initial reaction to this proposal? Is this different than your response to transplanting animal organs in "Transplanting Animal Organs, Part B"? Why?
2. What reasons support your response?
3. Does an underlying moral principle exist that supports your response? Is this a different underlying moral principle for this case versus the scenario in "Transplanting Animal Organs, Part B"?
4. Can you think of viable arguments against your position?

TRANSPLANTING HUMAN ORGANS, PART B

Should the government permit further research to develop headless human clones? What role should ethics play in such public policy discussions?

Break into new groups of five. Prepare testimony for a U.S. Senate hearing on the question of whether the government should permit further research to develop headless human clones. Each person plays one of the following stakeholder roles:

The ethicist: Identifies critical moral issues on both sides of the question
The medical researcher in the organ-transplant community: Explains issues of equity and medical utility
The representative from the insurance industry: Explains the costs of current medical procedures and how they affect health-insurance rates

The family member of a patient needing a transplant: Portrays the emotional frustrations associated with being unable to receive a timely transplant

The religious representative: Explains the theological resources offered by his or her tradition (for example, Judaism, Christianity, Islam).

NOTES

1. Reprinted from *The Ag Bioethics Forum*, vol. 10, no. 2, Nov. 1998.
2. In fact, most veterinary hospitals that perform transplants require the owner of the recipient cat to adopt the donor cat. We did not include this information in the case in order to force students to compare the relative value of the lives of the donor recipient.

16

FARMS

LOST IN THE MAIZE

Isabel Lopez-Calderon, Steven Hill, L. Horst Grimme, Michael Lawton, and Anabela M. L. Romano

Genaro Moura (GM) and Oswaldo Fernándes (OF) are farmers who grow their crops in neighboring fields. GM has planted a non-sweet forage maize that is genetically engineered to contain high levels of essential amino acids. He has followed all regulations but has not informed anyone that he is planting transgenic plants. This is because he is afraid that the environmental organization "Maize Liberation Front" will destroy his crop and organize a campaign against him.

His neighbor OF is opposed to genetic modification and uses only organic methods of agriculture. He is under a strict contract to sell his crop of sweet corn to the Gerber Baby Food Company. His crop must be certified as organic (under current regulations, organic food must be free of any transgenic material).

GM has lately noticed some sweetness in some of the corn kernels of his own crop. He suspects that some of the pollen from OF's field has drifted into his field and has pollinated his plants. This does not affect the value of his crop. But he realizes that pollen from his genetically modified plants may have also drifted into his neighbor's field. GM realizes that if foreign genes have been transferred to OF's maize, the entire organic crop will be rejected by the Gerber Baby Food Company.

GM decides *not* to inform OF that his crop may be contaminated with foreign genes.

Has GM made the right decision? Identify the *principles* behind GM's decision. Identify the interested parties and how they may be affected by GM's decision.

Later that week, GM reconsiders his decision. He decides to tell OF about the possible cross pollination. OF is upset but realizes that his own economic livelihood is threatened if it is revealed that his organic crop is contaminated by the GM maize.

What do you think OF should do? Justify your position.

OF decides to find out more about the genetically modified maize that GM has grown. He spends six hours on the Internet to research this topic. He finds that the maize has been approved by the Federal authorities and that it has been grown for more than three years without any health problems being reported. Although OF is opposed to genetically modified food, he concludes that in this case the risk of substantial contamination is very low. He decides not to inform the Gerber Baby Food Company about the problem.

Did OF make the right decision? Justify your answer.

OF performs tests that show that his crop is completely free of any genetically modified material. He informs the Maize Liberation Front about GM's crop, knowing full well that they will destroy it.

Did OF do the right thing? Justify your answer.

MAGNANIMOUS IOWANS

RICARDO SALVADOR, STEPHEN MOOSE, BRUCE CHASSY, AND KATHIE HODGE

Four students are delivering a presentation in a senior-level course on world food issues. The four students were born and raised on Iowa farms and are very impressed with the prowess of the agricultural producers of their home state and their country. They express the belief that the way that U.S. producers can contribute toward allaying global hunger is to produce surplus grain so that a portion of this is available to enter food aid channels in the form of donations to needy nations. Their paper, entitled "*America: The Land of Plenty*," begins[1]:

> For the last forty years the U.S. has been setting the standard for agricultural production all over the world. We have gone from a self-sustaining production system to one which has made us a leading exporter of agricultural products world wide. There have been many reasons for this change including improved technology and farming methods. The U.S. now has a system, which consistently overproduces. Most of this overproduction is exported throughout the world, while some of the surplus is used to combat hunger abroad through U.S. food aid programs.

Their paper concludes:

> The future of this world is uncertain. Nobody knows how large the population will grow or how many people will continue to die from hunger. But one thing that is certain is that the U.S. will continue to overproduce. U.S. farmers have adopted technology and farming methods that have caused yields to steadily increase year by year, and these farmers have been known to be very reluctant to change. We can only hope that the U.S. continues to use this abundance to help feed the hungry throughout the world.

During the presentation, team members display historical U.S. production and export curves for four major grain crops, including maize. Melissa, a member of the class, notes that for the years 1999 and 2000, the volume of U.S. maize exports decreased noticeably and asks for an explanation. Ben, speaking for the group of presenters, explains that the major export markets for U.S. grain, European nations and Japan, have had reservations about the transgenic grains produced in the United States and that, specifically in year 2000, the presence of StarLink™ corn[2] in the U.S. shipping pipeline deterred export sales. Mark, another member of the class, asks the obvious question: "So, what is happening with all that extra grain?" Ben replies that most of it is redirected as feed for U.S. livestock. Then Nick, a member of the presenting team, points out that during the course of his research, he read that a portion of GMO corn without a buyer was being donated as food aid. On hearing this, Martine, a member of the class who is a wealthy Haitian studying in the United States, exclaims, "But that is unethical!" Ben looks at her directly and asks incredulously: "Wouldn't it be worse to let them starve?"

DISCUSSION QUESTIONS

1. Do rich countries have a moral obligation to donate food to poorer countries?
2. Is it ethical to offer food that hasn't been approved for human consumption in your own country as food aid to another country?
3. Do different standards of food safety apply when the trade-off involved is that starving people could be saved?
4. Is it moral to withhold grain when others might eat it and live?
5. Under what conditions would using this grain as food aid be acceptable?
6. Is it ethical for a foreign government to accept StarLink™ corn on behalf of its people?
7. Are the ethics of trade and aid different? Note that in trade, parties purchase products from one another. In aid, parties exchange donations or support with the intention of benefiting one another.

8. In the example provided, what are the comparative benefits derived by the donor and recipient of aid?

NOTES

1. For the text of the full paper, including a summary of U.S. food aid programs, see http://www.agron.iastate.edu/courses/agron342/reports/s01/surplus.htm

2. A variety of corn modified with a bacterial protein that is toxic to Lepidopteran insects. When this case took place, StarLink™ had received federal approval as a livestock feed but not for human consumption. This was because the transgenic protein was found to be indigestible and this indicated the need for subsequent tests to determine potential allergenicity in humans. For further information, see http://www.anzfa.gov.au/documents/fs053.asp

EXERCISES

EXERCISE 1.A

A good way to learn how to analyze assumptions is to practice putting yourself in your neighbor's shoes. The following exercise asks you to figure out, and then to defend, someone else's argument. It also asks you to discover unstated premises in the other person's argument.

Your instructor may wish to copy this page, handing out one copy to each student at the beginning of the exercise. Each student then answers the first question. When finished, pass your paper to the person on your right, per instructions. (Note to instructor: You may have to be creative in finding a way to link all students together across your classroom's rows and spaces.)

WRITE AND PASS

Assume that Emily's case has no unusual excusing conditions. Under normal circumstances, most people think it is wrong to cheat. However, our reasons for this conclusion may vary. Some think cheating is a violation of duties to one's fellow students, or the teacher, or, perhaps, the institution. Others may think cheating has bad consequences, such as the tarnishing of the reputation of Emily or her teacher, or a decline in respect for Emily's institution were the institution widely believed to condone cheating. Others think cheating is wrong for all these reasons and others.

1. Assuming that Emily's case has no unusual excusing conditions, do you think it would be morally wrong for Emily to cheat?

Answer yes or no and provide one reason to defend your answer. Using at least one complete sentence, write your answer here:

_____

Pass this paper to the person on your right.

2. Write down one good reason that supports the claim stated in (1).

_____

Pass this paper to the person on your right.

3. Together, the two preceding statements form an argument (that is, a claim and a reason to support the claim). Every argument makes certain assumptions. (For example, from the claim that "All scientists are smart," it follows that "Lewontin is smart," but only if one assumes that Lewontin is a scientist.)

List one assumption made in this argument.

_____

Pass this paper back *to the person on your left*. Allow him or her to review what you have written on it. Finally, pass this paper back to its original owner. Do you agree with the way your argument was developed? With the assumptions identified in it?

EXERCISE 1.B

DISCUSSION QUESTIONS

Imagine that you are Emily. Should you hand in the sheet of paper with Doug's name on it?

Write out answers to the following questions.

1. **Issues and points of conflict:** What exactly is the issue that Emily must decide? What values are in conflict? List as many as you can think of.
2. **Interested parties:** Who are the stakeholders, that is, humans who might be affected by Emily's decision? List as many as you can think of.
3. **Potential consequences:** What things (for example, relationships, reputations, social groups, cultural institutions) might be affected by Emily's decision? List as many as you can think of.

4. **Obligations:** What duties does Emily owe to people? List as many as you can think of.

How would you respond to the following questions that Emily is asking herself?

1. Don't different cultures have different ways of doing things?
2. Who is to say what's right and wrong?
3. What's the relationship between ethics and the law?
4. What's the relationship between ethics and religion?
5. Is ethics, unlike science, completely subjective? Are there any right and wrong answers in ethics? If not, why do most people think cheating is ethically wrong? If there are, then why do many people think that no method exists to determine the right answers?

EXERCISE 1.C.

Shock Treatment for Naïve Relativism (or Three Things Everyone Everywhere Absolutely Positively Should Not Do, Ever)

List three actions that you believe are clearly ethically wrong. You must describe the actions in such a way that no one in your classroom will disagree with you.

If you think this will be a difficult assignment, you're right, but only if you describe the actions too generally or ambiguously. To avoid trouble, describe actions in very specific terms. If you say *cheating*, or *killing*, or *abusive behavior* is clearly ethically wrong, someone is sure to be able to find a convincing counterexample.

So describe the three actions in careful detail, staying away from obviously hot issues where disagreement is likely to occur. Avoid abortion, euthanasia, gay rights, genetically modified foods, and animal rights. A safe bet is to choose an outrageous action and define it so narrowly that everyone will see at a glance why it is immoral. Give your actors names, describe their motives, explain the consequences of their actions, and rule out all exceptions that might make the action morally acceptable.

Here's an example to get you started:

It is wrong for Emily to write Doug's name on work she has done in order to prevent Doug from failing the assignment unless she has promised her instructor that she will appear to cheat in order to help the instructor carry out an experiment.

Here is a template you might want to use in stating the actions:

It is morally wrong for _name_ to _particular action_ in order to _motive_ unless _exceptions_.

Following this strategy should help you to articulate at least three actions that are clearly ethically wrong. When finished, write three actions that are clearly ethically right. Here are some more examples.

It is morally wrong for _Dirk Smith_ to _pour gasoline on his niece_ and light a match _in order to kill her_ unless she is a Buddhist and has asked him to help her immolate herself as a protest against an unjust war that has killed all other family members.

1. _____

2. _____

3. _____

It is morally right for _Dirk Smith_ to _help his niece obtain information about relatives_ lost in the war _in order to try to save them from harm_ unless by doing so he will jeopardize their safety.

1. _____

2. _____

3. _____

If we collect all of our claims, we will have begun a fairly substantial list of particular moral judgments on which we agree.

EXERCISE 2.A

DISCUSSION QUESTIONS

1. Would you consider yourself religious?
2. If you answered the first question _Yes_:
   - Do you feel comfortable discussing your religious beliefs with others in general? Why or why not?
   - Do you feel comfortable discussing your religious beliefs in university classrooms? Why or why not?

3. If you answered the first question *No*:
   • Do you feel comfortable discussing the religious beliefs of others in general? Why or why not?
   • Do you feel comfortable discussing the religious beliefs of others in university classrooms? Why or why not?

4. Do you agree with Rich that discussion of religion should not take up much time in university classes devoted to ethics? Why or why not?
5. If you know what the Divine Command Theory is, please explain it.

EXERCISE 3.A

Decide whether each of the following examples is or is not an argument. Briefly explain your judgment. If the example *is* an argument, underline the reason that is given for the conclusion. If the example is *not* an argument, indicate whether it is one of the following: an intuition; an emotional appeal; a mere description of some state of affairs or someone's beliefs; a question; or an exclamation forcefully stating one's personal belief or opinion.

A. You have argued that moral rights do not come from God, the Congress, or evolution, so where do they come from?
B. Mountain biking is good for the soul because it connects you with nature.
C. Wendell Berry believes that our community is disintegrating because it has lost the necessary understanding of the relations among materials and processes, principles, body and spirit, city and country, life and death, and civilization and wilderness.
D. Agrarianism represents a great good and requires our allegiance.
E. Soil and water are crucial to agriculture. Since whatever is crucial to agriculture ought to be preserved, soil and water ought to be preserved.
F. I have this intuition that Congress should do all it can to save the family farm.
G. Living on a family farm is the best way to live because it teaches self-reliance and appreciation for plant and animal life. Therefore, Congress should pass legislation to save the family farm.
H. If farmers can profit in the short term from depleting soil and water resources, then they have an interest in exploiting the land in this way. Future generations may not need soil and water if future generations can find alternative ways of feeding themselves. Taken together, these reasons lead to this conclusion: Farmers have no moral duty to farm in an ecologically sustainable way.

EXERCISE 3.B

Make up two really terrible arguments, the worst or silliest you can think of. Write them here. Make sure to include two premises, and a conclusion that clearly does not follow from those premises.

1. _____
2. _____
   _____
3. _____
   _____

1. _____
2. _____
   _____
3. _____

EXERCISE 3.C

Indicate which, if any, of the arguments in A–H in Exercise 3.A are **moral** arguments.

EXERCISE 3.D

For each of the following arguments, state the final conclusion in your own words. Is the conclusion explicit or implicit? Put brackets around any clue words that indicate conclusions and identify the clues that indicate the final conclusion. Where possible, underline the portion of the passage that comes the closest to stating the final conclusion.

1. If you want effective relief, buy Brand X. Brand X contains the ingredient doctors recommend most.
2. This object must be a diamond since it will scratch glass.
3. This object is a diamond. Therefore, this object will scratch glass.
4. By voting themselves a hefty pay raise, members of Congress proved that they are not interested in fighting the budget deficit.
5. Pit bulls are dangerous dogs. According to the Humane Society of the United States, in the four years since July 1983, pit bulls have been responsible for twenty of the twenty-eight deaths that occurred after dog bites in the nation, including five in 1987. The breed accounts for no more than one percent of all dogs in the nation.

6. Linus Pauling told his audience that vitamin C must be taken in doses much higher than those recommended by the FDA. This contrasts with the way therapeutic drugs act. He said: "A large enough dose, no matter how useful the drug, can be deadly. Vitamins, however, are natural substances, and mankind has become accustomed to them through the ages, so one can't take too much vitamin C."

7. Chimpanzees learn language much more slowly than people and require special tutoring. So, with chimps we can get a better perspective on both the factors that facilitate learning and the factors that interfere with learning. For example, we can completely control their training. We can make the chimps proficient in some areas of language but not in others; we can systematically emphasize certain aspects of their language learning.

EXERCISE 3.E

In each of the following examples, decide whether reasons (premises) are being given to support the truth of a conclusion. Identify the premises (if any), being careful to distinguish them from introductory remarks, restatements, or elucidations of a position, mere persuasion, and disclaimers. Briefly justify your answer.

1. Quite simply, this eiderdown comforter is the showpiece of our collection! Like a rare antique, eiderdown is of superior quality and is coveted by connoisseurs around the world.

2. Pigs are quickly replacing dogs as laboratory animals because their use provokes less of an outcry from the public.

3. Since people who can handle poison ivy with no ill effects can lose their immunity at any time, they should avoid unnecessary contact with the plant.

4. If we are to regain our position as a scientifically advanced nation, we must increase aid to elementary schools, for lack of basic education at the earliest stages can never be overcome.

5. Humans are higher than animals, but humans should not exploit animals. In the Christian tradition, the lower creation should serve the higher creation, yet God does not want humans to kill animals because Christ's death puts an end to the need for blood offerings. As a higher life form, God condescended to a lower life form, humans, in the person of Jesus Christ, so humans, a higher life form, should condescend to a lower life form, animals, by loving the animals.

6. Stealth's invisible. Enemy radar can't see it. And it's the newest electronic marvel to come off the drawing board. Now you may be thinking that there's not much in common between a Stealth Bomber and an automated cassette deck. After all, a Stealth Bomber can't fly backward. But wait, before you decide: This automated auto-reverse deck has a "radar avoidance system" called dbx. No, it's not

an MX missile. But if the Stealth Bomber is invisible to radar, wait until you hear how "invisible" tape hiss will become to your ears with this dbx deck. [From DAK Industries Inc. Winter 1986 catalog.]

7. People cooking live lobsters believe that dunking arthropods in boiling water does not cause them pain. This common view of pain in invertebrates has now been challenged, at least with regard to spiders. Honeybee venom and wasp venom injected into the leg of some types of spider cause the spider to detach the affected leg. Because the response is so swift, the venom has little chance to reach the spider's body. Spiders that do not discard their legs when stung in the leg usually die. Thus, discarding the leg has definite survival value. [Adapted from *The Science Almanac: 1985–1986 Edition*, ed. by Bryan Bunch (Garden City, NY: Anchor Books, 1984), 169.]

EXERCISE 3.F

INCOMPLETE ARGUMENTS[1]

Here are some incomplete arguments. Your task is to add the necessary premise (or premises) that will make the premises of the argument support the conclusion. Do not concern yourself with whether you agree with the premises or conclusions. Your only job is to add the missing premise that will make the premises support the conclusion.

(1)   Premise: Nonhuman animals suffer, have thoughts, and feel pain.
     Conclusion: Therefore, killing nonhuman animals is morally wrong.

Missing premise: _____

(2)   Premise: It's morally wrong to treat human beings as mere objects.
     Conclusion: So, genetically engineering human beings is morally wrong.

Missing premise: _____

(3)   Premise: The state ought to license all activities that can cause great amounts of harm.
     Conclusion: So, the state ought to require a license for all agricultural biotechnology.

Missing premise: _____

(4)   Premise: It is biologically natural for humans to eat animal flesh.
       Conclusion: Therefore, it is morally permissible for humans to eat animal flesh.

Missing premise: _____

(5)   Premise: For transnational corporations to patent genes taken from developing countries is a form of theft.
       Conclusion: For this reason, it is morally wrong for transnational corporations to patent genes taken from developing countries.

Missing premise: _____

(6)   Premise: It is our moral duty to provide food for future generations.
       Conclusion: It follows that it is our moral duty to genetically engineer crops.

Missing premise: _____

(7)   Premise: It is morally wrong to engage in activities that undermine the natural order of things.
       Conclusion: Hence, genetic engineering is morally wrong.

Missing premise: _____

(8)   Premise: Making transgenic animals fails to maximize the balance of happiness over unhappiness.
       Conclusion: Thus, it is ethically unacceptable to make transgenic animals.

Missing premise: _____

NOTE

1. Adapted by Gary Comstock from an exercise written by Michael Bishop of the Philosophy and Religious Studies Department, Iowa State University.

EXERCISE 3.G

ARGUMENTS FOR DIAGRAMMING

1. Using the numbers indicated, diagram the following arguments.

A. [1] If we are to regain our position as a scientifically advanced nation, we must increase aid to elementary schools, for [2] lack of basic education at the earliest stages can never be overcome.

B. [1] The lower creation was made to serve the higher creation. [2] Humans are the higher creation, and [3] animals are the lower creation. [4] Therefore, humans may eat animals.

C. [1] People cooking live lobsters believe that dunking arthropods in boiling water does not cause them pain. This common view of pain in invertebrates has now been challenged, at least with regard to spiders. [2] Honeybee venom and wasp venom injected into the leg of some types of spider cause the spider to detach the affected leg. Because the response is so swift, the venom has little chance to reach the spider's body. [3] Spiders that do not discard their legs when stung in the leg usually die. [4] Thus, discarding the leg has definite survival value. [5] Although this behavior in itself does not prove that some spiders feel pain, the components of the venom associated with leg detachment suggest that these spiders do feel pain. [6] Melittin, histamine, phospholipase A, and serotonin, found in the venoms, are known to cause human pain.

D. [1] Animal liberationists insist that we have a moral obligation to efficiently relieve animal suffering. [2] The misery of wild animals is enormous. [3] In the natural environment, nature ruthlessly limits animal populations by doing violence to virtually every individual before it reaches maturity. [4] The path from birth to slaughter, however, is nearly always longer and less painful in the barnyard than in the woods. [5] Thus, the most efficient way to relieve the suffering of wild animals would be to convert our national parks and wilderness areas into humanely managed farms. [6] It follows, therefore, that animal liberationists cannot be environmentalists since they must be willing to sacrifice the authenticity, integrity, and complexity of ecosystems for the welfare of animals.

2. Supplying your own numbers, diagram the following arguments:

A. Living on a family farm is the best way to live. It would be nice if most Americans could live on family farms. Therefore, the U. S. Congress ought to provide funds so that all Americans can live on family farms.

B. The personality of the farmer is basically healthy. The reason is that he is self-reliant and independent, committed to fairplay, due process, and democratic ideals. But a darker side is characterized by scapegoating, violence, and ideologies that would bring fewer gains to farmers than losses to consumers and society as a whole.

C. Farmers have an interest in depleting soil and water resources if they can profit in the short term from exploiting the land in this way. Future generations may not need soil and water if they can find alternative ways of feeding themselves. Taken together, these two reasons lead to this conclusion: Farmers have no moral duty to farm in an ecologically sustainable way.

D. It is possible to question whether future generations indeed have a right to food. First, the question of which individuals will make up future generations is unclear. The reason is that the choices we make today affect which individuals are alive tomorrow. Second, utilitarians think that individuals have rights only when doing so produces the greatest good for the greatest number. Third, the greatest good for the greatest number might be obtained by giving everything to the present generation and not worrying about future generations.

3. Consider the following argument:

1. Humans are rational creatures.
2. Animals are not rational creatures.
3. Rational creatures are higher than nonrational creatures.
4. Higher creatures may use lower creatures.
5. Humans may use animals.
6. In the course of nature, rational creatures use nonrational creatures for their purposes.
7. In the course of nature, humans raise, kill, and eat animals and use them in research.
8. God created the course of nature.
9. What God creates cannot be wrong.
10. Whatever happens in the course of nature cannot be wrong.
11. It cannot be wrong for humans to raise, kill, and eat animals and use them in research.

Which of the following diagrams most accurately portrays the logic of this argument?

a.

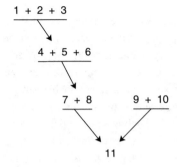

b.

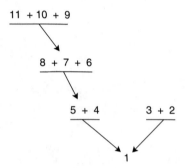

c.

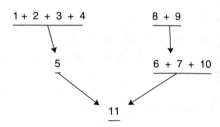

FIGURE A.1, CONTINUED

d.

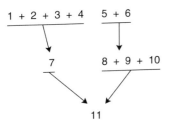

e.

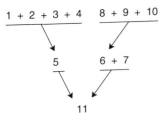

4. Consider the following argument:
   1. Animals have value.
   2. Whatever has value has intrinsic value.
   3. Humans may not claim to be the only measure of good as regards animals.
   4. If humans may not claim to be the only measure of whether something's good, then that thing has intrinsic value.
   5. Animals have their own needs, interests, and patterns of behavior.
   6. Whatever has its own needs, interests, and patterns of behavior has intrinsic value.
   7. Animals have intrinsic value.
   8. It is morally wrong to cause anything with intrinsic value avoidable death or injury through deprivation or starving.
   9. Using animals for food always causes them avoidable injury or death.
   10. It is morally wrong to use animals for food.

Which of the following diagrams most accurately portrays the logic of this argument?

a.

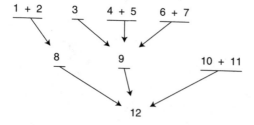

b.

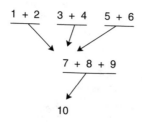

c.

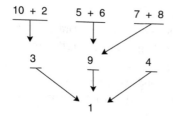

d.

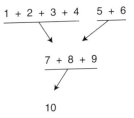

e.

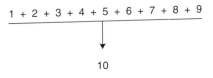

EXERCISE 4.A

DISCUSSION QUESTIONS

1. Do you think ethicists have a method they follow that is as objective as the method that scientists follow?
2. If you answered "yes," explain analogies you see between ethics and science.
3. If you answered "no," explain any disanalogies you see between ethics and science.
4. Dennis gave three reasons for thinking ethics is not like science. Explain whether you agree or disagree with each argument, and why.
   A. People have conflicting definitions of ethical terms.
   B. People make conflicting ethical judgments.
   C. We cannot establish the reliability of our ethical judgments without using circular arguments.

EXERCISE 5.A

DISCUSSION QUESTIONS

1. Imagine that you are Emily. For the moment, forget any doubts you may have about whether Marie is an environmental nutcase. The immediate question you must answer is very specific: Do you think that Marie is correct, that it is possible to harm a natural entity like a river? Is it possible to harm a natural entity even if one does not harm any humans in the process? Explain your answers.
2. The American environmentalist Aldo Leopold wrote that a thing is morally right when it tends to support the stability, integrity, and beauty of the land and it is wrong when it tends otherwise. What do you think Leopold meant? Do you agree with him? Apparently, Marie has read Leopold. Do you think Leopold would support her view about the moral standing of the Chatham?
3. How would you know when an action was tending to support the stability, beauty, and integrity of an ecosystem?
4. To what extent, if any, should the town council shape public policy to take account of what Marie calls her "natural" argument? To what extent, if any, is it possible for us to shape public policy on the basis of duties to what Leopold called "the land"?

EXERCISE 5.B

THE SPOTTED OWL

First, consider the statement: "We ought to preserve the old-growth forest in the Pacific Northwest because it is the last habitat for the spotted owl, an endangered species." On a scale of one through five, with one representing strong disagreement, three neither agreement nor disagreement, and five strong agreement, choose a number that best represents your attitude. If you chose one or two, write down a sentence or two explaining why you disagree. If you chose three, explain what additional information or arguments might move you to one side or the other of the scale. If you picked four or five, indicating strong agreement, complete the following statement:

We should protect endangered species like the spotted owl because _____

_____.

One obvious purpose of this exercise is to articulate and identify some of the many ways in which we justify concerns for the environment (what sorts of reasons were given in the fill-in-the-blank exercise), and also reasons that may come into conflict with those concerns (the sorts of considerations articulated by those who chose one or two). However, there is an additional twist, hidden in the "because" of the original statement. In fact, the Endangered Species Act was, and continues to be, invoked as an argument to justify protection of the old-growth forests: Does preservation of a particular species, such as the spotted owl, drive our concern for the habitat that it needs to survive, or does our concern for the habitat, the ecosystem, drive our concern about the perilous status of the spotted owl? Further answers to the fill-in-the-blank part of this exercise often turn back to a stronger emphasis on the value of the ecosystem, with the plight of the owl viewed as a symptom of a more basic problem. Although this distinction may seem purely academic, it has important implications for environmental policy. Suppose, for example, that spotted owls thrive and reproduce quite well in zoos, meaning that we could preserve the species and continue logging. In such a case, should we still be concerned about the habitat? Why or why not?

This brings us back to the questions of intrinsic value. Is it the individual owls, the survival of a unique species, the ecosystem of which it is currently a part, or some combination of these that is purported to have intrinsic value? To make progress toward answering, or even understanding these questions, we need to survey some of the options listed previously. Before we do that, however, an initial understanding of the concept of intrinsic value and its relation to claims such as "nature has rights" is essential.

EXERCISE 6.A

DISCUSSION QUESTIONS

1. Imagine that you are Emily. For the moment, forget your doubts about whether any relief agency actually forwards the majority of the money you give to the intended recipients. The immediate question you must answer is very specific: Do you think Emily is right to assume that each of us has the opportunity to do something about the situation? Explain your answer.
2. If Emily's factual assumption is right, is our inaction the moral equivalent of murder?
3. Suppose that letting someone die is not the moral equivalent of murder. It may nevertheless be very wrong. In addition to an obligation not to kill, we may have a separate obligation to help. Do we have a general duty to help all people, including total strangers in distant lands?

4. What might be the limits of a duty to help? According to some philosophers, there are no limits: We should continue to benefit others until further efforts would burden us as much as they would help the others. Do you find this plausible?

EXERCISE 7.A

DISCUSSION QUESTIONS

1. Do you think that animals such as cattle and chickens are capable of feeling pain? Do you think that they have a sense of their own future and preferences about what happens to them? What sort of evidence leads you to answer "yes" or "no" to each of these questions regarding cattle and regarding chickens? What animals are *not* capable of each of these things? What makes you think they aren't?

2. How much merit is there in the idea that a kind of farm that takes animal interests into account and satisfies them more fully is better than a farm that does not? Explicitly distinguish the normative premises that support your conclusion from the empirical ones.

3. A majority of Americans are not vegetarians and do not accept animal rights arguments. How important, from a moral point of view, is this fact to Doug's family's decisions about the place of cows in their future farm operation?

4. It would be difficult, economically and socially, for dairy farmers to not slaughter cows. Given the state in which veal calves live out their short lives, do you think we need scientific research to make it possible that only heifers would be born on dairy farms? If so, who should pay for this?

5. Should we shape public policy so that cows are put under as little stress as possible on dairy farms? Why or why not? Should we outlaw veal production? Why or why not?

EXERCISE 8.A

DISCUSSION QUESTIONS

1. Imagine that you are Emily. For the moment, forget any doubts you may have about whether environmentalists are right or wrong in their belief that it is possible to harm nature even if you do not harm any humans, present or future, in the process. Forget, too, your questions about the relative numbers of farmers

versus city dwellers who will be helped or hurt by the town council's proposal. The immediate question you must answer is very specific: Do you think that there is some scientific process by which Gordon can figure out for himself which party, the farmers, the townsfolk, or the environmentalists, is most justified in its claim to have rights to the water? Explain your answer.

2. In general, Americans tend to think that contested policy issues are not settled on rational grounds but rather by politicians trying to satisfy the most politically powerful group. Assume that there are twenty times as many residents of Springdale as there are farmers, and three times as many residents of Springdale as there are environmentalists. How important are these facts to deciding what is the correct answer to Gordon's moral dilemma? How much do the numbers count?

3. Libertarians believe that others should not interfere with our basic liberty and property rights. Libertarians also believe that property owners should enjoy the authority to utilize their property according to their wishes, whatever their wishes may be, as long as they do not harm or interfere with others. If the farmers adopted a libertarian position in ethics to defend their claims to the water in Chatham River, how would their argument go?

4. Utilitarians believe the right action is always the one that will maximize the greatest benefits for the greatest number. Utilitarians also believe that it is permissible to sacrifice the interests of the few if so doing will promote the greatest good of the many. If Gordon were to adopt a utilitarian position in ethics to defend Springdale's appropriation of Chatham River water, how would his argument go?

5. To what extent, if any, should the town council shape public policy to take account of the interests of people who do not live in town?

EXERCISE 9.A

DISCUSSION QUESTIONS

1. For the moment, forget your doubts about whether it is possible to make transgenic Egg Machines. The immediate question you must answer is very specific: Do you think Dr. Krista is right to assume that Egg Machines might be a good idea from a moral point of view? Who would be the affected stakeholders? Write down as many individuals as you can. Be prepared to defend your answer.

2. If Krista's factual assumptions are correct that the Egg Machines are feasible and they would lack any feelings or consciousness, should we adopt public policies that would provide poultry breeders with funds to engage in molecular engineering

with the goal of producing nonsentient, animal-derived organisms capable of producing industrial quantities of foods? Who might benefit from this policy? Who might lose?

3. Do you find the imaginary scene in the warehouse repulsive? If so, why? If not, why not? What role do you think our emotions should play in deciding which agricultural vision to pursue?

4. Do you think poultry gene splicers are "playing God" with nature in a way that may be objectionable on theological grounds? Do you think it is morally impermissible to make transgenic animals at all? If so, why? If not, why not? What role do you think religious beliefs should play in secular discussions of ethics and public policy?

5. Who should own the products of genetic engineering? The communities of the women farmers in developing countries that hand selected chickens for breeding over hundreds of years? The corporations that invest millions in research to change a gene or two, patent the product, and then sell it? The governments that sponsor the basic research at universities on which the private sector builds its applied research? The taxpayers who fund the government's basic research? How should our basic social institutions be set up so that all stakeholders benefit fairly?

EXERCISE 10.A

DISCUSSION QUESTIONS

1. Imagine that you are Emily. For the moment, forget any doubts you may have about whether you would want to live the way Karen and her husband want to live. Forget, too, your questions about which style of farming is most likely to make the three families the most money. It appears that either style of farming will make some money, even though the three smaller operations will probably make less than the one larger operation. The immediate question you must answer is as follows: Do you think that there is merit in Karen's and her husband's idea that three smaller family farms are better in a moral sense than one larger family farm? Explain your answer.

2. In general, farms are getting larger and larger, not smaller and smaller. List five reasons that the family should opt for one large farm; next, list five reasons they should not make that choice. Then list five reasons that the family should opt for

the three smaller farms and five reasons that they should not make that choice.

3. Using your best critical reasoning skills, assess the reasons you have just listed. Identify those that you think Emily should throw out as bad reasons and those you think she should pursue with Doug as good reasons.

4. Should we shape public policy so that it would be easier for entering farmers to make the choice Roy's daughter is recommending to her dad and brother?

5. It is difficult, economically and socially, for farmers to make the choice Roy's daughter is recommending. Colleges of Agriculture generally do not hold up the small farm as the farm to be emulated. Should we hire agriculture professors who would teach their students to aspire to have small farms?

EXERCISE 10.B

QUESTIONS

1. Consider Marty Strange's characterization that follows of family farming. To what extent is this notion linked to a specific model of the family (traditional or nontraditional)?

> In a family farming system, farms rely on family labor and management skills. The family lives its life in harmony with its workplace. There is no division between home and work. Children grow up learning to farm by apprenticeship. Formal education is not eschewed; in fact, it is valued as a means of increasing the human skills on the farm. But the practical aspects of farm management and decision making, of work and reward, and of problem solving are learned by doing. Most important, responsibility is shared by all family members old enough to assume any. (Strange, 1988, 34)

2. To what extent is a case for family farming affected by the kind of farm involved? For example, does the case for family farming explored in this chapter change if it does not concern a dairy farm like Roy's? Imagine that the farm is devoted to any of the following: cash grain, tobacco, cotton, vegetables, fruit, nuts, nursery or greenhouse, other crops; beef, hogs, sheep, poultry, other livestock. Does the region (such as Northeast, Appalachia, Southeast, Delta, Corn Belt, lake states, northern plains, southern plains, Pacific) matter?

3. Adam Smith, the great eighteenth century Scottish philosopher and champion of the limited free market, was critical of specialized labor. His views reinforce Berry's as cited in this chapter. Assess the following claim by Smith:

> In the progress of the division of labor, the employment of the far greater part of those who live by labor, that is, of the great body of people, comes to be combined to a few very simple operations, frequently to one or two. But the understandings of the greater part of men are necessarily formed by their ordinary employments. The man whose whole life is spent in performing a few simple operations of which the effects too are, perhaps, always the same, or very nearly the same, has no occasion to exert his understanding, or to exercise his invention in finding out expedients for removing difficulties which never occur. He naturally loses, therefore, the habit of such exertion, and generally becomes as stupid and ignorant as it is possible for a human creature to become. The torpor of his mind renders him, not only incapable of relishing or bearing a part in any rational conversation, but of conceiving any generous, noble, or tender sentiment, and consequently of forming any just judgment concerning many even of the ordinary duties of private life.. His dexterity at his own particular trade seems, in this manner, to be acquired at the expense of his intellectual, social, and martial virtues. But in every improved and civilized society this is the state into which the laboring poor, that is, the great body of the people, must necessarily fall, unless government takes some pains to prevent it. (734–735)

4. Along with assessing a position such as Smith's, you may wish to develop an educational program (or perhaps a course or series of courses) that would introduce agricultural knowledge and technology to the general public. Marty Strange writes: "Only an informed and alert public can defend itself against the misuse of either technologies or regulations" (Strange, 1988, 290). Address Strange's proposal that follows. In your region of the country, is there any extant program now in place (or in development) that would secure what Strange describes as public-public relations?

> Better to focus on the broader educational needs of society to understand technology than to train only the brightest to use it. Instead of private-public partnerships that corrupt the research process, why not public-public relationships between agricultural universities and public schools in which the universities help students grasp the meaning of technologies and the alternative paths to technical development? (Strange, 1988, 290)

5. To what extent do you think that the debate over family farming should be affected by aesthetics (judgments of beauty and ugliness)? To what extent can the case for or against the family farm be bolstered through aesthetic experience or through art, literature, theater, music, film, or poetry? In what respects may any of these have not merely an emotional, persuasive force but raise important reasons and arguments for debate? You may wish to consider some of the following works: *Remembering*, by Wendell Berry; *The Grapes of Wrath*, by John Steinbeck; *A Thousand Acres*, by Jane Smiley; *Founding Farms; Portraits of Five Massachusetts Family Farms*, by Michael Gery. From antiquity, you may wish to engage Virgil's extended poem on farming, *Georgicas* (first century BC).

6. Family farmers have sometimes been led to protest government policies. See, for example, Dianna Hunter's *Stories of the Minnesota Farm Advocates*. Sometimes protests have involved destroying crops and livestock to affect the market and general population, sometimes they have involved protest marches and the like. When do you think it is ethically permissible for a farmer to destroy livestock in protest of a government policy? Imagine, for example, that Doug's father is upset at the price control of milk.

7. If Berry is right about virtue, how should this influence a philosophy of agricultural education?

8. How strong do you think familial obligations are? Imagine that Doug's parents are good and kind. He is divided about what occupation to pursue. He would enjoy farming but also some other occupation. His parents ask him to farm. Is he obliged to do so?

9. Utilitarianism was presented as an impartialist ethic that highlights the overall good. One objection to this is that it does not seem to leave sufficient space for other goods and rights, such as the good of integrity. In one well-known exchange, Bernard Williams held that a utilitarian would have to sacrifice the integrity of an individual to promote the greater good (Williams, 1973). To what extent do you think that utilitarianism would be able to take seriously family agrarian claims about the integrity of personal identity and land use?

10. List some of the virtues that you think are integral to family farming. How may the case for family farming differ from or be in league with the case for protecting small businesses?

11. Some philosophers have advanced wager arguments designed to tip the scales to one side or the other. Thus, the French philosopher Blaise Pascal (1623–1662) held that when in doubt about whether God exists, and given the choice to believe or disbelieve, it would be wiser to believe because the opportunity to net good would be greater and the risk of ill lessened. Can a wager be formulated in the case of family farms? Imagine that the case for and against

preserving family farming is equally balanced except that the loss of family farming involves a risk of losing an important, good component of our heritage and national identity. It is not known to incur such a loss, but the loss is a live possibility. Under such conditions, is it better to be safe than sorry and to preserve family farms?

EXERCISE 15.A

CASE STUDY RESPONSE FORM: INDIVIDUAL[1]

Your Name: _____

1. What ethical question are you trying to answer? (Your question should include a normative word, such as *should* or *ought*. For example: "Should I support the production of headless cats?")

_____

2. Give your first, intuitive answer to your question.

_____

3. List three reasons for your answer.

_____

_____

_____

4. List three facts that support your answer. (For example, "Headless cats do not feel pain"; "Chances are that a homeless cat will be euthanized.")

_____

_____

5. List one moral principle that supports your answer. (For example, "We ought to do good"; "We ought not to cause unnecessary pain.")

_____

_____

_____

6. Using at least one factual premise that you just wrote and at least one moral principle, construct a valid moral argument that leads logically to your answer. Write a factual premise on line A; write a moral principle on line B; and write your conclusion on line C.

A. _____

B. _____

C. _____

7. List two potential objections to your argument.

A. _____

B. _____

8. Describe how you would respond to each objection.

A. _____

B. _____

EXERCISE 15.B

CASE STUDY RESPONSE FORM: GROUP[1]

Your Name: _____

1. What ethical question are you trying to answer? (Your question should include a normative word, such as *should* or *ought*. For example: "Should I support the production of headless cats?")

_____

_____

_____

2. Give your first, intuitive answer to your question.

_____

3. List three reasons for your answer.

_____

_____

_____

4. List three facts that support your answer. (For example, "Headless cats do not feel pain"; "Chances are that a homeless cat will be euthanized.")

_____

_____

_____

5. List one moral principle that supports your answer. (For example, "We ought to do good"; "We ought not to cause unnecessary pain.")

_____

6. Using at least one factual premise that you just wrote and at least one moral principle, construct a valid moral argument that leads logically to your answer. Write a factual premise on line A; write a moral principle on line B; and write your conclusion on line C.

A. _____

B. _____

C. _____

7. List two potential objections to your argument.

A. _____

B. _____

8. Describe how you would respond to each objection.

A. _____

B. _____

NOTE

1. Case Study Response Forms were developed by Gary Comstock and originally appeared in the *Iowa State University Bioethics Institute Handbook, May 1998.*

NOTES FOR INSTRUCTORS

CHAPTER 1: ETHICS

EMILY THE STUDENT

The ethical problem embodied in Emily's case is a problem of conflict or conscience for an individual. Our teaching objectives in it are to help students articulate what is wrong with cheating and to help them realize that there is a broad range of judgments they share about what is morally right and wrong. Salient features are that, all other things being equal, cheating is morally wrong and students think that it is wrong. They think it is wrong for anyone in a situation similar to Emily's in all of the relevant details. They are, therefore, moral objectivists. Ironically, however, many of these same students think of themselves as naive relativists, and they will deny that there are any particular moral judgments universally shared across times and cultures. The main objective in discussing this case is to help them begin to overcome their naïve relativism and to acknowledge the fact that there are many particular moral judgments on which we all agree.

"Write and Pass" is a pedagogical exercise to help students gain clarity about their own views, to learn how to compose an argument, and to gain confidence in their innate ability to examine the assumptions of arguments.

Suggestion: Use the "Write and Pass" exercise immediately after the students read "Case 1: Emily the Student."

Like all of the pedagogical exercises found at the end of this book (in Part 3), "Write and Pass" may be adapted for use with other case studies.

"Discussion Questions" is a graded exercise intended to help students identify ethical issues and points of conflict, use their moral imaginations to recognize potentially interested parties, learn to envision potential consequences, and articulate obligations.

Suggestion: Use the "Discussion Questions" exercise after the students do the "Write and Pass." Following are some notes for grading the exercise.

EVALUATION CHECKLIST[1]

1. **Issues and points of conflict** (<u>5</u> points total)
   One correct answer earns <u>3</u> points.
   Two correct answers earns <u>4</u> points.
   Three correct answers earns <u>5</u> points.
   Emily's duty to be honest versus her duty to look out for her friend, Doug ____
   Emily's duty to respect her fellow students versus her desire to help Doug ____
   Emily's duty to treat Dr. Wright with respect versus her duty to pursue her own interests   ____

   Subtotal   ____

2. **Interested parties** (<u>3</u> points total: 1-2 = <u>1</u>, 3-4 = <u>2</u>, 5+ = <u>3</u>)
   Emily   ____
   Emily's family   ____
   Doug   ____
   Dr. Wright   ____
   Other students in the class   ____
   Other students at the university   ____
   The university itself   ____
   The general community   ____

   Subtotal   ____

3. **Potential consequences** (<u>4</u> points total: 1-2 = <u>1</u>, 3-4 = <u>2</u>, 5-7 = <u>3</u>, 8+ = <u>4</u>)
   to Emily's reputation   ____
   to Emily's career   ____
   to Emily's relationship with Doug   ____
   to Emily's relationship with Dr. Wright   ____
   to Emily's family   ____

to other students in the class	____
to the university	____
to future students	____
Subtotal	____

4. **Obligations** (5 points total: 1 = 1, 2 = 2, 3 = 3, 4 = 4, 5+ = 5)

to be honest	____
to obey the university's rules	____
to respect Doug; to respect herself	____
to respect Dr. Wright	____
to do no harm	____
to act in a way that she could approve of other students acting	____
to consider her own needs and interests	____
Subtotal	____
Total	____

16–17 = A
14–15 = B
11–13 = C
9–12 = D

Note

1. Based on a form developed by Muriel J. Bebeau, et al., *Moral Reasoning in Scientific Research* (Poynter Center for the Study of Ethics and American Institutions, Indiana University, 1995).

CHAPTER 2: RELIGION

RICH THE ATHEIST

The ethical problem embodied in Rich's case is a problem of conflict or conscience for an individual. Our teaching objectives in it are to help students articulate their views about the relevance of religious beliefs to ethics, to help them realize that there is at least one ethical theory that is deeply religious, and to raise questions for them about the acceptability of that theory. Salient features are that, all other things being equal, religious beliefs ought to be a topic for discussion in university classrooms; yet, we live in a pluralistic democratic society in which religion can be a conversation stopper.

Postscript to Chapter 2: Dealing with Religion in Secular Life Science Ethics Discussions

Science instructors may experience some uneasiness in dealing with students' religious convictions during a discussion of ethics at a secular institution. Here are some tips.

1. Acknowledge the validity of religious convictions.

    Religion has been and continues to be an important source of many peoples' ethical values, concepts, and ideals. No matter what culture we study, religious communities have been the repositories and incubators of moral values and principles. Even if we think that religion is to be abhorred for the values it has taught, we should begin by honoring the religious student's perspective. We can do this by acknowledging the historical role of religion in teaching values.

    We might respond by saying: "Good point. Matters of faith and God and religious instruction play a central role in forming our values, don't they? Thanks for reminding us not to overlook religion as we think through this issue."

    Then we must go on to point out that there are difficult problems in bringing religion into secular ethics conversations. Such as:

2. Appeals to religion can be used to stop rather than encourage conversation.

    "Abortion is wrong because it says so in the New Testament (or Hebrew Bible or Qu'ran or Upanishads)."

    "Euthanasia is wrong because God (or the mu'azzin or Pope or rabbi) opposes it."

    When someone tries to use a religious appeal to end a conversation, we can point out that this is a *faulty* appeal to authority. However, we must be careful to explain why. Suppose that the authority is scripture, tradition, the head of the church, or the deity. Appeals to these authorities are faulty *not* because scripture, tradition, and God are necessarily faulty authorities. They are faulty appeals because *the structure of the argument* is faulty. The appeals beg the question of whether the authorities are incorruptible and infallible. What if someone appeals to the authority of Charles Manson or Adolph Hitler? Participants in ethics conversations must, at the very least, remain free to ask whether the authority is reliable, and free to pursue questions in this area.

3. Ask the student why he or she thinks God (or the Bible) would command something.

   The best way to convert religious appeals from conversation stoppers to conversation starters is to ask students to think about God's *reasons* for doing something. "Religion is important in these conversations. Now, let's try to figure out *why* God would make it wrong to have an abortion or engage in homosexual behavior." This will help religious students begin to figure out the philosophical grounds of their values.

   What if a student balks at this point, refuses to entertain the question, protesting that it would be impious or dangerous to inquire into God's motives? There is every reason to answer such students straightforwardly. They have articulated an issue that lurks in the minds of many less outspoken students.

4. Ask whether there is unanimity about issues of morality *within* the student's tradition.

   All traditions, even fundamentalist traditions, are at least mildly pluralistic. There is always a Mr. Jones who sits in the back pew and, although he is in all other respects a member in good standing of the congregation, is fondly known for all for his contrary opinions. Ask your student to identify such a person. For example: "Is there anyone in your tradition who holds a different view of the morality of abortion?" Assign them the task of asking their Mr. Jones about his views, and have them carefully write down his reasons for diverging from the tradition's teachings.

   If the student continues to refuse to name any dissenters, you may be dealing with one of two kinds of student. It is important to figure out which kind. The first type of student is good-willed but misinformed. Such students genuinely don't know that there are dissenters in their tradition. They may need your help to identify contrary-minded individuals whom the student can trust.

   The second kind of student is more difficult. These students' lack of information is compounded by the fear that modern science is going to destroy their faith. There is no guaranteed method for dealing with these students, but I suggest the following.

5. Assign the students the task of having their spiritual mentor (pastor or priest) help them find literature written by people in their tradition that fairly represents ethical views conflicting with theirs.

   The problem with the fearful student is that they do not trust their university instructor, probably not always without good reason, with what seem to them spiritual questions. Try to get these students in touch with someone they trust. When these students find such a mentor, the mentor may be able to assist the students in realizing that they can be simultaneously pious and intellectually active.

6. If the appeal is to the Bible, ask whether the verse has a wider context and whether there are any larger, overall, Scriptural themes that seem to be in tension with the verse.

A strong case can be made that the overall themes of the Hebrew, Christian, and Muslim scriptures are righteousness, justice, love, compassion, and mercy. Try to draw this point out of other students' comments, but be ready to point it out if no one steps forward.

7. Ask whether there is unanimity about issues of morality *among* religions.

For example: "Does anyone know whether Christianity, Buddhism, and Hinduism have the same view of the Bible (or "God's will" or "duties to animals")?

8. Ask whether the United States should make public policy on the basis of religious revelations given to one religious tradition.

Point out that although students come from various religious traditions, and some from no religious tradition at all, they actually agree about a fairly wide range of values. Mention the long lists of rights and wrongs they "wrote on the walls" during the Shock Treatment for Relativism exercise. Remind them that the best way to make progress in ethics is not to start with values that are specific to their tradition, because those values may be very contentious. It is to start from values we agree about and then work from there toward the thorny questions.

There is also the matter of public policy. Unlike most other nations, the United States does not have a state religion. The country's founders wanted to separate church and state in order to protect the free exercise of religion. People here have the freedom to follow whatever religion suits their conscience. Then you might ask your students the following sorts of questions: "Is this a good way to proceed?" "Does anyone have reasons to support their view about whether we should continue to try to make public policy on nonreligious grounds, without privileging the views of any single religion?" "Do you think we should decide policy matters on religious grounds or on grounds that might be shared both by believers *and* atheists?"

9. Suggest that students pursue religious issues in courses devoted to that subject.

This is a natural way to end discussions if you've done #1 through #8. Students will appreciate knowing that there are places they can go to pursue the issues further. Many secular campuses have courses explicitly called "Religious Ethics."

10. Summarize the discussion.

We can often avoid hard feelings simply by bringing closure to the discussion. Here are some ideas: "As we have noted, religion is an important source of ethical values and belongs in rational ethics discussions. But there are problems. First, within a single tradition, there are varying interpretations of what God commands. Second, there are varying interpretations of morality among the various religions. Third, in the United States we honor religious liberty by trying to make public policy without privileging the authority of any one religion. How do we address the problems? First, we can ask what reasons God has for commanding something. Second, we can try to identify values that are shared widely across many religious traditions. And third, we can explore these issues in more depth in classes devoted to this topic."

CHAPTER 3: REASONING

KAREN THE ETHICIST

The problem embodied in the students' conversation is the problem of knowing how to proceed when thinking about ethics. Our teaching objectives in it are to help students articulate their doubts about ethical thinking being rigorous or methodical; help them to see some ways in which ethical thinking is rigorous and methodical; and assist them in learning how to think critically and carefully in this area. The emphasis on good arguments and objective criteria can also be used to set the stage for Chapter 4, and to show why ethics is not "just merely" a matter of opinion.

CHAPTER 4: METHOD

DENNIS THE RELATIVIST

The problem embodied in Dennis's case is a problem of ethical theory. Our teaching objectives in it are to help students articulate their worries about the objectivity of ethical judgments, to help them realize that others have these same worries, and to raise questions about whether students must simply accept naïve relativism. Salient features are that questions about the objectivity of ethics are legitimate, concerns about ethics being unscientific are widespread, and yet there are good reasons to think that the skeptical, relativist, challenge can be met.

CHAPTER 5: ENVIRONMENT

MARIE THE ENVIRONMENTALIST

The ethical problem embodied in this case is a problem of deciding on the moral standing of nonhuman entities. The teaching objectives are as follows: to explore the various theories regarding the relative weight to be attached to the moral status of humans, animals, ecosystems, and future generations. The salient facts or features that create the problem are the following: (1) the fact that farmers, individual animals, ecosystems, and future generations all have an interest in using natural resources such as water and soil; (2) there may not be enough natural resources to serve all these interests; and (3) it may be that "natural resources" should not be understood as "resources" at all, but rather as intrinsically valuable things. If it is appropriate, the last point could lead into a more theoretical discussion of the intrinsic/instrumental values: how one would justify the claim that something has intrinsic value, and how the distinction is relevant to moral reasoning.

CHAPTER 6: FOOD

DHRUVA THE DESTITUTE

The ethical problem embodied in this case is that of deciding on a general policy or principle. The teaching objectives are to explore the various theories regarding the relative weight to be attached to our duties to provide food aid to those close to us and those far away from us. The salient facts or features that create the problem are (1) the fact that many people die of hunger each year, and (2) farmers produce sufficient food each year to feed everyone.

CHAPTER 7: ANIMALS

MISHA THE COW

The ethical problem embodied in this case is a problem of deciding on a general policy or principle. The teaching objective is to explore the various theories regarding the moral standing of farm animals. The salient facts or features that create the problem are (1) the fact that many farm animal species exhibit complex behaviors and apparently possess reasonably sophisticated mental states, and (2) the moral protections afforded humans of comparable behaviors and states are not afforded to farm animals.

CHAPTER 8: LAND

GORDON THE LAWYER

The ethical problem to be embodied in this case is that of deciding on a general policy or principle. The teaching objectives are to explore the various theories regarding the relative weight to be attached to the property rights of farmers, consumers, and nonhuman entities, such as bodies of water, soil, or wildlife habitat areas. The salient facts or features that create the problem are (1) the fact that farmers, consumers, and wildlife all need water and soil, and (2) in some areas there is not enough water, soil, and/or habitat to go around.

CHAPTER 9: BIOTECHNOLOGY

Additional discussion questions that may be used after students have read the chapter:

1. Would such a product as the Egg Machine ever be made? On what do you base your answer?
2. Can you give a definition of *natural* that implies that Egg Machines are unnatural but that the use of airplanes and selective breeding is not?
3. Do you think there *is* a morally relevant difference between Egg Machines and Tissue Culture? If so, what is it? Or do you think that they should be seen as morally on par—that if one of them is acceptable, then the other must be as well?
4. Do you think that genetic engineering in agriculture would make us more likely to apply genetic engineering to *humans*? Why or why not?
5. Is there something morally bad about always being at the ready to make things out of bits of living nature, or viewing it in terms of *commercial* potential? Why or why not?
6. It was said that the Raw Materials view would encourage the following: Confronted with animal suffering in agricultural contexts, someone with this view would tend to try to change the animals so that they don't feel the suffering, rather than change the conditions so that the animals aren't under the stress. Is there anything wrong with this? If so, what?
7. It's often said that most Americans don't think that much about where their food comes from, and also that, given the suffering of animals involved, there are further pressures to not think about it, since this is a disturbing thought if focused on consciously. How would Egg Machines and Krista's scenario concerning Tissue Culture affect these tendencies of ours?

CHAPTER 10: FARMS

Roy the Dairy Farmer

The ethical problem embodied in this case is that of deciding on a course of action related to career and family. The teaching objectives are as follows: to articulate the moral virtues and vices traditionally associated with family farming, to discuss the costs and benefits of larger scale industrial agriculture, and to comment on the relevance of religious considerations to these issues.

The salient facts or features that create the problem are (1) the fact that there are many virtues in a way of life spent on a medium-sized, owner-operated dairy farm, yet (2) such farms are quickly disappearing as the structure of the dairy industry moves increasingly toward a system dominated by a few very large farms.

CHAPTER 11: ENVIRONMENT

Rare Plants

For this case study, I usually divide the class into small groups of three to five people and hand out the exercise. The students read the information and then discuss the questions within the group. Each student writes his or her own answers so that there is room for disagreement. This exercise typically takes place in a forty-minute period. Depending upon time available and the nature of the class, additional reading outside class may be beneficial. The full text of the Convention on the International Trade of Endangered Species can be accessed at http://www.cites.org/. For an interesting perspective on CITES and its effects, read "Orchid Fever" by Eric Hansen. Further discussion of the ethics of specimen-based research may also be fruitful (for example, the sale of fossils is an interesting topic).

CHAPTER 12: FOOD

Infant Deaths in Developing Countries

You may also wish to introduce issues related to post-illness quality of life. Many older flu victims, for example, are prone to subsequent illnesses including stroke and pneumonia.

EDIBLE ANTIBIOTICS IN FOOD CROP

Teaching objectives:
1. To have the students research and identify the issues, pros and cons.
2. To identify the facts or features that create the problem.
3. To identify risks, costs, benefits and detriments. Identify who bears risk, who benefits and who pays.
4. To recognize varying levels of certainty in the arguments regarding risks and benefits. To be able to compare use of the precautionary principle to current regulatory methods.
5. To be able to evaluate competing arguments and identify acceptable alternatives and/or compromises.

CHAPTER 13: ANIMALS

VETERINARY EUTHANASIA

The exercise is designed for use with veterinary students but has been used successfully with nonveterinary classes. It includes a pre-test, which may be administered before the discussion, and a post-test, which may be administered after the discussion. The students' progress can thus be measured by comparing their scores on the pre- and post-tests.

CHAPTER 14: LAND

HYBRID CORN

This case study consists of three parts. It is very useful for the case to limit yourself to the facts as they are presented and to imagine actually being there. However, introducing and discussing the precautionary principle prior to the case study is very useful. In addition, the article "Biotechnology in the 1930s: The development of hybrid maize" by Don Duvick in the January 2001 issue of *Nature Reviews* should also be required reading prior to the case study. The case study probably works best if the participants have ample time to think about each section and answer the questions before moving on to the next session. The instructor should pay close attention to the answers the students give and clearly point out to students if and when they change their answers over time. Often, the initial answers about hybrid corn are positive and then change to negative when introduced to biotechnology and Bt Corn.

CHAPTER 15: BIOTECHNOLOGY

GOLDEN RICE

The following sections are for instructors to use as a guide to this case study exercise.

Purpose

This case is designed to explore the ethical ramifications of biotechnology. The goal is for students to begin to appreciate the complexity of evaluating new technologies, and to practice making and evaluating moral arguments about a concrete topic in bioethics. Students will be asked to consider all salient facts and to appeal explicitly to moral values in supporting their groups' positions on golden rice. Important details about golden rice, as well as indications of how golden rice may advance or threaten some moral values, are available in the background materials. Moreover, the groups in this exercise are real groups, so this exercise can be used as a research project, in which students are assigned to learn as much as they can about their own groups in preparation for the in-class exercise. (The exception is "People from Developing Nations." However, the constituency of this group is real enough that students can easily research the condition within developing nations and the interests of people in such nations as preparation for the exercise.)

Procedure

The exercise has been used successfully with university faculty in a single hour-and-a-half session. With college students, however, we recommend a minimum of four one-hour class periods.

Class Period #1

The instructor gives an introduction to golden rice, handing out the one-page description of the exercise. Students are assigned to one of the five groups. We recommend that the instructor divide students randomly by, for example, asking them to count off from one to five. The group of students numbered "one" is assigned to the role of the WHO; the group of students numbered "two" is assigned to the role of Friends of the Earth; and so on.

The rest of this class period is spent giving instructions, meeting in groups, choosing a spokesperson for the group, and beginning to formulate strategy for the final presentation. The instructor also directs all students to read all the supporting documents, paying special attention to the documents supporting their group's position. The arguments provided in these materials are not exhaustive; other arguments can be made. It is up to the instructor's discretion whether to encourage students to do additional research on the Web or in their groups to discover other arguments. Instructors should carefully monitor the groups because some groups may need more assistance than others in extrapolating arguments from the information provided.

Class Period #2

Groups meet to discuss the readings, marshal the arguments for their position, and plan their final presentation. The WHO group discusses its decision and takes a preliminary vote, which it keeps secret from the other groups. The purpose of the vote is simply to inform members of the WHO how the respective members of the WHO are disposed. The WHO group also plans its behavior during final presentations. For example, it may wish to assign one student to be responsible for posing one question to the Friends of the Earth after the Friends have presented their arguments on the third day. Another student can pose a question to IRRI, and so on.

Class Period #3

Final presentations from each group. Each presentation must be no longer than five minutes. At the end of each presentation, the WHO is entitled to ask one question of each group, and the group's spokesperson must respond, taking no more than two minutes to do so.

After all four groups have made their presentations, the WHO recesses to another room. Taking no more than ten minutes, it discusses the arguments one last time and votes. It then returns to the room and announces its decision.

Class Period #4

This period is spent discussing the exercise and permitting students to vent feelings of frustration. Students in groups who lost the argument may feel disenfranchised. They may feel that the WHO did not adequately appreciate the weight of their arguments, did not understand the gravity of their concerns, and so on. The instructor can use this time for productive discussion of democratic institutions, the place of minority opinions, the difficulty of governing, the importance of open and transparent decision-making, and the like.

With the instructor's consent, additional information not included in the enclosed materials is permissible if it originates from any sources used in this project. However, participants must approach the instructor before the debate if they wish to use that information during the debate. At that time they must also present documentation showing the information's source so that the instructor is given time to determine its admissibility.

Background Materials

These materials are reproduced here to aid students in researching the arguments made by their respective groups. Students should be instructed to pay careful attention to specific factual claims as well as to any indications of the moral values that their groups endorse.

Panel: World Health Organization (WHO)

Objectives and functions[1]

WHO is defined by its Constitution as the directing and coordinating authority on international health work aim: "The attainment by all peoples of the highest possible level of health." The following are listed among its responsibilities:

- To assist governments, upon request, in strengthening health services;
- To establish and maintain such administrative and technical services as may be required, including epidemiological and statistical services;
- To provide information, counsel, and assistance in the field of health; to stimulate the eradication of epidemic, endemic, and other diseases;
- To promote improved nutrition, housing, sanitation, working conditions, and other aspects of environmental hygiene;
- To promote cooperation among scientific and professional groups which contribute to the enhancement of health;
- To propose international conventions and agreements on health matters; to promote and conduct research in the field of health;
- To develop international standards for food, biological, and pharmaceutical products; and,
- To assist in developing an informed public opinion among all peoples on matters of health.

Mission Statement[2]

The objective of WHO is the attainment by all peoples of the highest possible level of health. Health, as defined in the WHO Constitution, is a state of complete physical, mental, and social well-being and not merely the absence of disease or infirmity.

Group #1: Friends of the Earth (FOE)

Friends of the Earth is an international organization concerned with environmental and social issues. Friends of the Earth members view golden rice as a smokescreen used by biotechnology companies to convince the world that biotechnology is necessary to combat hunger and malnutrition, and to distract people from the risks of biotechnology. In a statement on golden rice, the group asks, "Is Golden Rice a triumph of biotechnology that could eradicate unnecessary suffering? Or is it merely a PR maneuver by a threatened industry that would thrust an unproven, unwanted, and perhaps even harmful technology upon the developing world?"[3]

One reason for the group's suspicion about golden rice is that vitamin A deficiency is usually correlated with general malnutrition. Presumably, general and widespread malnutrition can be addressed most effectively by addressing food security issues like ensuring that the poor have land on which to grow a varied diet or enough money to buy healthy foods. Golden rice therefore seems to Friends of the Earth like an excessively technical solution to a problem that might best be solved with traditional, low technology efforts to improve food security and combat poverty.

Friends of the Earth estimates that $100 million has been spent to develop golden rice.[4] Critics of golden rice point out that the charitable organizations that funded the development of golden rice might just as well have funded low-tech solutions to vitamin A deficiency like already-existing programs to distribute vitamin A capsules. While vitamin A capsules are neither problem-free nor a complete solution to malnutrition in the developing world, distribution programs are already in place, while golden rice is still in the research and testing phase. Moreover, the risks of capsule distribution are fairly well-known compared with the less-understood risks of biotechnology. Other options exist as well. Friends of the Earth reported that many agricultural and public education programs exist in areas where malnutrition is a problem, including an advertising campaign in Thailand to encourage people to grow a variety of vitamin-A rich foods, and the use of natural predators to control pests in food crops in Africa. Friends of the Earth sums up: "One must wonder how many other low-tech, sustainable, people-centered solutions to hunger and malnutrition go unfunded thanks to government and biotech industry obsession with the hugely expensive technology of genetic engineering."[5]

Group #2: MASIPAG: Philippine Farmer Scientist Partnership

Students in this group represent an indigenous group of farmers in the Philippines whose name translates into English as the Farmer Scientist Partnership for Development. MASIPAG believes all of the following claims: That golden rice is a technofix solution to a problem that requires a more fundamental restructuring of the global agricultural system. That golden rice only helps biotechnology companies and the governments friendly to them to continue the Green Revolution path, a path entailing that "malnutrition will even reach greater heights, as people will have more unbalanced diets based only on few foods."[6]

While those pushing golden rice have declared that the seeds will be distributed to poor farmers free of charge, MASIPAG believes that the technology will bear other costs. MASIPAG cites the case of Mr. Afsar Ali Miah, a Bangladeshi farmer, who lived through the Green Revolution and now observes that "Nothing comes in free anymore, without its consequence, especially if it is driven by profit motives." Ali Miah interprets the Green Revolution as follows:

At that time, the technology was started with all out support from the government and many farmers responded positively making use of the packaged technology of modern high-yielding varieties together with pesticides, and chemical fertilisers and a certain amount of credit. But when the uncertainty and fear of new was mitigated, the government slowly started withdrawing support and the farmers were left to deal with poor soil, lost seeds and declining diversity in the field, and dependency on pesticides and fertilisers. In the process, farmers lost control of their food system. According to Mr. Ali Miah, "Because of pesticides, people are no longer eating what little edible green leafy vegetables (and fishes) there are left in the fields anymore. If we allow this golden rice, and depend for nutrition on it, we might further lose these crops, our children losing knowledge of the importance of other crops such as green leafy vegetables."[7]

MASIPAG believes that the roots of Vitamin A deficiency are in the industrialization of agriculture. MASIPAG argues that as the diverse crops of yesteryear are replaced with monocultures, the diversity of nutrients will be increasingly narrowed, citing Ardhendu Chaterjee of the Development Resource and Service Center (DRCSC) in Calcutta, India:

The problem of malnutrition is linked not with rice per se, but with the way rice is produced now.[8] "In the past," writes Chaterjee, "integrated rice-fish-duck-tree farming was a common practice in wetlands. This does not only meet peoples' food, fodder and fuelwood needs, but it provides superior energy-protein output to that obtained from today's monoculture practice of growing high-yielding varieties. These fields also serve as the hatcheries for many fishes and aquatic organisms, which multiplied and spread to other wetlands. In the rainy season, these lowland rice fields often become connected to the water bodies like lakes and rivers. Agrochemicals applied in the paddy pollute these water-bodies and hence affect the entire food chain, thereby causing a decline in the overall fish, shrimp and frog supply—a resource freely available to the poor. Aquatic weeds which are rich in vitamin A are also becoming scarce." Sadly this is a scenario fast becoming common in most of Calcutta and over the whole Asian region.[9]

MASIPAG believes that there are alternative, better ways to provide vitamin A. The organization encourages integrated and sustainable forms of agriculture, including backyard or "kitchen" gardens, arguing that local, small-scale gardens can supply ample amounts of fruits and vegetables, foods that go a long way toward meeting micronutrient needs. MASIPAG notes that groups promoting gardens in West Bengal have had great success with this strategy.

After just two seasons of her garden, Kobita Mondall relates that, "We have already consumed all that we can, have given some to the neighbours and sold some in the market, and still we're getting something from our backyard." Kobita's garden consists of a 300 square foot plot near their home, planted with more than 30 kinds of fruits and vegetables. [10]

Hence, MASIPAG concludes as follows:

> While many doubt the ability of golden rice to eliminate vitamin A deficiency, the machinery is being set in motion to promote a GE strategy at the expense of more relevant approaches. The best chance of success in fighting vitamin A deficiency and malnutrition is to better use the inexpensive and nutritious foods already available, and in diversifying food production systems in the fields and in the household. The euphoria created by the Green Revolution greatly stifled research to develop and promote these efforts, and the introduction of golden rice will further compromise them. Golden rice is merely a marketing event. But international and national research agendas will be taken by it. The promoters of golden rice say that they do not want to deprive the poor of the right to choose and the potential to benefit from golden rice. But the poor, and especially poor farmers, have long been deprived of the right to choose their means of production and survival. Golden rice is not going to change that, and nor will any other corporately-pushed GE crop. Hence, any further attempts at the commercial exploitation of hunger and malnutrition through the promotion of genetically modified foods should be strongly resisted. [11]

Group #3: People from Developing Nations

Dr. Florence Wambugu is a scientist who has worked to bring the benefits of agricultural biotechnology to her home country of Kenya and to other countries in Africa. Dr. Wambugu herself developed a genetically engineered virus-resistant sweet potato. This development has significant potential to improve the nutritional status of Kenyan farmers, whose sweet potatoes are often shriveled and sparse due to the ravages of viruses.

Dr. Wambugu and others from developing countries argue that biotechnology can drastically improve agriculture in their homelands. They view the opposition to biotechnology in agriculture as a predominantly privileged kind of activism. In their view, American environmentalists are neither vitamin A deficient nor otherwise malnourished, so they tend to underestimate, or even totally ignore, the potential nutritional benefits of biotechnology. In a statement she published in the Washington Post, Dr. Wambugu claimed that the critics of biotechnology are insensitive to the needs of Africans: "These critics, who have never experienced hunger and death on the scale we sadly witness in Africa, are content to keep Africans dependent on food aid from industrialized nations while mass starvation occurs."[12]

Dr. Norman Borlaug, the recipient of the 1970 Nobel Peace Prize, succinctly endorses Dr. Wambugu's main point: "The affluent nations can afford to adopt elitist positions and pay more for food produced by the so-called natural methods; the 1 billion chronically poor and hungry people of this world cannot. New technology will be their salvation, freeing them from obsolete, low-yielding, and more costly production technology."[13]

In response to the critics of golden rice who argue that biotechnology will only benefit agribusiness corporations, Gregory Conko of the Competitive Enterprise Institute points out that it is a common phenomenon that new technologies may take some time to "trickle down" to the needy, but once they do, the benefits are real. "Wealthy consumers are usually first to benefit from innovations—from automobiles to antibiotics. Today, those once exorbitantly priced luxury items can be found across the globe and in use by many of modest means. The reason is that costs tend to fall over time due to economies of large-scale production, and once R&D expenditures are recouped." Applying this general analysis to biotechnology, he points out that we can expect more and more biotechnology products to benefit those in the developing world: "Once developed and commercialized, the technological knowledge used by for-profit endeavors is easily applied to far less profitable products. Many patented genetic discoveries are already being used to create extraordinarily promising plants solely for use in developing countries."[14] If this analysis is correct, there is no reason to be skeptical of the potential benefits of golden rice for the developing world.

Group #4: International Rice Research Institute

Students in this group will defend a nongovernmental organization involved with developing golden rice. IRRI is investigating the following claims: That golden rice is just the first of many biotechnologies that may assist IRRI's clients, who are among the poorest people of the world. That IRRI exists to deliver new technologies and options, free of charge, to developing country farmers.[15] That golden rice may help IRRI meet its objectives if it is first proved to be safe for people and the environment.

IRRI's mission statement reads as follows:

> IRRI is a nonprofit agricultural research and training center established to improve the well-being of present and future generations of rice farmers and consumers, particularly those with low incomes. It is dedicated to helping farmers in developing countries produce more food on limited land using less water, less labor, and fewer chemical inputs, without harming the environment.[16]

In January 2001, IRRI received its first research samples of golden rice. The sample came from the co-inventor, the German scientist, Dr. Ingo Potrykus. IRRI, aware of criticisms of the technology, read with interest Dr. Potrykus's interview with Michael Fumento of *American Outlook* magazine.[17] Here is the substance of that interview:

AO: Do you believe biotech companies have "overhyped" the value of golden rice?

Potrykus: I did not follow the advertisements of the industry, but it is difficult to overhype the value of golden rice.

AO: How many companies had to grant you licenses for golden rice to be distributed?

Potrykus: As our partner AstraZeneca [now its spin-off, Syngenta Crop Protection] took care of many IPRs [intellectual property rights], we ultimately needed free licenses from only four companies.

AO: Isn't it true that golden rice not only contains added iron but has been engineered to make the iron already present in rice more readily absorbed by the human body? Has Greenpeace or the Union of Concerned Scientists [UCS] made any mention of this?

Potrykus: This is true and the opposition has, so far, ignored this. However, the golden rice we can currently give out has only beta-carotene. For the iron traits we again first have to settle the [licensing problems].

AO: I have heard that research is already being conducted on a new and improved version, which will express vitamin A at a higher level. Is there any truth to that? Also, what about the claims that people must have a diet rich in fat and protein in order to absorb beta-carotene?

Potrykus: The golden rice that everybody is talking about is the first prototype, and we are, of course, continuously working on its improvement. It is true that uptake of beta-carotene requires fat (though not protein), but there is oil in rice endosperm [the nutritive, starchy mass in the center of grains] that will be studied to see whether it alone is sufficient for efficient uptake.

AO: To your knowledge, has Greenpeace, other advocacy groups, or any other biotech company, misrepresented your words on the nutritional value of golden rice?

Potrykus: Greenpeace has a strategy to convince people that golden rice provides so little beta-carotene that it is useless. This group and its allies base their argument on 100 percent of the recommended daily allowance [RDA], thus hiding the fact that far lower values are effective against mortality, morbidity, and blindness. The golden rice that the public will receive will provide true benefits at just 300 grams [10.5 ounces] per day.

AO: Greenpeace and the UCS claim that the timing of the announcement of golden rice was "suspicious," intended to give the agbiotech [agriculture biotechnology] multinationals a needed publicity boost. Can you refute this?

Potrykus: This is so stupid. When we initiated our work ten years ago, agbiotech definitely had no acceptance problems.

AO: Do you see golden rice as "the answer" to nutritional problems in the underdeveloped world where rice is a staple, or is it merely a tool to be used alongside others?

Potrykus: Golden rice is meant only to complement traditional interventions and to improve the vitamin A intake in poor populations. That said, it will probably be the cheapest and most sustainable solution.

AO: Do you see a role for golden mustard, golden canola, or other transgenic plants in providing more vitamin A and more nutrition in general to people in underdeveloped countries?

Potrykus: Of course I see a role for further food crops providing beta-carotene. We've already had discussions with scientists who want to introduce the trait into wheat, maize (white maize of Africa), cassava, sweet potato, banana, and so on. Naturally, the work with mustard and canola helps also. What I want is not only the addition of beta-carotene but nutritional improvement in general. That's why I have also added the iron trait, and I am working on a high-quality protein trait.

AO: Do you concur with the assertion that simply by raising nutrition levels of underdeveloped nations, we can help them become less poor, leading to overall better nutrition?

Potrykus: Yes, I certainly do.

AO: What do you think of Greenpeace's insistence that it reserves the right to take "direct action" against golden rice test plots?

Potrykus: If Greenpeace does this, they will be guilty of a crime against humanity.

AO: What measures were taken in the past to address vitamin A deficiency, and what were the problems with those alternatives? Do you think that Greenpeace's suggested plan of mass distribution of vitamin pills makes sense in terms of distributing them to hundreds of millions of people?

Potrykus: There is a need for distribution, fortification, dietary diversification, and education. All of these are important. These interventions have used an impressive amount of funds that have been spent over the last twenty years and have been very helpful. But we still have 500,000 blind children and millions of vitamin A deficiency deaths every year. The problem with vitamin A pill distribution is that it does not reach many of those who need it.

AO: Greenpeace has declared the rice to be "fool's gold." How do you respond to their accusation that it would take an incredible amount of golden rice consumption to give children the recommended daily allowance of vitamin A, plus prevent blindness?

Potrykus: This is not true. The golden rice that will finally be given out to the public will be effective on 300 grams of rice in the diet per day.

AO: In many parts of the world, people who eat rice value its whiteness. It has a special meaning to them. Will they eat rice that is not very white? Hasn't this been a barrier to previous efforts in adding supplements to rice?

Potrykus: This is a problem in some parts of the world, although probably not in India. People will have the freedom to decide whether they want healthy children or white rice. We are, however, working on a solution for the color problem, and we believe that we know how to solve it.

AO: Critics insist that $100 million was spent researching golden rice, but others claim that this figure was total Rockefeller Foundation spending on rice research over ten years to hundreds of scientists. Can you clarify this?

Potrykus: The total cost for golden rice development was $2.6 million, spent over ten years in the lab of Peter Beyer and myself. These funds were from the Rockefeller Foundation, the Swiss Federation, the National Science Foundation, and the European Union. The investment was approximately one-fourth of 1 percent of the money spent on traditional interventions. Compared to the $100 million plus Greenpeace spends per year, this was a very small investment.

Notes

1. http://www.who.int/aboutwho/en/objectiv.htm
2. http://www.who.int/aboutwho/en/mission.htm
3. Friends of the Earth, Link Magazine, April/June 2000. http://www.foei.org/publications/link/93/e93goldenrice.html
4. Friends of the Earth, "Golden Rice and Vitamin A Deficiency," http://www.foe.org/safefood/rice.html
5. Ibid.
6. Genetic Resources Action International (GRAIN), "Grains of delusion: Golden rice seen from the ground," February 2001. http://www.grain.org/publications/delusion-en-p.htm. This document was researched, written, and published as a joint undertaking between BIOTHAI (Thailand), CEDAC (Cambodia), DRCSC (India), GRAIN, MASIPAG (Philippines), PAN-Indonesia and UBINIG (Bangladesh). MASIPAG, Los Baños, Laguna, Philippines.
7. GRAIN, "Grains of delusion: Golden rice seen from the ground," February 2001.
8. MASIPAG communication with Ardhendu Chaterjee, Director, DRCSC, 21 July 2000.
9. GRAIN, "Grains of delusion: Golden rice seen from the ground," February 2001.
10. Ibid.
11. Ibid.
12. The Washington Post Company, Sunday, August 26, 2001, B07. http://www.washingtonpost.com/ac2/wp-dyn/A59811-2001Aug24
13. Norman Borlaug, "Ending World Hunger. The Promise of Biotechnology and the Threat of Antiscience Zealotry", *Plant Physiology*, October 2000, Vol. 124, pp. 487-490. http://www.plantphysiol.org/cgi/content/full/124/2/487
14. Gregory Conko, "Hype, Not Hype, in the Golden Grains", AgBioWorld. http://www.agbioworld.org/biotech_info/topics/goldenrice/hype.html
15. IRRI, "Golden Rice: The Eyes of the World Are Watching." http://www.irri.org/ar2001/datta.pdf

16. http://www.irri.org
17. "Golden Rice: A Golden Chance for the Underdeveloped World," *American Outlook*, July-August 2001. http://www.fumento.com/

CHAPTER 16: FARMS

LOST IN THE MAIZE

The ethical problem addressed in this case is that of conflict or conscience for an individual. The teaching objective for this case is to practice articulating alternative courses of action.

Additional features that one might want to add to the case:

1. GM and OF are close relatives.
2. The genetically modified maize has not been approved for human consumption (only for animal feed).
3. GM's crop is worth fifty times that of OF's.

Following are some notes to help you assist your students in responding to the worksheets.

A. Question 1: Did GM make the right decision?

A) Write your answer here (Yes or No):

B. Pass the paper to your neighbor on the left.

[Solicit answers and read/tally]

B) Write down one reason that supports the decision passed to you ("Yes" or "No"). Important: You have to write about the decision in front of you, not the one you wrote yourself.

C. Pass your paper to your neighbor on the left.

[Solicit answers and read out—illustrate different kinds of arguments]

C) Look at the answer and supporting arguments now in front of you. What kind of ethical principle do you think explains the answers to parts A and B? Write down a few words to sum up your thoughts.

[Solicit answers and have students read their answers out loud. Use their answers to illustrate different principles: legal, moral, ethical, economic justice, religious, ecological justice]

[After discussion, have students pass papers back to original source]

After several sleepless nights, GM reconsiders his decision and decides to tell OF about the possible cross-pollination . . . . It is also possible that the test will show that there has been *no* contamination of OF's organic crop.

[Ask: *Are there any factual questions about this part of the case?*]

A. *Work in pairs to answer questions 2, 3, and 4. Write down your answers in a few words.*

B. *Question 2: Should OF get his crop tested for the presence of genetically modified maize? Why?*

C. *Question 3: If tests shows that there IS genetically modified material present in OF's corn, should he tell Gerber? Why?*

OF has the tests performed. They show that his crop is completely free of any genetically modified material. He informs the Maize Liberation Front about GM's crop, knowing full well that they will destroy it.

Question 4: Should OF have informed the Maize Liberation Front, given that he has not suffered any economic harm himself? What ethical principle might justify OF's behavior?

Organizational Aspects

Ask students to write down "what is ethics?" Introduce the concept of doing the right thing. Apply to bioethics. Give examples of problems.

Introduce this case. Give out page 1 first. Get several students to read the paragraphs up to question 1. Ask whether there are any factual questions about the case—anything not understood and so forth. Now give out page 2.

Question 1: Write & Pass. Solicit answers. Discuss. Now give out page 3.
Questions 2–4: Pair, Write, Share.

Topics for Discussion

1. Even before planting, what are GM's obligations to OF? Do they extend beyond what is legally required? If GM tells OF, isn't there an assumption that there is something wrong with the genetically modified crop? Yet it was approved. So his

bringing it up would make it seem as though he agreed that there was a problem, even though he thinks there is none.

2. What does it mean for something to be *free* of any transgenic material? How might contamination occur? Is it possible to detect such contamination and, if so, at what levels? Do very low levels (for example, 0.1 percent) represent a threat? Should levels of contamination be set? What would have to be done to ensure that there is *no* contamination? Compare this case with pesticide residues that are detectable on organic crops.

3. In reality, Gerber would test every batch itself, so the company might discover contamination in its supply. However, testing procedures are patchy and many products labeled as GM-free or organic have been shown to contain some detectable genetically modified material.

4. Insurance might cover OF's losses and GM's liability. In fact, the major re-insurers refuse to insure claims against loss caused by genetically modified crops. A perceived liability can prevent a technology being deployed, whether or not the risk is real. The insurance company is employing the "precautionary principle," but is this appropriate?

5. What different ways could genetically modified maize "contaminate" organic maize? How would you ensure that organic maize is absolutely free of any contamination?

6. Some students may cite the precautionary principle. This is a relatively new concept in the legal system, though it has been adopted to some extent in Europe. In addition to actual harm, it covers anything that has the potential to cause harm. How do you apply this principle? Would it preclude deployment of new technologies?

Additional issues that could be added to the case:

- The genetically modified maize has not been approved for human consumption (only for animal feed).
- OF's crop is worth fifty times that of GM's.
- OF's family has been farming on this site for two hundred years. GM has been there for only five years.

MAGNANIMOUS IOWANS

This case study faithfully describes an event that took place in a course on world food issues at Iowa State University during spring of 2001. The study is intended to provide a focal point for discussion on the ethics and morality of giving, particularly at the international level. A key issue involved in the case study is transgenic technology. Sufficient empirical data to explain and illustrate the technical issues involved in this case are included herein. We recommend that instructors emphasize at all times (when presenting the study, when charging students with discussion objectives, and when moderating the resulting discussion) that the study is not about the ethics or morality of transgenic organisms per se, but rather about the ethics and morality of giving. A relevant level of discussion regarding how transgenic technologies affect the decisions to be made in this case will of course be necessary, but we advise instructors to be alert and ready to curb discussions that veer away from the principal focus of the case.

We also note that the study could be used in at least two additional ways:

1. If an instructor in fact wishes to deal with the morality of transgenic technologies, this case study may be an effective way to introduce the topic by showing how transgenic organisms are already influencing moral decisions. However, in such an event, a greater amount of time and a larger body of supporting materials would have to be provided to students.

2. This case study may also be used for an audience of educators. There is an aspect of the study that will be self evident and involves the ethics of the educational process. We don't provide a full set of questions to guide discussion on this dimension of the case, but suggest the following initial line of questioning: What should an instructor do, given the classroom circumstance obtaining at the end of this case description?

INDEX

ISBN 0-8138-2835-X

9 780813 828350